PRIMATE AUDITION
ETHOLOGY AND NEUROBIOLOGY

METHODS & NEW FRONTIERS IN NEUROSCIENCE

Series Editors
Sidney A. Simon, Ph.D.
Miguel A.L. Nicolelis, M.D., Ph.D.

Published Titles

Apoptosis in Neurobiology
Yusuf A. Hannun, M.D., Professor of Biomedical Research and Chairman/Department of Biochemistry and Molecular Biology, Medical University of South Carolina
Rose-Mary Boustany, M.D., tenured Associate Professor of Pediatrics and Neurobiology, Duke University Medical Center

Methods for Neural Ensemble Recordings
Miguel A.L. Nicolelis, M.D., Ph.D., Professor of Neurobiology and Biomedical Engineering, Duke University Medical Center

Methods of Behavioral Analysis in Neuroscience
Jerry J. Buccafusco, Ph.D., Alzheimer's Research Center, Professor of Pharmacology and Toxicology, Professor of Psychiatry and Health Behavior, Medical College of Georgia

Neural Prostheses for Restoration of Sensory and Motor Function
John K. Chapin, Ph.D., Professor of Physiology and Pharmacology, State University of New York Health Science Center
Karen A. Moxon, Ph.D., Assistant Professor/School of Biomedical Engineering, Science, and Health Systems, Drexel University

Computational Neuroscience: Realistic Modeling for Experimentalists
Eric DeSchutter, M.D., Ph.D., Professor/Department of Medicine, University of Antwerp

Methods in Pain Research
Lawrence Kruger, Ph.D., Professor or Neurobiology (Emeritus), UCLA School of Medicine and Brain Research Institute

Motor Neurobiology of the Spinal Cord
Timothy C. Cope, Ph.D., Professor of Physiology, Emory University School of Medicine

Nicotinic Receptors in the Nervous System
Edward D. Levin, Ph.D., Associate Professor/Department of Psychiatry and Pharmacology and Molecular Cancer Biology and Department of Psychiatry and Behavioral Sciences, Duke University School of Medicine

Methods in Genomic Neuroscience
Helmin R. Chin, Ph.D., Genetics Research Branch, NIMH, NIH
Steven O. Moldin, Ph.D, Genetics Research Branch, NIMH, NIH

Methods in Chemosensory Research
Sidney A. Simon, Ph.D., Professor of Neurobiology, Biomedical Engineering, and Anesthesiology, Duke University
Miguel A.L. Nicolelis, M.D., Ph.D., Professor of Neurobiology and Biomedical Engineering, Duke University

The Somatosensory System: Deciphering the Brain's Own Body Image,
Randall J. Nelson, Ph.D., Professor of Anatomy and Neurobiology,
    University of Tennessee Health Sciences Center

New Concepts in Cerebral Ischemia
Rick C. S. Lin, Ph.D., Professor of Anatomy, University of Mississippi Medical Center

DNA Arrays: Technologies and Experimental Strategies
Elena Grigorenko, Ph.D., Technology Development Group, Millennium Pharmaceuticals

Methods for Alcohol-Related Neuroscience Research
Yuan Liu, Ph.D., National Institute of Neurological Disorders and Stroke, National Institutes of Health
David M. Lovinger, Ph.D., Laboratory of Integrative Neuroscience, NIAAA

In Vivo Optical Imaging of Brain Function
Ron Frostig, Ph.D., Associate Professor/Department of Psychobiology,
    University of California, Irvine

PRIMATE AUDITION
ETHOLOGY AND NEUROBIOLOGY

Edited by
Asif A. Ghazanfar, Ph.D.
Max Planck Institute for Biological Cybernetics
Tübingen, Germany

CRC PRESS

Boca Raton   London   New York   Washington, D.C.

Cover illustration credits

Photographs: Cotton-top tamarin, Primate Cognitive Neuroscience Lab, Cambridge, MA: Geertrui Spaepen; rhesus monkey, Cayo Santiago, Puerto Rico: Cory Miller; chacma baboon, Kruger National Park, South Africa: Spook Skelton.

Vocalizations: Cotton-top tamarin long call and rhesus monkey shrill bark, courtesy of Primate Cognitive Neuroscience Lab, Cambridge, MA; baboon "wahoo" call, courtesy of Dorothy Cheney and Robert Seyfarth. Brain outlines are based on photographs from the Comparative Mammalian Brain Collection at the University of Wisconsin–Madison and Michigan State University.

Library of Congress Cataloging-in-Publication Data

Primate audition : ethology and neurobiology / edited by Asif A. Ghazanfar.
      p. cm. -- (Methods and new frontiers in neuroscience)
Includes bibliographical references and index.
ISBN 0-8493-0956-5
  1. Auditory cortex. 2. Auditory perception. 3. Comparative neurobiology. 4. Primates--Physiology. I. Ghazanfar, Asif A. II. Methods & new frontiers in neuroscience series

QP383.15 .P75 2002
156'.215--dc21
                                                                                                 2002067481

This book contains information obtained from authentic and highly regarded sources. Reprinted material is quoted with permission, and sources are indicated. A wide variety of references are listed. Reasonable efforts have been made to publish reliable data and information, but the author and the publisher cannot assume responsibility for the validity of all materials or for the consequences of their use.

Neither this book nor any part may be reproduced or transmitted in any form or by any means, electronic or mechanical, including photocopying, microfilming, and recording, or by any information storage or retrieval system, without prior permission in writing from the publisher.

All rights reserved. Authorization to photocopy items for internal or personal use, or the personal or internal use of specific clients, may be granted by CRC Press LLC, provided that $1.50 per page photocopied is paid directly to Copyright Clearance Center, 222 Rosewood Drive, Danvers, MA 01923 USA. The fee code for users of the Transactional Reporting Service is ISBN 0-8493-0956-5/03/$0.00+$1.50. The fee is subject to change without notice. For organizations that have been granted a photocopy license by the CCC, a separate system of payment has been arranged.

The consent of CRC Press LLC does not extend to copying for general distribution, for promotion, for creating new works, or for resale. Specific permission must be obtained in writing from CRC Press LLC for such copying.

Direct all inquiries to CRC Press LLC, 2000 N.W. Corporate Blvd., Boca Raton, Florida 33431.

Trademark Notice: Product or corporate names may be trademarks or registered trademarks, and are used only for identification and explanation, without intent to infringe.

Visit the CRC Press Web site at www.crcpress.com

© 2003 by CRC Press LLC

No claim to original U.S. Government works
International Standard Book Number 0-8493-0956-5
Library of Congress Card Number 2002067481
Printed in the United States of America 1 2 3 4 5 6 7 8 9 0
Printed on acid-free paper

Methods & New Frontiers in Neuroscience

Our goal in creating the **Methods & New Frontiers in Neuroscience** series is to present the insights of experts on emerging experimental techniques and theoretical concepts that are or will be at the vanguard of the study of neuroscience. Books in the series cover topics ranging from methods to investigate apoptosis to modern techniques for neural ensemble recordings in behaving animals. The series also covers new and exciting multidisciplinary areas of brain research, such as computational neuroscience and neuroengineering, and describes breakthroughs in classical fields such as behavioral neuroscience. We want these to be the books every neuroscientist will use in order to get acquainted with new methodologies in brain research. These books can be given to graduate students and postdoctoral fellows when they are looking for guidance to start a new line of research.

Each book is edited by an expert and consists of chapters written by the leaders in a particular field. Books are richly illustrated and contain comprehensive bibliographies. Chapters provide substantial background material relevant to the particular subject; hence, they are not only "methods" books. They contain detailed tricks of the trade and information as to where these methods can be safely applied. In addition, they include information about where to buy equipment and about Web sites that are helpful in solving both practical and theoretical problems,

We hope that as the volumes become available, the effort put in by us, by the publisher, by the book editors, and by the individual authors will contribute to the further development of brain research. The extent to which we achieve this goal will be determined by the utility of these books.

Sidney A. Simon, Ph.D.
Miguel A.L. Nicolelis, M.D., Ph.D.
Series Editors

Preface

Neuroethological research has been critical for our understanding of brain function and how natural selection shapes brain design for complex behaviors. The value of this approach is evident in the auditory model systems that are currently popular: echolocation in bats, song learning and prey localization in birds, and mate choice in frogs. Even in human neurobiological studies, speech perception and production represent the paradigm example of a specialized system in the cerebral cortex. It is therefore surprising that few researchers interested in the neural substrate of non-human primate auditory processing have adopted a similar naturalistic approach. With the advent of new signal-processing techniques and the exponential growth in our knowledge of primate behavior, the time has arrived for a neurobiological investigation of the primate auditory system based on principles derived from ethology.

A neuroethology of primate hearing may also yield insights into human speech processing. Like speech, the species-specific vocalizations of non-human primates mediate social interactions, convey important emotional information, and in some cases refer to objects and events in the caller's environment. These functional similarities suggest that the selective pressures that shaped primate vocal communication are similar to those that influenced the evolution of human speech. As such, investigating the perception and production of vocalizations in extant non-human primates provides one avenue for understanding the neural mechanisms of speech and for illuminating the substrates underlying the evolution of human language.

Primate Audition: Ethology and Neurobiology is the first book whose primary purpose is to bridge the epistemological gap between primate ethologists and auditory neurobiologists. To do this, the knowledge of leading world experts on different aspects of primate auditory function has been brought together in a single volume. The book covers the state-of-the-art work on a variety of issues in primate auditory perception. Topics include the functional organization and anatomy of the primate auditory system, spatial localization of sounds and its neural basis, function and perception of conspecific and heterospecific vocalizations and their ontogeny, neural encoding of complex sounds, vocal production and its relationship to perception, and the acoustic cues guiding vocal recognition. This synthesis of ethological and neurobiological approaches to primate vocal behavior is likely to yield the richest understanding of the acoustic and neural bases of primate audition.

Acknowledgments

I am grateful to all the authors who have contributed to this book, for without them it would not exist. Thanks are also due to Roian Egnor, Tecumseh Fitch, Don Katz, Merri Rosen, and Laurie Santos for generously helping me edit a good number of the chapters. I would also like to express my gratitude to my former mentors (in the temporal order of their initial impact), Professors Mark Desantis, Matthew Grober, Miguel Nicolelis, and Marc Hauser. The ideas behind putting together this book are the result of their continuing and converging positive influence on me. Finally, I would like to thank the series editors, Miguel and Sid, and Barbara Norwitz, Pat Roberson, and their colleagues at CRC Press for pushing, sympathizing, and supporting when necessary and expertly guiding the book to its completion.

The Editor

Asif A. Ghazanfar, Ph.D., is a research scientist at the Max Planck Institute for Biological Cybernetics in Tübingen, Germany. Born in Pullman, WA, and raised in nearby Moscow, ID, he received his Bachelor of Science degree in philosophy at the University of Idaho. While earning his degree, he studied the neural and hormonal bases for sex reversal in a coral reef fish, the saddleback wrasse. In 1998, he earned his doctoral degree in neurobiology from Duke University in Durham, NC. His dissertation research uncovered some of the dynamic properties of single neurons and neural ensembles in the somatosensory corticothalamic pathway. Since then, he has combined his dual interests in ethology and neurophysiology by studying the natural vocal behavior of primates and its neural basis. As a postdoctoral fellow at Harvard University, he studied the acoustic bases for vocal recognition in three species of non-human primates. Using this ethological work as a foundation, he is currently investigating how behaviorally relevant acoustic features of species-specific vocalizations are processed in the auditory cortex of rhesus monkeys.

Contributors

James A. Agamaite
Laboratory of Auditory
 Neurophysiology
Department of Biomedical Engineering
Johns Hopkins University School of
 Medicine
Baltimore, Maryland

Michael Brosch
Leibniz-Institut für Neurobiologie
Magdeburg, Germany

Charles H. Brown
Department of Psychology
University of South Alabama
Mobile, Alabama

Julia Fischer
Max Planck Institute for Evolutionary
 Anthropology
Leipzig, Germany

W. Tecumseh S. Fitch
Department of Psychology
Harvard University
Cambridge, Massachusetts

Asif A. Ghazanfar
Max Planck Institute for Biological
 Cybernetics
Tübingen, Germany

Jennifer M. Groh
Department of Psychological and Brain
 Sciences
Center for Cognitive Neuroscience
Dartmouth College
Hanover, New Hampshire

Troy A. Hackett
Department of Psychology
Vanderbilt University
Nashville, Tennessee

Marc D. Hauser
Primate Cognitive Neuroscience Lab
Department of Psychology
Harvard University
Cambridge, Massachusetts

Siddhartha C. Kadia
Laboratory of Auditory
 Neurophysiology
Department of Biomedical Engineering
Johns Hopkins University School of
 Medicine
Baltimore, Maryland

Kristin A. Kelly
Department of Psychological and Brain
 Sciences
Center for Cognitive Neuroscience
Dartmouth College
Hanover, New Hampshire

Colleen G. Le Prell
Kresge Hearing Research Institute
University of Michigan
Ann Arbor, Michigan

Li Liang
Laboratory of Auditory
 Neurophysiology
Department of Biomedical Engineering
Johns Hopkins University School of
 Medicine
Baltimore, Maryland

Thomas Lu
Laboratory of Auditory
 Neurophysiology
Department of Biomedical Engineering
Johns Hopkins University School of
 Medicine
Baltimore, Maryland

Ryan Metzger
Department of Psychological and Brain
 Sciences
Center for Cognitive Neuroscience
Dartmouth College
Hanover, New Hampshire

Cory T. Miller
Primate Cognitive Neuroscience Lab
Department of Psychology
Harvard University
Cambridge, Massachusetts

O'Dhaniel A. Mullette-Gillman
Department of Psychological and Brain
 Sciences
Center for Cognitive Neuroscience
Dartmouth College
Hanover, New Hampshire

David B. Moody
Kresge Hearing Research Institute
University of Michigan
Ann Arbor, Michigan

John D. Newman
Laboratory of Comparative Ethology
National Institute of Child Health and
 Human Development
National Institutes of Health
Poolesville, Maryland

Kevin N. O'Connor
Center for Neuroscience
Section of Neurobiology, Physiology,
 and Behavior
University of California
Davis, California

Gregg H. Recanzone
Center for Neuroscience
Section of Neurobiology, Physiology,
 and Behavior
University of California
Davis, California

Lizabeth M. Romanski
Department of Anatomy and
 Neurobiology
University of Rochester
Rochester, New York

Laurie R. Santos
Department of Psychology
Yale University
New Haven, Connecticut

Henning Scheich
Leibniz-Institut für Neurobiologie
Magdeburg, Germany

Mitchell L. Sutter
Center for Neuroscience
Section of Neurobiology, Physiology,
 and Behavior
University of California
Davis, California

Xiaoqin Wang
Laboratory of Auditory
 Neurophysiology
Department of Biomedical Engineering
Johns Hopkins University School of
 Medicine
Baltimore, Maryland

Daniel J. Weiss
Department of Brain and Cognitive
 Sciences
University of Rochester
Rochester, New York

Uri Werner-Reiss
Department of Psychological and Brain
  Sciences
Center for Cognitive Neuroscience
Dartmouth College
Hanover, New Hampshire

Klaus Zuberbühler
Department of Psychology
University of St. Andrews
Fife, Scotland

Contents

Chapter 1
Primates as Auditory Specialists 1
Asif A. Ghazanfar and Laurie R. Santos

Chapter 2
Causal Knowledge in Free-Ranging Diana Monkeys 13
Klaus Zuberbühler

Chapter 3
Auditory Temporal Integration in Primates: A Comparative Analysis................... 27
Kevin N. O'Connor and Mitchell L. Sutter

Chapter 4
Mechanisms of Acoustic Perception in the Cotton-Top Tamarin........................... 43
Cory T. Miller, Daniel J. Weiss, and Marc D. Hauser

Chapter 5
Psychophysical and Perceptual Studies of Primate Communication Calls 61
Colleen G. Le Prell and David B. Moody

Chapter 6
Primate Vocal Production and Its Implications for Auditory Research 87
W. Tecumseh S. Fitch

Chapter 7
Developmental Modifications in the Vocal Behavior of Non-Human Primates... 109
Julia Fischer

Chapter 8
Ecological and Physiological Constraints for Primate Vocal Communication 127
Charles H. Brown

Chapter 9
Neural Representation of Sound Patterns in the Auditory Cortex of Monkeys ... 151
Michael Brosch and Henning Scheich

Chapter 10
Representation of Sound Location in the Primate Brain 177
Kristin A. Kelly, Ryan Metzger, O'Dhaniel A. Mullette-Gillman, Uri Werner-Reiss, and Jennifer M. Groh

Chapter 11
The Comparative Anatomy of the Primate Auditory Cortex 199
Troy A. Hackett

Chapter 12
Auditory Communication and Central Auditory Mechanisms in the Squirrel
Monkey: Past and Present 227
John D. Newman

Chapter 13
Cortical Mechanisms of Sound Localization and Plasticity in Primates 247
Gregg H. Recanzone

Chapter 14
Anatomy and Physiology of Auditory–Prefrontal Interactions in Non-Human
Primates 259
Lizabeth M. Romanski

Chapter 15
Cortical Processing of Complex Sounds and Species-Specific Vocalizations
in the Marmoset Monkey (*Callithrix jacchus*) 279
Xiaoqin Wang, Siddhartha C. Kadia, Thomas Lu, Li Liang, and James A. Agamaite

Index 301

1 Primates as Auditory Specialists

Asif A. Ghazanfar and Laurie R. Santos

CONTENTS

I. Introduction 1
II. Avoiding Predators 2
III. Finding Food 4
    A. Direct Auditory Cues 4
    B. Indirect Auditory Cues 5
IV. Finding (and Keeping) a Mate 6
    A. Copulation Calls 6
    B. Primate Songs 7
V. A Socioecologically Sensible Auditory Neuroscience 8
    A. Temporal Cues in Rhesus Monkey Calls 8
    B. Temporal Cues in Tamarin Long Calls 8
VI. Conclusion 10
References 10

I. INTRODUCTION

A brief survey of the animal kingdom quickly reveals an impressive array of sensory specializations, each engineered to solve a particular type of adaptive problem. Such a diverse collection of adaptive specializations is especially prevalent in the auditory domain. Acoustic engineers would likely be impressed by the diverse solutions different animals have found to the problems of sound localization, auditory discrimination, and vocal recognition. Echolocating bats, for example, have an auditory system exquisitely tuned for nocturnal prey catching. They are able to use the difference between the sound of an emitted vocalization and its subsequent echo to identify and localize moving targets in a dark environment. Similarly, the auditory systems of many anuran and avian species are specifically tuned for recognizing mating calls and other conspecific vocalizations.

    Unfortunately, when one thinks of the impressive auditory specializations of the animal kingdom, one rarely considers those of animals within the primate order. Although human primates clearly possess specialized mechanisms for processing speech, few consider the auditory systems of other primate species to be exemplary in any way. In fact, while many consider primate auditory systems to be important and effective models of *general* auditory processing, few auditory physiologists would describe these systems as particularly specialized for species-specific ecological problems or as masterfully designed as those of bats and anurans.

This species-general view of primate audition is incorrect and, more importantly, detrimental to a rich understanding of the structure and function of primate auditory mechanisms. We hypothesize that the design of primate auditory circuitry, like that of other taxa, reflects the specialized functions that these systems have evolved to carry out. In this chapter, we review the ways in which primates naturally use their auditory systems, focusing particularly on the problems of avoiding predation, locating food, and finding mates. We propose that a more thorough understanding of the adaptive nature of primate audition is necessary for neuroscientists to develop better questions about the structure and function of primate auditory mechanisms.

II. AVOIDING PREDATORS

One of the most persistent adaptive problems facing our primate ancestors was the task of avoiding potential predators. Today, most primate species serve as prey to at least one kind of predator, and many species are hunted by a number of different predator types. Red colobus monkeys (*Procolobus badius*), for example, are preyed upon by eagles, leopards, chimpanzees, and human poachers.[1] Predation rates can often be incredibly high, and in some species predators bring about more deaths than disease, injuries, and other causes. In vervet monkeys (*Cercopithecus aethiops*), for example, predation alone accounts for over 70% of all deaths.[2] With mortality rates as high as these, natural selection has exerted strong pressure on the evolution of antipredator tactics and, in particular, on mechanisms specialized for avoiding predators. In the auditory domain, these tactics can include the detection of predatory-specific auditory signals as well as recognition of the alarm calls of conspecifics and other sympatric species.

In response to hearing or seeing a predator, many primates produce alarm calls. More impressively, a number of primate species distinguish among various predators, producing acoustically different alarm calls for different classes of predators. Perhaps the best known antipredator tactic is the alarm-calling behavior of vervet monkeys. Vervet monkeys in the Amboseli National Forest are preyed upon by at least three different predator classes: eagles, leopards, and snakes.[3,4] Because each of these predators hunts in a different way, no single general-purpose antipredator tactic would be effective against all. When faced with an eagle, for example, a vervet monkey's safest response is to hide as close to the ground as possible. When faced with a leopard, however, the ground is the most dangerous place to be; instead, vervets must immediately move as high off the ground as possible. Vervets must distinguish among different predator classes and react in a predator-appropriate way. One way vervets manage to categorize different predators is through the use of three predator-specific alarm calls, one for aerial predators, one for leopards, and one for snakes.[3] Seyfarth et al.[4] played these different calls back to vervet subjects and found that individuals always reacted to the calls in a predator-appropriate way. When subjects heard playbacks of leopard calls, they ran into the trees; but when they heard eagle alarm calls, they hid under bushes on the ground. These results suggest that vervets naturally distinguish between both the different predator classes and the vocalizations associated with those classes.

Conspecific alarm calls are not the only kind of acoustic information relevant for predator detection. Many primates live sympatrically with other alarm-calling primate and nonprimate species that are preyed upon by similar classes of predators. As such, the ability to detect, learn, and respond adaptively to the alarm signals of heterospecific individuals would be strongly selected for during an organism's evolutionary history. There is now much evidence to suggest that primates do just this; many primate species can learn, and respond adaptively to, the alarm calls of other species with whom they are sympatric.[5–9] In southern India, for example, bonnet macaque (*Macaca radiata*) groups are found in association with Nilgiri langurs (*Trachypithecus johnii*), Hanuman langurs (*Semnopithecus entellus*), and sambar deer (*Cervus unicolor*), but the frequency of such associations varies between groups of bonnet macaques. All four species fall prey to leopards and produce alarm calls upon detecting a leopard. Ramakrishnan and Coss[7] compared the responses of bonnet macaques to playbacks of conspecific alarm calls with playbacks of alarm calls of the other three species. They found that no differences in the latencies to flee following conspecific vs. heterospecific alarm calls. Thus, bonnet macaques treated the heterospecific alarm calls with as much urgency as they would a conspecific alarm call. On a group-by-group basis, however, the responses to alarm calls of species to which bonnet macaques were not frequently exposed were significantly different from the responses to their conspecific alarm calls. This suggests that sufficient experience is necessary for bonnet macaques to learn the alarm calls of other species and argues against the idea that the responses are driven solely by the acoustic features of alarm calls in general (for similar results in Diana and Campbell's monkeys, see Chapter 2).

The ability to learn a secondary cue for danger, such as the alarm vocalizations of another species, may be a capacity specific to the auditory domain. While primates seem to readily learn the secondary auditory cues of danger (e.g., the alarm calls of other species or the sounds that predators themselves make), there is scant evidence that they understand secondary *visual* cues of danger. Cheney and Seyfarth[5] tested whether vervet monkeys knew enough about the behavior of leopards to understand that a carcass in a tree in the absence of a leopard represented the same potential danger as did the leopard itself. A stuffed carcass of a gazelle was positioned up in a tree. When vervets saw the carcass, none produced alarm calls or showed any increased vigilance behavior. Thus, despite all the experience vervets have with seeing leopards and their prey together in trees, they did not behave as though the carcass might mean that a leopard was nearby. A similar experiment was conducted for another predator class. Could vervets associate the distinctive tracks of a python with the possible presence of the python itself? Pythons prey upon vervets frequently, and vervets readily alarm call upon seeing a python. Yet, upon seeing fresh python tracks, vervets did not alarm call or increase their vigilance. In stark contrast, vervets readily respond to the alarm calls of starlings, a secondary auditory cue for the presence of terrestrial and aereal predators.[5,6]

Cheney and Seyfarth[5] have argued that vervets' inability to adaptively use secondary visual cues of predators may be related to their limited use of visual signals as secondary cues in their social interactions. In the auditory domain, however, vervets regularly use auditory signals to designate objects or events, and this may facilitate their use of auditory signals as cues when dealing with other species.

III. FINDING FOOD

Predator avoidance is not the only significant adaptive problem that all primates must face. In order to survive, primates, like other organisms, must forage for food. At a proximate level, a primate's size and food processing apparatus (i.e., its teeth and gut) constrain its food choices. Although the diet of a given primate can be diverse enough to include insects, fruits, leaves, vines, nectar, sap, and resin, most primates are easily classified according to the food type they most commonly eat.[10] The five main dietary classes are insectivores, gummivores (gums and sap), frugivores (fruits), gramnivores (seeds), and foliovores (leaves). These dietary variations among species can often be related to morphological and physiological differences. For example, within the Callichtrichids, marmosets (*Callithrix* and *Cebuella*) have long, forward-projecting lower incisors consistent with their gummivory (necessary to gouge holes in tree trunks), while tamarins (*Saguinus* and *Leontopithecus*), like most primates, do not have such a dental adaptation.[11] A more familiar food-related adaptation is cheek pouches, typically used by Cercopithecinae for their "retrieve-and-retreat" pattern of feeding. Such a tactic is useful in situations where competition from conspecifics is intense and for the enzymatic predigestion of food.[12,13] Specialized adaptations for feeding and foraging should be no less prevalent in the auditory system. In this section, we consider the relationship between perceptual specializations in the auditory domains and the foraging behaviors of different primates.

A. DIRECT AUDITORY CUES

The two categories of auditory cues that primates can use to detect food are direct and indirect.[14] Direct cues include the sounds of prey or self-generated acoustic cues that could be used to identify where prey might be found. Indirect auditory cues include the calls of conspecifics or other sympatric species that could lead a listener to the location of a food source.

The use of direct auditory cues can sometimes be related to the anatomy of the ears. For example, many insectivorous nocturnal primates have large, membranous outer ears that can be moved in several directions.[15] The prosimian aye-aye (*Daubentonia madagascarensis*), for example, uses its large ears to detect larvae that reside in wood cavities. By tapping on wood surfaces with their long middle finger, they seem to detect cavities made by insects by changes in the reverberation of sound.[16] Such detection is, without a doubt, aided by their large ears. Another prosimian, the galago (*Galago demidovii* and *Euoticus elegantulus*), locates scurrying or flying insects primarily by the sounds these prey make.[17] A cricket can escape detection as long as it remains quiet and immobile; however, as soon as it moves or makes a sound, a galago will move its ears in the appropriate direction to locate the prey and then grab it, often in mid-air. Indeed, experiments in which insects were placed behind a plywood screen revealed that galagos followed the movements of insects with precise head movements, almost as if they can see the prey.[17] Direct auditory cues are also used by chimpanzees (*Pan troglodytes*) when they hunt monkey prey.[18] In the dense forest, chimpanzees detect their preferred prey species, the red colobus monkey (*Procolobus badius*), by the colobus' frequent vocalizations. Red colobus monkeys are very vocal, especially compared to sympatric primates,[18] and, naturally, one of their antipredator strategies upon hearing chimpanzees is to reduce their vocalization rates.[19]

B. INDIRECT AUDITORY CUES

Food-associated vocalizations — indirect auditory cues — are produced by many New and Old World primates.[14] These calls can let receivers know the location and characteristics of a food source. Depending on the species, characteristics such as quantity, quality, and divisibility can be indicated by the probability and rate of food calling by the discoverer. For example, toque macaques (*Macaca sinica*) produce special food calls when they discover a new location of food.[20] It is the quantity (must be large), not the food type, that seems to elicit calls from the discoverers. In other words, the same food but in low quantities does not elicit food calls from toque macaques. Upon hearing food calls, dispersed receivers immediately run to the site and feed there. These calls, therefore, convey information about the presence of a food source, its location, and its quantity.

Rhesus macaques (*Macaca mulatta*) produce three different calls (chirps, warbles, and harmonic arches) associated exclusively with the discovery of preferred, rare foods (e.g., coconuts, berries, etc.).[21] The rate of calling is positively correlated with the hunger level of the caller. When conspecifics hear these calls, they typically approach the caller. Interestingly, all members of a given rhesus monkey troop do not call with equal probability upon discovering high-quality foods. Adult males are less likely to produce food calls compared to adult females, and among those who do call they are more likely to be residents of the group than peripheral males. Among females, those within large matrilines are more likely to produce food calls compared to females in smaller matrilines.[21] What happens when a rhesus monkey finds good food and fails to give a food call? If such cheating is discovered by a conspecific, the result is increased aggression from group members, regardless of the cheater's rank.[22]

In cotton-top tamarins (*Saguinus oedipus*), a New World monkey, food calling may serve slightly different functions and may depend on social conditions. Tamarin social groups include a stable, monogamous pair with one or more generations of offspring. They produce two calls in their repertoire that are exclusively associated with food-related activities: the C-chirp and the D-chirp.[23] C-chirps are produced as they approach a food source, and D-chirps are produced during feeding. In a series of experiments on captive cotton-top tamarins, Roush and Snowdon[24] tested the influence of food types, food quantity, food distribution, and audience effects on food calling rates. While neither food quantity nor distribution affects the calling rate, adult mate-pairs call at a higher rate than immature individuals or mate-pairs with offspring.[24] The presence or absence of a mate-pair in the vicinity of the food has no effect on food calling rates among discovers, suggesting that there is no audience effect on food calling. These findings for the cotton-top tamarin are in contrast to the results reported by Caine et al.[25] for the closely related red-bellied tamarin (*Saguinus labiatus*). In this species, increased calling rates were associated with increased food palatability and quantity. Furthermore, the visibility of conspecifics also influenced calling rates: When conspecifics were out of sight, food discoverers gave more food calls.[25]

For all four species described above, food calls ultimately serve as indirect auditory cues, attracting conspecifics to a food source; however, there are important differences between species, even within a genus. Within *Macaca*, toque macaques give food calls in response to quantity as opposed to food type, while rhesus macaques produce food calls upon the discovery of rare, high-quality foods. Within *Saguinus*, red-bellied tamarins adjust their food call rates according to the quantity and palatability of the discovered food and whether or not conspecifics are out of sight. Cotton-top tamarin food calling behavior, on the other hand, reveals no such distinctions.

IV. FINDING (AND KEEPING) A MATE

A. COPULATION CALLS

Although avoiding predators and locating food are essential tasks for survival, a primate's ultimate reproductive success requires finding high-quality mates. In a wide variety of mating systems, mate choice is based on an assessment of auditory signals presumed to correlate with fitness. For example, for one species of frog (*Physalaemus pustulosus*), the fundamental frequency of the male advertisement call (which correlates positively with body size) is used by females to select the biggest male as a mate.[26] In another species of frog (*Hyla versicolor*), the duration of the advertisement call is used as an indicator of male genetic quality: longer duration calls are preferred by females, and males with the longest calls sire the fittest offspring.[27]

In many Old World monkey and ape species, particularly those with multi-male mating systems, individuals (males, females, or both sexes) produce copulation calls.[28,29] These calls are produced immediately before, immediately after, or during copulation and serve as auditory cues for reproductive status. In rhesus monkeys, only males produce copulation calls and always during copulation.[30] Whether or not their production is under voluntary control, they could be used as indicators of male quality. Three pieces of evidence support this idea.[30] First, the number of calling males decreases with increased competition for estrous females. Second, males, independent of rank, who produce copulation calls receive more aggression from conspecifics. Finally, copulation-calling males have a greater mating success (in terms of number of copulations) than silent males. In essence, males who call are more fit because they can withstand the aggression of other males following their copulation calls. However, it is not known whether females show preferences among the males who do call or what acoustic features they might be using as cues.

Semple[31,32] has studied the function of female copulation calls in Barbary macaques (*Macaca sylvanus*). Following playbacks of their copulation calls, Barbary macaque females were mated sooner than following a control playback. Playbacks to male dyads revealed that only the higher ranking of the two would approach the sound source, while the other male stayed behind. These results suggest that these Barbary macaque copulation calls provide an indirect mechanism of female choice (because females end up with the higher ranking male more often) and promote sperm competition by reducing the interval between matings.[31] Of course, a male Barbary macaque looking for a suitable mate would be wise to select, and fight for, a mate who is at the peak stage of fertility. Indeed, male Barbary macaques can actually distinguish the reproductive states of conspecific females based on voice alone.[32] Playbacks of female copulation calls produced during late estrus (when she is most likely to ovulate) elicited stronger responses from males than calls produced during early estrus. The dominant frequency and/or the duration of the copulation call are two acoustic cues males may use to discern the reproductive states of females. Late estrus calls had longer durations and higher dominant frequencies than early estrus calls.[32] Taken together, the data from both rhesus and Barbary macaques suggest that these primates can and do use copulation calls to influence their reproductive success.

B. Primate Songs

Singing has evolved four times in the primate lineage and is represented by 26 species of primates.[33] All of them have monogamous mating systems. What is a primate song? Like birdsong, it is defined as a concatenation of different notes that together form a recognizable pattern over time; they can last for several tens of minutes. Both males and females can sing in some species, and sometimes they sing together in duets. Also like birdsong, primate songs are thought to have two main functions: to defend territories and to strengthen pair bonds. For strengthening pair bonds, there has been scant supporting evidence,[34] but recent studies of gibbon singing have provided some new insights.[35,36]

Geissman and colleagues studied the duet songs of one species of gibbon, the siamang (*Hylobates syndactylus*).[35,36] Duet songs are mostly produced by mated pairs and are exquisitely timed vocal exchanges between mates using components of their individual songs.[33] As a male sings his song, the female will insert, at regular intervals, long parts of her song. How does each member of the pair know when to start and stop singing? If such song coordination had to be learned (an investment of time and energy), then this would support the idea that singing (and duetting, in particular) can function to strengthen pair bonds.[37] To test this idea, the changes in duet structure were examined for two pairs of siamangs during a forced partner exchange.[35] When the new pairs formed, they had difficulties in synchronizing their duets and produced atypical (relative to their song behavior before the partner exchange) sequences of calls. Thereafter, the two newly formed pairs showed changes in their duetting behavior that can only be interpreted as a learning effort: one partner would adapt its singing behavior to compensate for the mate's song structure. For example, a male in one pair dropped the last syllable of one of his songs so that it did not overlap with a similar sounding syllable sung by the female.[35]

If duetting strengthens pair bonds, then one would expect that the intensity of duetting (number of duets per day and duration of duets) should be correlated with other measures of pair bond strength, such as grooming, interindividual distance, and behavioral synchronization (e.g., eating or resting at the same time). In siamangs, both the number of duets produced each day and the average duration of a duet were positively correlated with the frequency of grooming and negatively correlated with interindividual distance.[36] Thus, for siamangs at least, duetting seems to be important for strengthening pair bonds.

V. A SOCIOECOLOGICALLY SENSIBLE AUDITORY NEUROSCIENCE

In the previous sections, we provided a broad overview of three adaptive problems that primates face and ways in which different primate species use audition to solve those problems. Primate audition, then, can be seen not simply as a set of general perceptual mechanisms but also as a suite of specifically tailored, species-specific auditory systems which, like those of bats and birds, are built to solve particular ethological problems. As such, primate auditory mechanisms must be specialized to attend to particular acoustic features, namely those features that are relevant to particular adaptive problems. The task for those interested in auditory processing, then, is to identify first the acoustic features that are relevant for particular adaptive problems and use this information to examine how neural mechanisms are able to extract out such cues.[38] Toward this goal, recent playback experiments have investigated the acoustic features rhesus monkeys and cotton-top tamarins use to recognize conspecific vocalizations.

A. Temporal Cues in Rhesus Monkey Calls

To date, only two field playback studies have explicitly explored the acoustic features that primates use to distinguish conspecific from nonconspecific sounds.[39,40] Building upon their initial findings using the head-orienting bias as an assay,[41] Hauser and colleagues[39] tested the role of interpulse interval on call recognition in rhesus monkeys (*Macaca mulatta*). Playback experiments using temporally manipulated exemplars of three different call types — shrill barks, grunts, and copulation calls — demonstrated that expanding or contracting the interpulse interval beyond the species-typical range eliminated the right-ear orienting bias normally seen for shrill barks and grunts, but not for copulation screams. Sensitivity to the shape of the amplitude envelope in rhesus monkey shrill barks and harmonic arch calls was tested by using time-reversed calls and the head-orienting assay.[40] Reversing the calls changed their temporal structure without affecting their spectral content. Rhesus monkeys responded with left-ear leading head turns in response to time-reversed exemplars (Figure 1.1), demonstrating that temporal cues in the amplitude envelope are important for recognition.

B. Temporal Cues in Tamarin Long Calls

Like many primates, cotton-top tamarins produce loud, multisyllabic calls dubbed *long calls*.[42] Upon hearing long calls, conspecifics will reliably respond with their

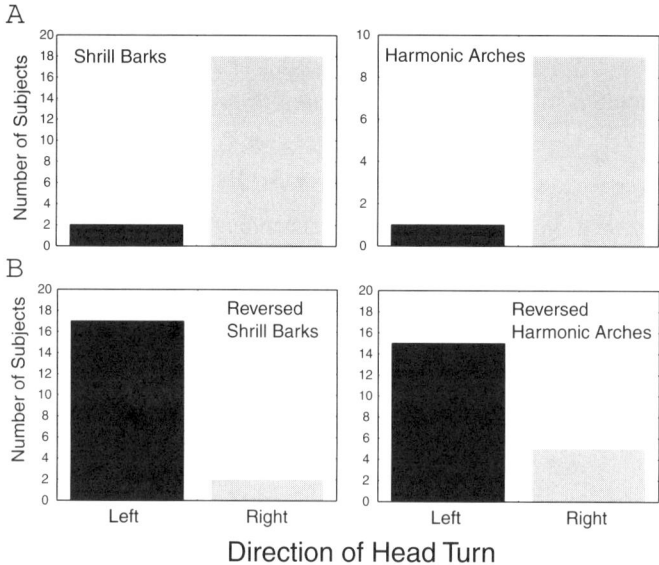

FIGURE 1.1 (A) Number of subjects orienting either left or right in response to normal calls. (B) Number of subjects orienting either left or right in response to reversed calls. On the *x*-axis, black bars represent left head turns, gray bars represent right head turns; *y*-axis, number of subjects.

own antiphonal long calls. Beyond the acoustic classification of these calls and descriptions of the behavioral context in which they are produced,[23] no experimental studies have examined how species-typical information is encoded in the structure of these signals until recently.[43,44] To isolate the mechanisms underlying the production of long calls in tamarins, combinations of naturally produced and experimentally manipulated long calls were used as playback stimuli. These stimuli were presented to individuals isolated from their group, and the relationship between signal design and the rate of antiphonal long call production by the test subject was measured. Tamarins antiphonally called preferentially to whole calls compared to signals composed of one of the two syllable types present in long calls.[43] Tamarins did not distinguish between normal calls and time-reversed or pitch-shifted long calls, but normal response rates did require the species-typical amplitude envelope.[44] Furthermore, evidence suggested that the number of syllables and the syllable rate may also influence antiphonal calling responses.[44]

Together, these experiments exploring the salient features of rhesus monkey and cotton-top tamarin vocal perception allow one to generate meaningful hypotheses for the neural mechanisms underlying primate communication. For example, in the case of the cotton-top tamarin, one might predict that neurons in the auditory cortex would be combination sensitive to the multisyllabic long call, yet insensitive to the order of syllables within the call. Another candidate mechanism would be neural selectivity to amplitude modulation. One might predict that decreasing the intersyllable interval, or altering the number of syllables in the tamarin long call not only may result in an increase in the number of antiphonal calls by subjects but also may result in an increase in the responsiveness of a subset of auditory neurons to such stimuli. In contrast, eliminating the species-typical amplitude envelope of tamarin long calls should result in a decrease in neural responsiveness.

VI. CONCLUSION

In conclusion, then, we argue that a rich understanding of the mechanisms underlying primate audition can only be achieved by focusing on the adaptive problems that shaped these mechanisms. By identifying the features that are relevant to solving these problems, scientists will be able to examine how primates represent and extract the relevant acoustic features. Such an ethological analysis will shed light on the neural mechanisms subserving these representations and may provide insight into the evolution of auditory capacities across the primate order.

REFERENCES

1. Stanford, C.B., *Chimpanzee and Red Colobus: The Ecology of Predator and Prey*, Harvard University Press, Cambridge, MA, 1998.
2. Cheney, D.L. and Seyfarth, R.M., *How Monkeys See the World*, University of Chicago Press, Chicago, IL, 1990.
3. Struhsaker, T.T., Auditory communication among vervet monkeys, in *Social Communication among Primates*, S.A. Altmann, Ed., University of Chicago Press, Chicago, IL, 1967.
4. Seyfarth, R.M., Cheney, D.L., and Marler, P., Monkey responses to three different alarm calls: evidence for predator classification and semantic communication, *Science*, 210, 801–803, 1980.
5. Cheney, D.L. and Seyfarth, R.M., Social and non-social knowledge in vervet monkeys, *Philos. Trans. Roy. Soc. London, B*, 308, 187–201, 1985.
6. Hauser, M.D., How infant vervet monkeys learn to recognize starling alarm calls, *Behavior*, 105, 187–201, 1988.
7. Ramakrishnan, U. and Coss, R.G., Recognition of heterospecific alarm vocalizations by bonnet macaques (*Macaca radiata*), *J. Comp. Psychol.*, 114, 3–12, 2000.
8. Zuberbühler, K., Interspecies semantic communication in two forest primates. *Proc. Roy. Soc. London B*, 267, 713–718, 2000.
9. Zuberbühler, K., Causal cognition in a non-human primate: field playback experiments with Diana monkeys, *Cognition*, 76, 95–207, 2000.
10. Richards, A.F., *Primates in Nature*, W.H. Freeman, New York, 1985.
11. Napier, J.R. and Napier, P.H., *A Handbook of Living Primates*, Academic Press, New York, 1967.
12. Murray, P., The role of cheek pouches in cercopithecine monkey adaptive strategy, in *Primate Functional Morphology and Evolution*, R.H. Tuttle, Ed., Mouton, The Hague, 1975, pp. 151–194.
13. Strier, K.B., *Primate Behavioral Ecology*, Allyn & Bacon, Boston, MA, 2000.
14. Dominy, N.J., Lucas, P.W., Osorio, D., and Yamashita, N., The sensory ecology of primate food perception, *Evolutionary Anthropology*, 10, 171–186, 2001.
15. Fleagle, J.G., *Primate Adaptation and Evolution*, 2nd ed., Academic Press, San Diego, CA, 1999.

16. Erickson, C.J., Percussive foraging in the aye-aye, *Daubentonia madagascarensis*, *Animal Behav.*, 41, 793–801, 1991.
17. Charles-Dominique, P., *Ecology and Behaviour of Nocturnal Primates*, Columbia University Press, New York, 1977.
18. Boesch, C. and Boesch-Achermann, H., *The Chimpanzees of the Taï Forest*, Oxford University Press, London, 2000.
19. Bshary, R. and Noë, R., Anti-predation behaviour of red colobus monkeys in the presence of chimpanzees, *Behav. Ecol. Sociobiol.*, 41, 321–333, 1997.
23. Cleveland, J. and Snowdon, C.T., The complex vocal repertoire of the adult cotton-top tamarin (*Saguinus oedipus oedipus*), *Z. Tierpsychol.*, 58, 231–270, 1982.
20. Dittus, W.P.J., Toque macaque food calls: semantic communication concerning food distribution in the environment. *Animal Behav.*, 32, 470–477, 1984.
21. Hauser, M.D. and Marler, P., Food-associated calls in rhesus macaques (*Macaca mulatta*). I. Socioecological factors, *Behav. Ecol.*, 4, 194–205, 1993.
22. Hauser, M.D. and Marler, P., Food-associated calls in rhesus macaques (*Macaca mulatta*). II. Costs and benefits of call production and suppression, *Behav. Ecol.*, 4, 206–212, 1993.
24. Roush, R.S. and Snowdon, C.T., Quality, quantity, distribution and audience effects on food calling in cotton-top tamarins, *Ethology*, 106, 673–690, 2000.
25. Caine, N.G., Addington, R.L., and Windfelder, T.L., Factors affecting the rates of food calls given by red-bellied tamarins, *Animal Behav.*, 50, 53–60, 1995.
26. Ryan, M.J., Female mate choice in a neotropical frog, *Science*, 209, 523–525, 1980.
27. Welch, A.M., Semlitsch, R.D., and Gerhardt, H.C., Call duration as an indicator of genetic quality in male gray tree frogs, *Science*, 280, 1928–1930, 1998.
28. Hauser, M.D., *The Evolution of Communication*, MIT Press, Cambridge, MA, 1996.
29. Semple, S., Female Copulation Calls in Primates, Ph.D. thesis, University of Sussex, U.K., 1998.
30. Hauser, M.D., Rhesus monkey copulation calls: honest signals for female choice?, *Proc. Roy. Soc. London B*, 254, 93–96, 1993.
31. Semple, S., The function of Barbary macaque copulation calls, *Proc. Roy. Soc. London B*, 265, 287–291, 1998.
32. Semple, S. and McComb, K., Perception of female reproductive state from vocal cues in a mammal species, *Proc. Roy. Soc. London B*, 267, 707–712, 2000.
33. Geissmann, T., Gibbon songs and human music from an evolutionary perspective, in *The Origins of Music*, N.L. Wallin, B. Merker, and S. Brown, Eds., MIT Press, Cambridge, MA, 2000.
34. Cowlishaw, G., Song function in gibbons, *Behaviour*, 121, 131–153, 1992.
35. Geissmann, T., Duet songs of the siamang, *Hylobates syndactylus*. II. Testing the pair-bonding hypothesis during a partner exchange, *Behaviour*, 136, 1005–1036, 1999.
36. Geissmann, T. and Orgeldinger, M., The relationship between duet songs and pair bonds in siamangs, *Hylobates syndactylus*, *Animal Behav.*, 60, 805–809, 2000.
37. Wickler, W., Vocal duetting and the pair bond. I. Coyness and partner commitment: a hypothesis, *Z. Tierpsychol.*, 52, 201–209, 1980.
38. Ghazanfar, A.A. and Hauser, M.D., The auditory behaviour of primates: a neuroethological perspective, *Curr. Opin. Neurobiol.*, 11, 712–720, 2001.
39. Hauser, M.D., Agnetta, B., and Perez, C., Orienting asymmetries in rhesus monkeys: effect of time-domain changes on acoustic perception, *Animal Behav.*, 56, 41–47, 1998.
40. Ghazanfar, A.A., Smith-Rohrberg, D., and Hauser, M.D., The role of temporal cues in conspecific vocal recognition: rhesus monkey orienting asymmetries to reversed calls, *Brain Behav. Evolution*, 58, 163–172, 2001.

41. Hauser, M.D. and Andersson, K., Left hemisphere dominance for processing vocalizations in adult, but not infant, rhesus monkeys: field experiments, *Proc. Natl. Acad. Sci.*, 91, 3946–3948, 1994.
42. Miller, C.T. and Ghazanfar, A.A., Meaningful acoustic units in nonhuman primate vocal behavior, in *The Cognitive Animal*, C. Allen, M. Bekoff, and G.M. Burghardt, Eds., MIT Press, Cambridge, MA, 2002.
43. Ghazanfar, A.A., Flombaum, J.I., Miller, C.T., and Hauser, M.D., The units of perception in the antiphonal calling behavior of cotton-top tamarins (*Saguinus oedipus*): playback experiments with long calls, *J. Comp. Physiol. A*, 187, 27–35, 2001.
44. Ghazanfar, A.A., Smith-Rohrberg, D., Pollen, A., and Hauser, M.D., Temporal cues in the antiphonal calling behaviour of cotton-top tamarins (*Saguinus oedipus*), *Animal Behav.*, in press, 2002.

2 Causal Knowledge in Free-Ranging Diana Monkeys

Klaus Zuberbühler

CONTENTS

I. Introduction 13
II. Philosophical and Empirical Approaches to Causality 14
III. The Evolution of Causal Cognition 14
    A. Physical Causation 15
    B. Social Causation 15
IV. Causal Problems in the Predation Context 15
    A. Understanding Predator Behavior 16
    B. Understanding Other Monkeys' Alarm Calls 18
    C. A Simple Syntactic Rule 19
    D. Understanding Nonprimates' Alarm Calls 22
V. Causal Understanding in Non-Human Primates 24
Acknowledgments 25
References 25

I. INTRODUCTION

Causal reasoning is a defining competence of the human mind that calls for a detailed account of the evolutionary emergence of this particular cognitive ability. So far, research on causal understanding in non-human primates has suggested that a primate's understanding of events is probably not based on a generalized notion of causality but rather on memories of individualized antecedent–consequent patterns. This conclusion is primarily based on the performance of non-human primates in tool experiments, which is somewhat problematic because most primates do not habitually use tools in the wild. However, when causal problems are posed in the social domain (for example, through the behavior of conspecifics), then primates seem to perform better. This chapter reviews a series of recent playback experiments conducted on free-ranging Diana monkeys in their natural forest habitat in Western Ivory Coast. The experiments were designed to investigate this primate's ability to assess behavioral causation in the context of predation. Results suggest that, in some instances, the monkeys responded to experimental event patterns as if they had a generalized notion of causality. In other cases, however, their responses can be explained more parsimoniously — that is, as the result of previously learned antecedent–consequent patterns.

II. PHILOSOPHICAL AND EMPIRICAL APPROACHES TO CAUSALITY

In humans, causal reasoning is reliably triggered by surprising, unexpected events, which typically induce a search for the corresponding antecedent events as well as possible consequences. When humans engage in such reasoning, they implicitly assume that the outside world has a causal structure whose relevant participants can be identified. This reasoning relies on memories of how the same or similar objects or events have co-occurred in space and time previously, and how the order was sequenced. Thus, David Hume and other philosophers have long suspected that this form of thinking, reasoning about causes, is firmly grounded in the laws of association; consequently, they have proposed that causality is a purely mental phenomenon that arises in the human mind in response to some highly specific event patterns.[1] Causality, thus, cannot be established by either rational analysis or higher metaphysics; habit or custom sufficiently accounts for the sensation that everything that begins must have a cause and that similar causes must have similar effects. Causality, in other words, is a psychological phenomenon and thus must be explained psychologically.[2]

Research on causal cognition has traditionally distinguished two kinds of causality, although they may be linked to the same mental process.[3] The first kind is reliably invoked when subjects observe certain physical or impulse phenomena, such as one billiard ball bumping into another.[4] A second kind is social in nature and arises when animate beings behave or interact in specific ways.[5] This line of research has identified the conditions required for an individual to perceive behavior as internally caused. Intention has been identified as the main underlying force, in some ways the social analog to the invisible physical forces that are perceived as impulse causality. In either case, a sensation of causality does not arise from the objects themselves but from their unique underlying relationship. The human mind thus seems to possess a special mental concept, the "causality concept," which gets activated when event patterns of a certain kind are detected. The causality concept allows humans to predict future effects from current constellations and to infer their associated antecedent events.

III. THE EVOLUTION OF CAUSAL COGNITION

From an evolutionary perspective, it is of interest whether non-human primates have a similar propensity to grant special status to events that humans judge as causal. Does the primate mind similarly perceive repeated, contiguous sequences of events as belonging to a special category, and does this category become activated when encountering certain novel problems, when making behavioral decisions, or when judging other subjects' behaviors? Can a primate understand antecedent–consequent sequences as the outcome of unobservable, mediating forces; or, alternatively, is primate understanding situationally bound and limited to specific and individually experienced stimulus constellations, as Tomasello,[6] for example, has put it?

A. Physical Causation

A review of the primate literature on causal cognition generates a largely negative picture.[6] In several studies, non-human primates behaved as if they did not really understand why things happened, both in the physical and in the psychological domain. When captive chimpanzees and capuchins (the only two primates that habitually use tools in the wild) were tested on their understanding of physical causality, individuals did not exhibit a very rich understanding of the physical forces involved.[7,8] Subjects were unable to predict the effects of gravity, a powerful and omnipresent force in the physical world, when pushing a piece of food through a transparent tube with a stick. Instead, they behaved as if they only knew the results of the actions of a familiar tool. A recent review by Povinelli[9] of his own work draws a similar picture of the causality-blind, tool-using chimpanzee. However, based on their spontaneous abilities and behavior in the wild, it is clear that more research is needed; chimpanzees regularly innovate novel tool techniques to obtain food, which requires at least some understanding of the physical forces involved.[10]

B. Social Causation

In the psychological domain, humans typically identify *intention* as the main causal force that drives the behavior of other individuals. Intention is assigned widely and quickly, sometimes even to inanimate objects such as computers or cars. Non-human primates clearly understand others as animate and separate beings with their own behavior, which can be manipulated with communicative signals,[11] but they probably do not understand each other as intentional beings.[3] Studies aimed at testing for primates' abilities to respond to the intentions of others have either yielded ambiguous results or were difficult to interpret because they involved enormous training efforts.[12–14] One field experiment conducted on free-ranging baboons (*Papio cynocephalus ursinus*), however, has shown that these animals may perceive or even recognize cause-and-effect relations in the context of social interactions.[15] In this species, dominance relationships are partially mediated by two kinds of vocalizations, the *grunts* given by a female to lower ranking group members and the *fear barks* given to higher ranking ones. Through the use of a playback experiment, the study showed that causally inconsistent call sequences — a higher ranking animal responding with fear barks to a lower ranking animal's grunts — elicited stronger responses in recipients than control sequences that were made causally consistent.

IV. CAUSAL PROBLEMS IN THE PREDATION CONTEXT

Primates do not possess specialized devices or weapons to protect themselves against predators. Instead, their success in avoiding predation largely depends on accurately interpreting and responding to cues that predict predator attacks. Because most predators' hunting success depends on the prey being unaware of their presence, primates and their predators find themselves in an arms' race of detection and concealment abilities. As a result, primates often have to deal with incomplete cause-and-effect sequences in the predation context, and individuals with improved reasoning abilities are likely to have a selective advantage over others. This makes predation particularly interesting for research on the evolution of causal understanding.

Instead of tool-using or mind-reading paradigms, the remainder of this chapter focuses on primate problem-solving abilities investigated by using acoustic stimuli related to predator presence. The author has conducted experiments in the Taï National Park of Western Ivory Coast with wild groups of Diana monkeys (*Cercopithecus diana*), using their own alarm calls, as well as those of sympatric Campbell's monkeys (*Cercopithecus campbelli*) and of other species. The Taï rainforest is characterized by extremely poor visual conditions; consequently, monkeys often find themselves in complete visual isolation from other group members. Reliance on acoustic information about ongoing events thus becomes crucial under these circumstances. Diana monkeys are very attentive to acoustic stimuli and use an elaborate vocal system to protect themselves against predation. Group members constantly exchange a specific call, the *clear call*, which informs other group members whether or not important changes in the environment have occurred, similar to the watchman's song in some bird species.[16,17]

Taï monkeys are hunted by leopards (*Panthera pardus*), crowned-hawk eagles (*Stephanoaetus coronatus*), chimpanzees (*Pan troglodytes*), and human poachers. In response to these predators, the monkeys react with predator-specific defense strategies, which strongly rely on vocal behavior. When detecting a leopard, for example, Diana monkeys break into a medley of loud and conspicuous alarm calling, and often the entire group approaches the predator in the lower forest canopy. This behavior communicates to stealth-hunting leopards that they have been detected and that further hunting will not be successful. Analyses of the ranging behavior of radio-tracked leopards have shown how effective primate alarm calls are in deterring leopards. Minutes after a monkey group had detected and alarm called to a hiding leopard, the leopard typically moved on and left the area.[18]

A. Understanding Predator Behavior

The chimpanzees in the Taï forest and in other places in Africa are notorious monkey predators,[19] but they elicit a different response from Taï monkeys than do leopards or eagles. In this case, the monkeys flee rapidly and hide silently somewhere in the upper forest canopy. Again, this response is highly adaptive as chimpanzees follow monkeys into the trees to hunt them in a seemingly cooperative manner.[20] Interestingly, the monkeys' response to humans is very similar.[21] Conspicuous vocal behavior in the presence of these two primate predators simply increases an individual's risk of being singled out and killed; however, Diana monkeys must also take into account the fact that chimpanzees themselves occasionally fall prey to leopards.[22] When encountering a leopard, chimpanzees produce loud and conspicuous alarm screams, also termed *SOS screams*,[19] which draw other group members to the site. Acoustically, the chimpanzee alarm screams differ from other types of chimpanzee screams, such as screams given during social conflicts. Taï chimpanzees do not seem to give these alarm screams to other predators, such that this call essentially functions as a leopard alarm call. When hearing chimpanzee alarm screams, therefore, Diana monkeys have the opportunity to infer the presence of a leopard, in addition to knowing that chimpanzees are present.

To investigate whether the monkeys are in fact able to attend to the referential information inherent in these calls, recordings of chimpanzee social screams (indicating the presence of chimpanzees) and alarm screams (indicating the presence of chimpanzees and a leopard) were played back to different groups of Diana monkeys. Diana monkeys responded to the chimpanzee social screams as if a group of chimpanzees were present (Figure 2.1). The monkeys silently disappeared and did not give any leopard alarm calls. Individuals tested with leopard growls responded as if a leopard were present, giving loud alarm calls and in some cases approaching the hidden speaker, seemingly looking for the leopard. Finally, in about half of all the groups tested, individuals responded to chimpanzee alarm screams as if a leopard were present, while the other half responded as if chimpanzees were present. This finding was unexpected and called for further investigation.

Chimpanzees spend most of their time in a fairly small core area within their 20- to 30-km² territory.[24] Further analysis of our data showed that Diana monkey groups living in the core area of a chimpanzee territory were more likely to give leopard alarm calls to chimpanzee alarm screams than groups living in the periphery.

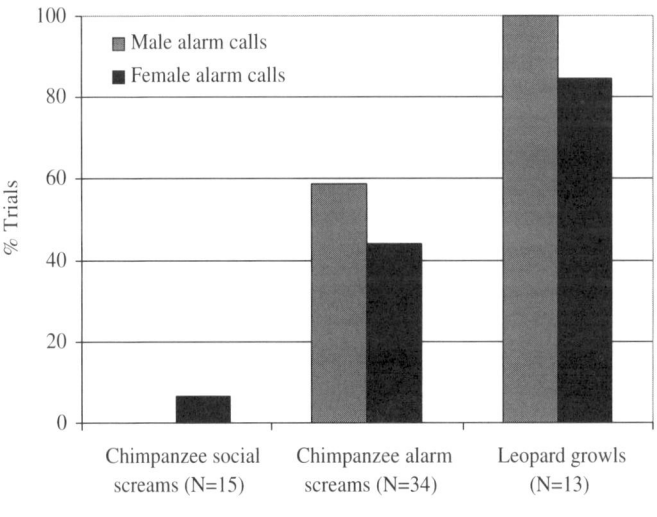

FIGURE 2.1 Alarm-call response of different groups of Diana monkeys to chimpanzee social screams, chimpanzee alarm screams, and leopard growls. Bars represent the relative number of Diana monkey groups in which at least one leopard alarm call was given by the adult male (dotted bars) or one of the adult females (black bars). (Data from Reference 23.)

Presumably, peripheral groups encountered chimpanzees much less often and were therefore less familiar with the chimpanzees' vocal repertoire. These groups responded to the chimpanzee alarm screams as if a chimpanzee group was present. Peripheral groups simply did not understand that a leopard could have caused the chimpanzee alarm screams. To test the causal knowledge hypothesis in more detail, Diana monkey groups were primed with a brief recording of a leopard growl, a chimpanzee alarm scream, or a chimpanzee social scream. Five minutes later, the same group heard a probe stimulus (a leopard growl) from the same location. For those groups that had heard the leopard as a prime stimulus before, this meant redundant information, simply indicating that the leopard was still present. For those groups primed with social screams, however, leopard growls represented unexpected and novel information and monkeys were therefore expected to react strongly. Finally, for the groups primed with chimpanzee alarm screams, leopard growls represented either novel information (if they did not understand the meaning of the calls) or redundant information (if they understood the meaning of the calls). When a Diana monkey heard a chimpanzee giving alarm calls to a leopard, the monkey was presented with a problem whose solution required some knowledge of the relationship "leopard → chimpanzee alarm scream," which, from the human perspective, is a cause-and-effect relation in the biological domain. Subjects perceived only the second part of the causal chain (the chimpanzees' alarm calls) and had to infer its cause in order to respond properly. Knowledgeable humans presented with a similar problem use causal reasoning to infer that the chimpanzees must have detected a leopard. When hearing chimpanzees' alarm screams, some Diana monkeys responded as if a leopard were present while others behaved as if the calls signaled only the presence of chimpanzees. Apparently, some subjects sensed the relation between chimpanzees' alarm screams and leopards, while others did not. It appears that these later groups took the chimpanzee alarm screams as a sign of the presence of a chimpanzee, unable to understand that these calls also indicated the presence of a leopard, as illustrated by the alarm call behavior of the males (Figure 2.2).

B. Understanding Other Monkeys' Alarm Calls

Chimpanzees are not the only species that provide Diana monkeys with vital information about ongoing events in the environment. Diana monkeys frequently associate with other primate species, including the Campbell's monkeys, and form large mixed-species groups, presumably to better protect themselves against predation.[25] Campbell's and Diana monkeys are hunted by the same predators, and the single males of both species produce predator-specific alarm calls.[21,26,27] A playback experiment confirmed that when hearing recordings of the predator-specific Campbell's monkey alarm calls, the Diana monkeys responded as if the corresponding predator were present. Adult females, for example, produced the same kind of vocal behavior, regardless of whether they heard the actual predator or the corresponding male Campbell's monkey alarm calls (Figure 2.3).

Experiments using a prime-probe technique akin to the previous experiments showed that when hearing Campbell's monkeys' eagle or leopard alarm calls, Diana monkeys attend to the meaning associated with these calls.[28] For example, when the

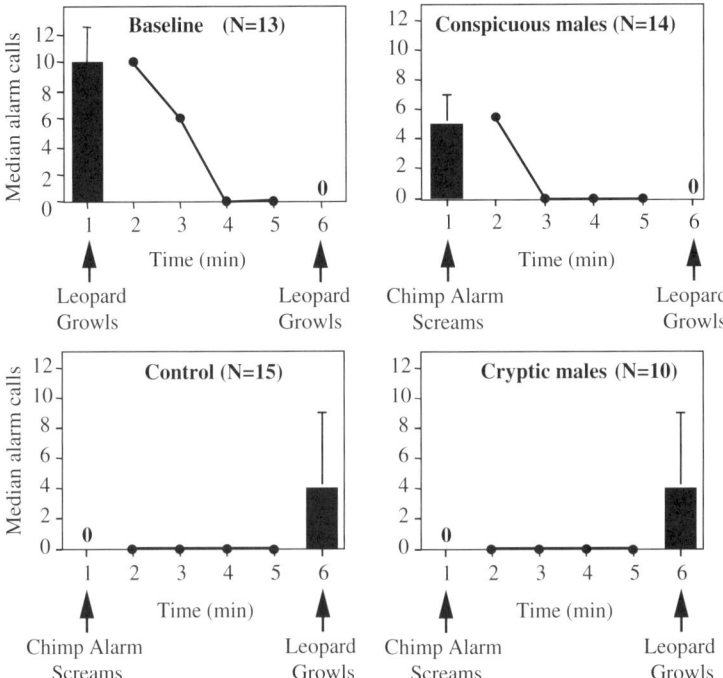

FIGURE 2.2 Alarm-call responses of male Diana monkeys to leopards after being primed with leopard growls, chimp alarm screams, or chimp social screams (median number of leopard alarm calls and third quartile). (Data from Reference 23.)

monkeys heard a series of Campbell's monkeys' eagle alarm calls, they did not show much response to playbacks of subsequent eagle shrieks, even though this stimulus was acoustically novel and normally highly efficient in eliciting vocal responses. In other words, the Diana monkeys behaved as if they understood that only the Campbell's monkeys' eagle alarm calls could indicate the presence of a crowned-hawk eagle (Figure 2.4) and were not surprised when hearing the calls of the eagle from the same direction.

C. A Simple Syntactic Rule

In addition to the two alarm calls described before, male Campbell's monkeys possess a third type of loud call, a brief and low-pitched *boom* vocalization. This call is always given in pairs separated by some seconds of silence and typically precedes an alarm call series by about 25 seconds. Boom-introduced alarm call series are given to a number of disturbances, such as a falling tree or large breaking branch, the far-away alarm calls of a neighboring group, or the presence of a predator. Unlike when callers are surprised by a close predator, the contexts in which Campbell's monkeys produce boom-introduced alarm calls do not involve a direct threat. The Campbell's monkeys' alarm calls following the booms show some structural similarities to the Campbell's monkeys' leopard alarm calls, but they might consist of a

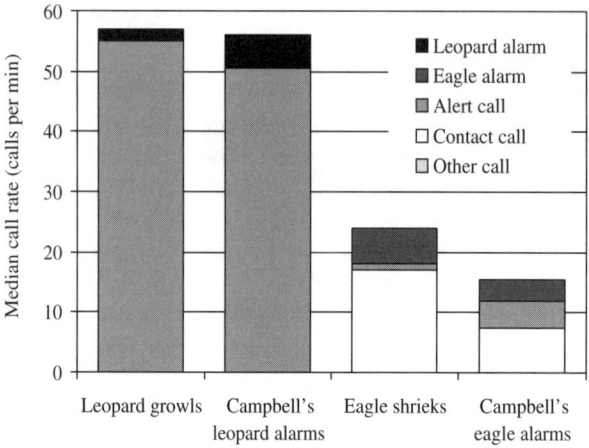

FIGURE 2.3 Vocal behavior of female Diana monkeys after hearing vocalizations of a leopard (16 groups), a crowned-hawk eagle (17 groups), or Campbell's monkey alarm calls to leopard (14 groups) or crowned-hawk eagle (12 groups). (Data from Zuberbühler, K., *Proc. Roy. Soc. London B*, 267, 713–718, 2000.)

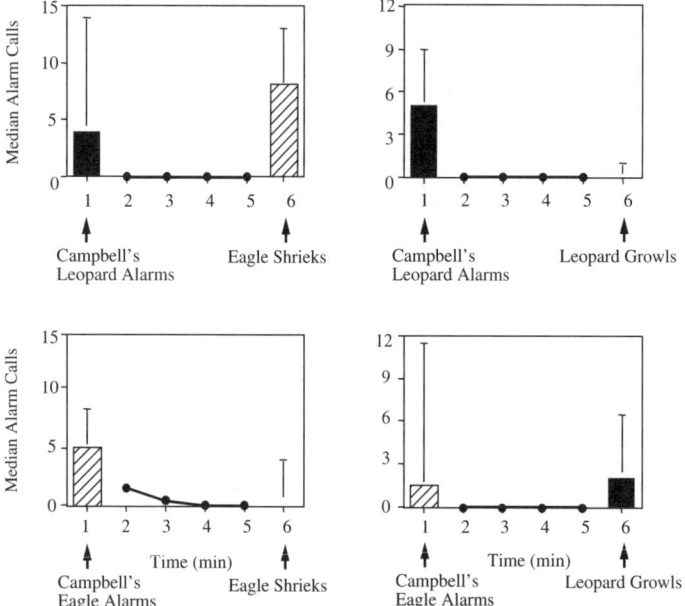

FIGURE 2.4 Alarm-call responses of female Diana monkeys to leopards and crowned-hawk eagles after being primed with Campbell's monkeys' eagle or leopard alarm calls (median number of calls and third quartile). Solid bars indicate female leopard alarm calls; striped bars indicate female eagle alarm calls. (Data from Reference 28.)

distinct class. When hearing boom-introduced Campbell's monkey alarm calls, Diana monkeys do not respond with their own alarm calls, which contrasts sharply with their vocal response to normal (that is, boom-free) Campbell's monkey's leopard or eagle alarm calls (see results in Figure 2.3).

To investigate whether or not monkeys understood the semantic changes caused by the presence of boom calls, the following playback experiments were conducted. In two baseline conditions, different Diana monkey groups heard a series of five male Campbell's monkey alarm calls given to a crowned-hawk eagle or a leopard. Subjects were expected to respond strongly — that is, to give many eagle or leopard alarm calls, as in the previous experiments (Figures 2.3 and 2.4). In the two test conditions, different Diana monkey groups heard playbacks of the exact same Campbell's monkey alarm call series, but this time two booms were artificially added 25 seconds before the alarm calls. If Diana monkeys understood that the booms acted as modifiers to affect the semantic specificity of subsequent alarm calls, then they should give significantly fewer predator-specific alarm calls in the test conditions compared to the baseline conditions.

Results of this experiment replicated the natural observations. Playbacks of Campbell's monkeys' eagle alarm calls caused the Diana monkeys to produce their own eagle alarm calls, while playbacks of Campbell's monkeys' leopard alarm calls caused them to give leopard alarm calls. Playback of booms alone did not cause any noticeable change in Diana monkey vocal behavior, but the boom calls had a significant effect on how the monkeys responded to subsequent Campbell's monkey alarm calls: Campbell's monkeys' leopard alarms no longer elicited leopard alarm calls, while Campbell's monkeys' eagle alarms no longer elicited eagle alarm calls in Diana monkeys. This study showed that the Campbell's monkey booms affected the way the Diana monkeys interpreted the meaning of subsequent Campbell's monkey alarm calls (Figure 2.5). These booms seemed to indicate that whatever message followed did not require the normal immediate antipredator response.

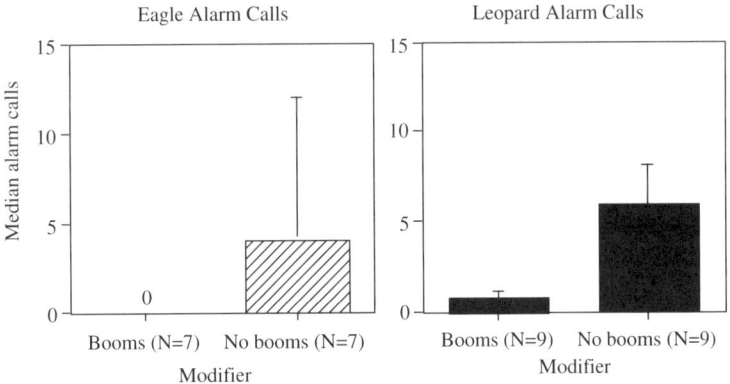

FIGURE 2.5 Alarm-call response of female Diana monkeys to Campbell's monkeys' eagle and leopard alarm calls, with or without preceding Campbell's monkeys' boom calls (median number of calls and third quartile). Solid bars indicate female leopard alarm calls; striped bars indicate female eagle alarm calls. (Data from Reference 29.)

Experimentally adding booms before an alarm call series created structurally more complex utterances with different meanings than the alarm calls alone. Judging from the Diana monkeys' response to these playback stimuli, the booms have actively modified the meaning of the subsequent alarm call series and transformed them from highly specific predator labels, requiring immediate antipredator responses, into more general signals of disturbance that do not require any direct responses.

D. Understanding Nonprimates' Alarm Calls

The playback stimuli used so far shared highly specific referential properties in which an acoustically distinct alarm call type was associated with one particular type of predator (e.g., Campbell's monkeys' eagle alarm call and the crowned-hawk eagle). Once learned, these relations do not require much flexibility on behalf of the monkeys, as one specific call type always refers to the same external event. However, there are a number of alarm signals that have more vague and ambiguous referents, which require the monkeys to take into account additional information before being able to respond adaptively. One good example is provided by the predator alarm call of the ground-dwelling crested guinea fowl (*Guttera pulcheri*). These birds forage in large groups and when chased produce conspicuously loud and rattling sounding alarm calls that can be heard over long distances.[30] Guinea fowls are not hunted by chimpanzees but may be taken by leopards and human poachers. Diana monkeys respond strongly to these alarm calls as if a leopard were present.

The following playback experiment compared the responses of different groups of Diana monkeys to recordings of guinea fowl alarm calls and leopard growls. To address concerns about Diana monkeys giving a leopard-type response to *any* loud stimulus from the ground, recordings of the alarm calls of the helmeted guinea fowl (*Numida meleabris*), a closely related species that does not occur in the forest, were also played back as well as recordings of human speech sounds. Results confirmed that Diana monkeys responded to crested guinea fowl alarm calls as if a leopard were present. Recordings of helmeted guinea fowls, however, elicited an unspecific response in Diana monkeys that resembled the Diana monkeys' response to human speech (Figure 2.6). These results were consistent with the idea that Diana monkeys treat Guinea fowl alarm calls as indicative of the presence of a leopard, even though guinea fowls appear to give these calls to any predator chasing them.

Diana monkeys seem to know that several predators can elicit these calls, but that in most cases the calls are caused by the presence of a leopard. To test whether Diana monkeys were able to take this information into account, different groups of Diana monkeys were primed either with leopard growls or human speech to simulate the presence of either of these predators. After a period of five minutes, recordings of guinea fowl alarm calls were played from the same location. If the monkeys were able to link the guinea fowl alarm calls to the presence of either of these two predators, the following should be found: First, groups primed with human speech should remain silent in response to subsequent Guinea fowl alarm calls because (1) humans were the likely cause of the birds' alarm calls, and (2) the appropriate response to humans is to remain silent. Second, groups primed with leopard growls should give alarm calls, because (1) the leopard was the likely cause of the birds' alarm calls, and (2) the appropriate response to leopards is to behave conspicuously.

Results showed that Diana monkeys were able to take these causal relations into account when responding to guinea fowl alarm calls (Figure 2.6). When primed with human speech, Diana monkeys remained mostly quiet to subsequent guinea fowl alarm calls, a behavioral pattern that was not found in groups that were primed with

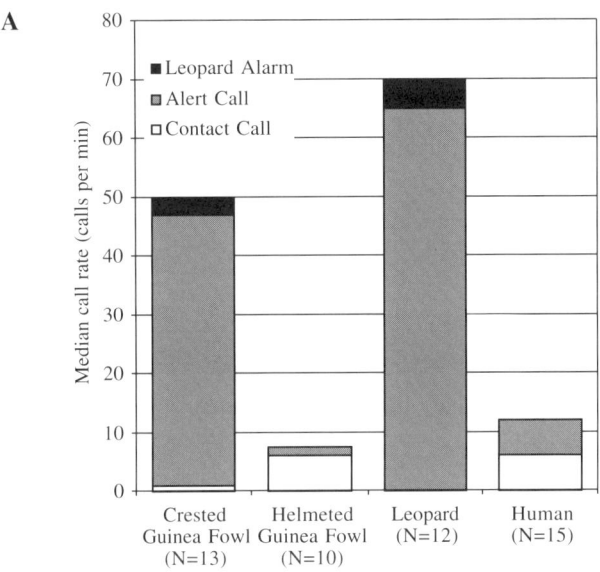

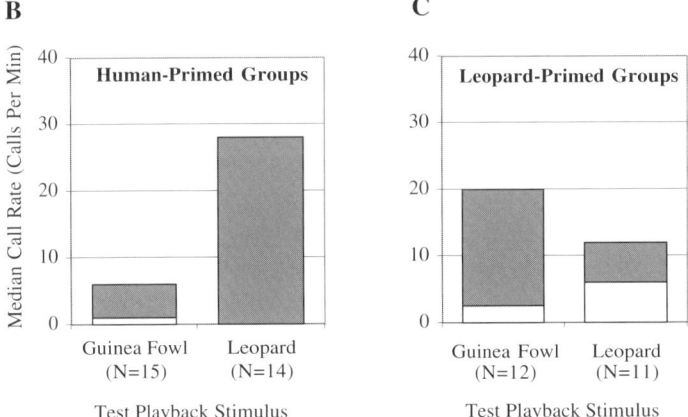

FIGURE 2.6 Top: (A) Vocal response of female Diana monkeys to vocalizations of leopards and humans, and guinea fowl alarm calls. Bottom: Vocal responses of female Diana monkeys to guinea fowl alarm calls and leopard growls after being primed with either human speech (B) or leopard growls (C). (Data from Reference 31.)

leopard growls. This difference did not arise because human-primed groups behaved cryptically to any subsequent stimulus. When leopard growls were played to human-primed groups, the monkeys responded by giving many alarm calls (Figure 2.6). The data suggest that when responding to guinea fowl alarm calls, the Diana monkeys responded to the differing reasons why the birds gave the calls, and not the calls themselves.

V. CAUSAL UNDERSTANDING IN NON-HUMAN PRIMATES

In sum, these experiments have shown that Diana monkeys, and by extension non-human primates, have a remarkable ability to respond adaptively to problems that, according to human judgment, have a causal structure. When hearing an alarm call, the monkeys respond not just to the acoustic features of the alarm calls but also to their associated meanings. Stimuli with very distinct meaning, such as a Campbell's monkey's leopard alarm call lost their effectiveness in eliciting behavior if the information was redundant (that is, if the monkeys were already informed about the leopard's presence) or if the calls were syntactically modified by boom calls. Moreover, the alarm calls did not simply act as a surrogate or substitute for a predator class, as evidenced by their highly flexible responses to guinea fowl alarm calls. In this case, the monkeys seemed to have assigned a default meaning to the calls, which probably coincided with the most likely scenario (i.e., presence of a leopard), but the monkeys were able to flexibly reassign a novel cause to these calls if the available evidence pointed elsewhere. It is clear that humans making causal judgments would behave in an analogous way.

None of the experiments described here, however, can clearly identify the exact mental mechanisms underlying the monkeys' behavior. Although the monkeys acted in many cases as if they attributed cause to events, it is entirely possible that individuals did so because they had a situationally bound understanding that was limited to the existing stimulus constellations, rather than a human-like conception of a cause-and-effect relationship mediating between the stimuli. Having said this, the human concept of causality is not that well understood, either. Attempts have been made to understand why humans are so inclined to explain every event in causal terms even though the real world seems to consist of nothing but covariations of various strengths. One explanation, suggested by Premack and Premack,[5] argues that our predisposition to explain everything in causal terms is due to our early understanding of intention. From an early age, human children are able to read other people's intentions. In addition, they have a natural tendency to comply with these intentions and to carry out the desired actions. With increasing age, children learn to inhibit this natural disposition, probably because they learn to distinguish their own intentions from those of others. Hence, intention is experienced as a most powerful force, ultimately responsible for all observed behavior. According to current theory, non-human primates are not able to distinguish their own intentions from those of a conspecific.[6] Without this ability, primates will be prevented from forming anything like a mental concept of causality, simply because they will not be able to identify behavior as the product of intentions and therefore might always be left to perceive the world at the level of individually learned stimulus–response associations, without ever really forming a sense of cause and effect, which so forcefully governs the human mind.

ACKNOWLEDGMENTS

My gratitude goes to Jennifer McClung for her comments and thoughts, as well as to the members of the Taï Monkey Project, the Centre Suisse de Recherches Scientifiques, the administration of the Taï National Park, and the Ivorian Ministry of Research and Higher Education for helping me with all aspects of the field work in Ivory Coast.

REFERENCES

1. Hume, D., *An Enquiry Concerning Human Understanding*, Hackett Publishing, Indianapolis, IN, 1748/1977.
2. Runes, D.D., *Dictionary of Philosophy*, Littlefield, Savage, MD, 1983.
3. Visalberghi, E. and Tomasello, M., Primate causal understanding in the physical and psychological domains, *Behav. Proc.*, 42, 189–203, 1998.
4. Michotte, A., *The Perception of Causality*, Methuen, Andover, MA, 1946/1963.
5. Premack, D. and Premack, A.J., Intention as psychological cause, in *Causal Cognition: A Multidisciplinary Debate. Symposia of the Fyssen Foundation*, D. Sperber and A. James Premack, Eds., Claredon, New York, 1995, pp. 185–199.
6. Tomasello, M., *The Cultural Origins of Human Cognition*, Harvard University Press, Cambridge, MA, 1999, p. 22.
7. Visalberghi, E. and Limongelli, L., Lack of comprehension of cause–effect relations in tool-using capuchin monkeys (*Cebus apella*), *J. Comp. Psychol.*, 108, 15–22, 1994.
8. Limongelli, L., Boysen, S.T., and Visalberghi, E., Comprehension of cause–effect relations in a tool-using task by chimpanzees (*Pan troglodytes*), *J. Comp. Psychol.*, 109, 18–26, 1995.
9. Povinelli, D.J., *Folk Physics for Apes: The Chimpanzee's Theory of How the World Works*, Oxford University Press, London, 2000.
10. Köhler, W., *The Mentality of Apes*, Routledge and Kegan Paul, London, 1925.
11. Tomasello, M. and Zuberbühler, K., Primate vocal and gestural communication, in *The Cognitive Animal*, A. Collin, M. Bekoff, and G.M. Burghardt, Eds., MIT Press, Cambridge, MA, 2002.
12. Heyes, C.M., Anecdotes, training, trapping and triangulating: do animals attribute mental states?, *Animal Behav.*, 46, 177–188, 1993.
13. Povinelli, D.J., Comparative studies of animal mental state attribution: a reply to Heyes, *Animal Behav.*, 48, 239–241, 1994.
14. Call, J. and Tomasello, M., A nonverbal false belief task: the performance of children and great apes, *Child Develop.*, 70, 381–395, 1999.
15. Cheney, D.L., Seyfarth, R.M., and Silk, J.B., The responses of female baboons (*Papio cynocephalus ursinus*) to anomalous social interactions: evidence for causal reasoning?, *J. Comp. Psychol.*, 109, 134–141, 1995.
16. Uster, D. and Zuberbühler, K., The functional significance of Diana monkey "clear" calls, *Behaviour*, 138, 741–756, 2001.

17. Wickler, W., Coordination of vigilance in bird groups: the "watchman's song" hypothesis, *Z. Tierpsychol.*, 69, 250–253, 1985.
18. Zuberbühler, K., Jenny, D., and Bshary, R., The predator deterrence function of primate alarm calls, *Ethology*, 105, 477–490, 1999.
19. Goodall, J., *The Chimpanzees of Gombe: Patterns of Behavior*, Harvard University Press, Cambridge, MA, 1986.
20. Boesch, C. and Boesch, H., Hunting behavior of wild chimpanzees in the Taï National Park, *Am. J. Phys. Anthropol.*, 78, 547–573, 1989.
21. Zuberbühler, K., Noë, R., and Seyfarth, R.M., Diana monkey long-distance calls: messages for conspecifics and predators, *Animal Behav.*, 53, 589–604, 1997.
22. Boesch, C., The effects of leopard predation on grouping patterns in forest chimpanzees, *Behaviour*, 117, 220–242, 1991.
23. Zuberbühler, K., Causal knowledge of predators' behaviour in wild Diana monkeys, *Animal Behav.*, 59, 209–220, 2000.
24. Herbinger, I., Boesch, C., and Rothe, H., Territory characteristics among three neighboring chimpanzee communities in the Taï National Park, Cote d'Ivoire, *Int. J. Primatol.*, 22, 143–167, 2001.
25. Noë, R. and Bshary, R., The formation of red colobus Diana monkey associations under predation pressure from chimpanzees, *Proc. Roy. Soc. London B*, 264, 253–259, 1997.
26. Zuberbühler, K., Referential labelling in Diana monkeys, *Animal Behav.*, 59, 917–927, 2000.
27. Zuberbühler, K., Predator-specific alarm calls in Campbell's monkeys, *Cercopithecus campbelli*, *Behav. Ecol. Sociobiol.*, 50, 414–422, 2001.
28. Zuberbühler, K., Interspecific semantic communication in two forest monkeys, *Proc. Roy. Soc. London B*, 267, 713–718, 2000.
29. Zuberbühler, K., A syntactic rule in forest monkey communication, *Animal Behav.*, 63, 293–299, 2002.
30. Seavy, N.E., Apodaca, C.K., and Balcomb, S.R., Associations of crested guineafowl, *Guttera pucherani*, and monkeys in Kibale National Park, Uganda, *Ibis*, 143, 310–312, 2001.
31. Zuberbühler, K., Causal cognition in a non-human primate: field playback experiments with Diana monkeys, *Cognition*, 76, 195–207, 2000.

3 Auditory Temporal Integration in Primates: A Comparative Analysis

Kevin N. O'Connor and Mitchell L. Sutter

CONTENTS

I. Introduction 27
II. Methods and Models of Investigation 28
III. Data Analysis 31
IV. Summary and Conclusions 39
Acknowledgments 41
References 41

I. INTRODUCTION

All sounds unfold in time. It seems likely that this fundamental characteristic of acoustic information transmission would place pressure on the design and behavior of auditory systems. Generally speaking, an inverse relationship exists between both the amount and quality of information that can be transmitted and the speed of transmission; in other words, it requires time to transmit a large quantity of high-quality information. Information theory informs us that, under noiseless conditions, the logarithm of the maximum number of discrete signals transmitted increases as a linear function of time.[1] The relationship between the quality of information transmission and time can be described by Fourier analysis; in general, the trade-off between time and frequency is such that the longer the duration of a signal, the more certain we are of its spectral content.*

These principles imply that, in order to maximize information transmission, senders gain by prolonging signal delivery, and receivers by waiting and listening to a signal for as long as possible. In opposition to these strategies, however, are the undoubted competitive advantages to transmitting and deciphering signals quickly and efficiently.

---

* Fourier analysis (or synthesis) employs a set of sine and cosine (basis) functions that are infinite in extent, while all real-world signals have finite duration. The implications and relevance of these facts to communication and hearing, particularly the limitations finite signals possess for precisely specifying frequency, have been explored extensively by Gabor.[2,3]

These conflicting pressures may drive auditory systems toward a design that optimizes the temporal integration of information, such that a signal is received and processed as quickly as possible without unduly sacrificing its content. If true, then we might expect that ecological selective pressures would play a strong role in determining the performance of auditory systems, including their temporal dynamics. In this case, it would not be surprising to find considerable systematic phylogenetic variation in rates of temporal integration, possibly related to the structure of species' acoustic signals. An alternative view is that the dynamics of temporal processing may be strongly constrained by the inherent structure of auditory systems. Although optimizing selective forces are and have been present, their effects are severely limited by the physical and physiological design of biological systems that transduce and process auditory information. If so, then, even though evolution would be expected to play some general optimizing role in system design, variation in temporal integration across species should be small and unrelated to the finer structure of species' acoustic communicative signals.

In this chapter, we will perform a quantitative analysis of data taken from studies across a range of species, both primates (including humans) and nonprimates, in order to obtain an objective cross-species comparison of temporal integration. A first step will be to determine whether the variation found across species appears to be systematic or random. If variation is systematic, the next step will be to determine if the dynamics of temporal integration are related to species phylogeny or whether other factors might account for this variation. This endeavor depends largely on the quality and validity of the data upon which it is based. It should also be mentioned that even if a relationship exists between communication and temporal integration, uncovering this fact might not be at all straightforward. Methodological differences across studies such as means of sound generation and/or behavioral measurement could introduce variability across studies, making the detection of systematic structure in the data a difficult or impossible task.

II. METHODS AND MODELS OF INVESTIGATION

Psychophysical methods have been, by far, the most popular approach for studying the sensory capabilities of both human and non-human subjects. A major reason for this popularity is that careful control over the generation and calibration of physical stimuli and the measurement of behavior is possible. For the study of temporal integration, most investigators have examined the relation between the duration of a pure tone (noise stimuli have been used much less often) and the intensity at which it is just detectable (i.e., determining the threshold intensity of a tone as a function of stimulus duration). It is well established that absolute sensory thresholds decline as a function of stimulus duration, up to a point. This decline is nonlinear: it is initially very steep as a function of duration, but the rate of descent decreases progressively until a constant threshold is reached at durations of about 0.5 to 1 sec (for acoustic stimuli). Thresholds are, in other words, a negatively accelerated, decreasing function of signal duration, up to some limit.* Much attention has been

---

* After reaching a minimum, a slight increase in threshold is often found with signals of increasing duration, an effect that has been attributed to sensory adaptation.

given to describing this rate of decline. The earliest and most popular quantitative description has been the simple function known as Bloch's law:

$$I \cdot t = k \qquad (3.1)$$

which states that a trade-off or reciprocity between stimulus intensity (I) and time (t) exists, and their product is a constant. Though the simplicity of this relation is appealing, it carries with it the assumption that integration is perfect, an assumption that is not necessarily obvious or warranted. Another source of its appeal is that it seems to be applicable to a wide range of phenomena, including visual behavior such as the detection of visual stimuli near threshold[4] and the perceived brightness of short-duration stimuli.[5] Photographers will note that this function is also descriptive of the relation between how long a shutter is open on a camera and the size of the aperture, factors that act reciprocally to determine the total exposure.

At first it would appear that Bloch's law describes a simple general law of temporal integration. The problem is that, as far as its application to biological systems is concerned, it is not quite correct. Early work clearly showed that Bloch's law was accurate only for stimuli of relatively brief durations, typically no longer than 100 to 200 msec. This is to be expected if there is some non-zero threshold for long-duration stimuli, a fact first pointed out for auditory detection by both Hughes[6] and Garner,[7] who assumed that sound energy below threshold for continuous tones did not undergo integration. The second problem concerns the perfect reciprocity between time and intensity level implied by the function. Rewriting Eq. (3.1) as:

$$I = k \cdot t^{-1} \qquad (3.2)$$

it is clear that the rate at which intensity changes as a function of time is determined by the exponent in the equation — here, a negative one. It seems unlikely, however, that nervous systems would be able to operate without some transduction loss or loss due to inefficiency of transmission and that, because of such potential losses, time and intensity could not be treated equivalently. If so, then integration should be less than perfect and the exponent of Eq. (3.2) should be between –1 and zero. We can account for these limiting factors by altering Eq. (3.2) to the form:

$$I_T = k \cdot t^{-m} + I_\infty \qquad (3.3)$$

where I_T is intensity at threshold,* k is a scaling parameter or constant of proportionality, t is time, m is a parameter controlling the rate of integration, and I_∞

---

* Investigators working in audition almost invariably express stimulus intensity in decibel units. This has been the convention when studying variants of Bloch's law, such as Eq. (3.3), and we follow this custom here. However, because the decibel scale is logarithmic, it has often been used as if it were equivalent to an arbitrary log-transform of an intensity scale, which it is not — in fact, most investigators use decibels as a (log) measure of sound pressure level relative to a specific standard: "zero" sensation level or absolute threshold (e.g., 0.0002 dyne/cm²). We believe that the use of an arbitrary log-intensity scale is more appropriate when using power functions such as Eqs. (3.1) and (3.3). This would avoid confusion and the often (we think) unintentional implications of a fixed-reference level scale.

represents the threshold for stimuli of infinitely long duration (a minimum threshold below which stimulus energy cannot be integrated). The scaling factor determines the total extent of threshold decline as a function of stimulus duration and takes account of our expectation that auditory systems are likely to display differential sensitivity to acoustic stimuli depending on parameters such as spectral content and amplitude variation, independent of possible differences in rates of integration.

Equation (3.3) is a power function, sometimes referred to as the *power model* of temporal integration, and because it is derived from Bloch's law it is based on a notion of integration of energy. An alternative model for auditory temporal integration proposed by Munson[8] and Plomp and Bouman[9] is referred to as a *leaky integrator* with (negative) exponential growth to a limit, analogous to the growth of charge on a resister–capacitor circuit. These authors implied that their model represented the growth of sensation within the nervous system, and the model's connection with exponential growth of neural excitation was made explicit by Zwislocki.*[10] The exponential model has intuitive appeal and is free from the "perfect integrator" implications of the Bloch's law-based power model. It can be expressed as:

$$I_T = k \exp(-t/\tau) + I_\infty \qquad (3.4)$$

where τ is the time constant for the rate of integration, and the remaining variables are defined as for Eq. (3.3).

All studies have used either the power or exponential functions to model temporal integration (for a recent review and discussion, see Eddins and Green[11]). Most of these studies have been concerned with estimation of the integration rate parameter (either m or τ) and have paid little attention to scaling parameter, which accounts for the total drop in threshold as a function of duration. As we discuss below, the focus on integration rate at the expense of scaling may have resulted in neglecting an important piece of information. The great majority of these studies have estimated one or the other of these rate parameters by employing a logarithmic transform or plot to convert the nonlinear functions to linear form.** The problem with this approach is that when I_∞ is non-zero (as it typically is), then these functions cannot be manipulated into linear form by a simple logarithmic transformation. Instead of a simple linear function (with negative slope), one obtains a "concave-upward" function having an initial linear decline at small durations, which flattens to a constant level at long durations, a consequence of the minimum, asymptotic threshold level (I_∞).

---

* Although the physical assumptions of these models (i.e., that the power model represents integration of energy and the exponential model integration of neural activity) are often remarked upon in the literature (with some investigators stating their preference for the "growth of sensation" idea), these models can also be considered just as two possible formal descriptions of temporal integration, without implying any particular physical or physiological representation. In this chapter, we consider the models to be formal descriptions not necessarily implying any specific physical mechanism.

** Very few temporal integration studies have employed a nonlinear analysis, despite the ready availability of nonlinear curve-fitting techniques for those studies conducted in more recent years. This may reflect the use of linear techniques originally, before the advent of inexpensive and convenient nonlinear analysis tools, and the desire among investigators to stick to familiar methods yielding ready comparisons.

Investigators have usually dealt with this problem by subtracting an estimated value of I_∞ from threshold data before performing a linear fit, or (equivalently) by performing the fit using only an estimated linear portion of the data set. These strategies necessitate estimating the precise location of the transitional point between the declining and constant portions of the function, which depends on the value of I_∞. The presence of curved "elbows" in some published "linearized" threshold data plots demonstrates that these subjective estimates are prone to error. Given this subjectivity and the possibility that different investigators may have employed different criteria and standards in the use of these methods, a direct comparison of the estimated rates of temporal integration between studies may not be valid.

This brief background reveals that there are at least two difficulties with directly proceeding with the analysis objectives stated earlier. The first is that the widespread effort to employ linear techniques on what is really a nonlinear problem may have led to inaccuracies due to subjective parameter estimation. This makes comparisons between studies somewhat suspect. The other problem is that studies almost invariably employ either one or the other model to derive parameters of integration rate, making the task of direct comparison between studies using different models impossible without further analysis. Our solution to these problems, given below, has been to reanalyze the available data using nonlinear curve-fitting techniques, making use of previously published data on temporal integration.*

III. DATA ANALYSIS

As mentioned above, most investigators have estimated rates of temporal integration by linearly transforming data (by creating log-log plots) and obtaining the slope of the best-fitting linear regression line as a measure of integration rate. For the power model, the slope gives the value of m; for the exponential model the slope is equivalent to the reciprocal of the time constant.** This is true, of course, only if the data have a linear form; in terms of Eqs. (3.3) and (3.4), then, the term I_∞ (the limiting or minimum threshold level) has been set to zero. The results of previous studies using both human and non-human subjects typically find m to be around one,[11–14] or slightly lower, and τ to center about 100 to 200 msec,[9,13–21] although the reported values show considerable variation.

In an initial effort to compare the results of a nonlinear analysis on their rhesus monkey (*Macaca mulatta*) temporal integration data with the previous linear methods for estimating temporal integration rates, O'Connor et al.[22] estimated m and τ in Eqs. (3.3) and (3.4) from previously published tables of threshold as a function

---

* Some of the results from these analyses have appeared previously in O'Connor et al.[22] That paper drew on the results of eight prior studies of temporal integration in different species to comparatively assess the results of our own experiments on temporal integration in rhesus macaques. In the current work, we analyze the data from 12 species taken from 14 studies.

** For the power model, the slope in decibels per $\log_{10}$ unit of time has usually been used to estimate m, on the assumption that a drop in 10 dB for every decade in time (or 3 dB for every doubling in time) implies an exponent of negative one.

of stimulus duration obtained from several different species.*[23] Surprisingly, we found an interesting and important difference in the performance of the models. The scaling parameter k in the exponential model was found to be directly proportional to the *range* or total drop in threshold found both within and across studies, whereas k was unrelated to this range when it was estimated using the power function (as were the parameters m and τ in both models). This implies that the scaling parameter — when it is in the exponential model — describes the influence of some experimental variable responsible for the total drop in threshold occurring between the shortest and longest stimulus durations.

The relevance of this result for modeling temporal integration can be understood by examining the details of Eqs. (3.3) and (3.4). In these models, the total amount of integration is equivalent to the total drop or range in threshold — the difference between the highest threshold found, generally for the shortest duration tone, to the lowest threshold which is usually found for the longest tone. The scaling parameter k defines this total decline in threshold measurement across stimulus duration. The rate of change at which this drop is attained is controlled by the parameters m or τ. At larger (absolute) values of m, or for shorter time constants, integration is more rapid and Eqs. (3.3) and (3.4) adopt steeper, more concave forms. Not surprisingly, we found the traditional measure of integration rate — slope measured in dB/decade — also to be proportional to the scaling parameter k (the total drop in threshold) but unrelated to the time constant τ (the rate of threshold decline). This suggests that the traditional measure of temporal integration is more descriptive of the *total* decline in threshold rather than the nonlinear *rate* of this decline.

According to the analysis of O'Connor et al.,[22] then, the exponential model provides a better description of temporal integration than the power model because it is able to account for two independent sources of variation in threshold decline, one a linear component or range characterized by the scaling parameter k, the other a nonlinear component descriptive of the decline within this range, the time constant τ. It seems likely that the exponential model will be more useful in explaining experimental results, and so we use it here for our comparative analyses.

To perform these analyses we adopted the techniques described in O'Connor et al.[22] Briefly, we used Fay's[23] extensive databook of vertebrate hearing which presents summaries of detection thresholds as a function of tone duration for different species (which were obtained from tables or estimated from figures in the original publications). We determined the rate parameters for the exponential model by performing a nonlinear curve fit of Eq. (3.4) to the data obtained for each species.** To aid

---

* In Fay,[23] threshold values for temporal integration are reported relative to thresholds at the longest durations tested (for each species and tone frequency); that is, the value at the longest threshold is subtracted from every threshold value, inclusively. In terms of using Eqs. (3.3) and (3.4), then, the assumption is made that I_∞ is equal to threshold for the longest duration stimulus (rather than for a continuous tone) and is set to zero.

** We analyzed data from the following 12 species: blue monkey (*Cercopithecus mitis*);[24] grey-cheeked mangabey (*Cercocebus albigena*);[24] rhesus monkey (*Macaca mulatta*);[22,25] human (*Homo sapiens*);[25,15] cat (*Felis catus*);[26] mouse (*Mus musculus*);[18] dolphin (*Tursiops truncatus*);[16] horseshoe bat (*Rhinolophus ferrumequinum*);[27] brown bat (*Myotis oxygnathus*);[27] chinchilla (*Chinchilla laniger*);[17,28] parakeet (*Melopsittacus undulatus*);[20,29] and field sparrow (*Spizella pusilla*).[2]

comparison with previous studies, we also performed linear regressions on threshold-log time plots for tone durations up to 200 msec.*

An important first question to ask is whether any systematic differences in the independent variables chosen for study might cause differences in temporal integration to be spuriously attributed to species factors. One such variable is tone frequency. Many studies have reported higher rates of integration for low frequencies. One reason proposed for this result is the broadening of the frequency spectrum that occurs as tones becomes shorter in duration, an effect that is more pronounced at low frequencies.[7] The argument is that the auditory system becomes less sensitive or efficient with a spread in spectral energy beyond a critical band, resulting in an increase in threshold at very short durations with an attendant increase in the estimate of integration rate.** A perusal of the literature reveals a wide range in the set of frequencies chosen for study, the highest found for the echolocating mammals. If integration rate is strongly related to tone frequency, and this relationship is preserved across species, then rates of integration could only be directly compared after somehow accounting for frequency differences between studies.

It seemed likely that earlier studies were detecting a negative relationship between tone frequency and the total decline in threshold (from the shortest to the longest duration tones), rather than between frequency and rate *per se*. To establish this relationship, we first computed correlation coefficients between temporal integration and frequency within each species. We compared frequency with the rate parameters in the exponential model and, to facilitate comparison with previous studies, with the linear slope measure (in dB/decade in time). Distributions for these correlation coefficients are presented in Figure 3.1. The rate measure yielding the greatest number of negative correlations is the slope parameter, shown in the left panel. The negative relationship found between slope and frequency is in agreement with previous analyses. Both other rate measures also tend to show negative correlations; in the case of the time constant, however, this means that *greater* rates of integration (corresponding to smaller time constants) are associated with *higher* tone frequencies. This result may seem surprising, but recall that our analysis showed that the linear slope measure is related to the *scaling* parameter in the exponential model, which accounts for the total decline in threshold, rather than to the actual *rate* of integration as given by the time constant.[22] Figure 3.1 shows that the scaling parameter, like the slope, tends to be negatively correlated with frequency, although more weakly. Therefore, it appears that the negative relationship found between slope and frequency is largely accounted for in the exponential model by the scaling parameter, while the time constant expresses an independent (and opposite) relationship between nonlinear integration rate and frequency.

Few of these correlations are statistically significant (these are shown in Figure 3.1 by the hatched bars), and for those species where significant correlations

---

* Separate curve fits and regressions were performed for each tone frequency tested (subjects were typically tested at more than one frequency) for each species (the summaries in Fay[23] are averaged over subjects). The majority of these fits (93%) were statistically significant ($p < 0.05$). Only results from significant fits are included in this report.
** It is not at all obvious that this argument is correct. According to the principle of loudness summation one would predict a *drop* in threshold as energy spreads beyond the bounds of a critical band.

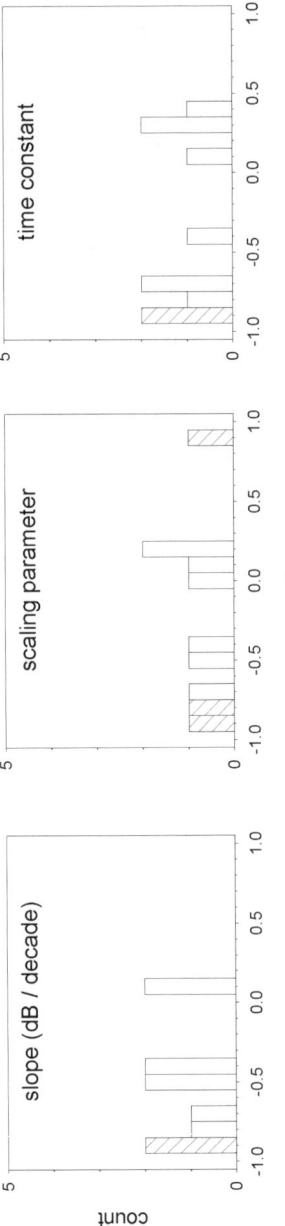

FIGURE 3.1 Distributions of Pearson correlation coefficients computed between three measures of temporal integration rate and tone frequency: the slope of the best-fitting linear regression for threshold-log time plots (durations < 200 msec) (left panel), and the scaling parameter k (center panel) and time constant τ (right panel) estimated for the exponential model (Eq. (3.4)). Statistically significant ($p < 0.05$) correlations are indicated by the hatched bars.

are found complete consistency is not evident across measures. In the case of the slope measure, rhesus monkeys (but only one study[25]) and chinchillas show significant negative correlations. For the rate parameters in the exponential model, the only species to show significant correlations are humans and dolphins. In the case of the scaling parameter k, however, one human study[16] showed the typical negative correlation, while the other human study[25] displayed a positive correlation. The remaining species, dolphins, showed a significant negative correlation between k and frequency. Overall, then, there does not appear to be a very strong relationship between integration rate and frequency when comparisons are made within species, although this conclusion must be tempered by the fact that a very limited number of frequencies are usually tested within each study.

Of more concern than the correlations within species is the relationship between rate of integration and tone frequency across species. Correlation coefficients* computed across species were weak and nonsignificant in the case of the slope and scaling parameters. The time constant, however, showed a modest yet significant negative correlation with frequency ($r = -0.387$; $df = 45$; $p < 0.02$), a finding in agreement with the within-species comparisons discussed above. This is illustrated in Figure 3.2A, which plots the time constant estimated for each tone frequency across species. The regression line indicates that for every tenfold increase in frequency, the estimated time constant drops by about a factor of two. Although there is considerable variation over species, these data do indicate that studies testing lower tone frequencies are more likely to find larger time constants for the rate of integration when using the exponential model. Given this result, it is certainly advisable to make species comparisons with care when those comparisons involve estimates of temporal integration based on disparate frequencies.

As mentioned above, the ability to detect very brief tones may not be described adequately by traditional models of temporal integration. Several investigators have noted that slopes for integration are often steeper for very brief tones (less than ~10 msec) than for moderate durations. Very brief tones, in other words, tend to be less detectable than one would expect on the basis of the exponential or power integration models. To determine whether rates of integration may be related to the shortest duration signals chosen for study, we computed the correlation coefficient between the rate parameters and the minimum durations used across species (equivalent durations were generally used within each study). Neither the slope measure nor the scaling parameter were significantly correlated with the shortest duration tone, but the time constant was ($r = 0.625$; $df = 45$; $p < 0.01$). Stimulus sets with brief minimum durations were more likely to yield small time constants than those with longer minimum durations. This relationship is displayed in Figure 3.2B. The regression line indicates a trend for the time constant to increase by a factor of almost ten as the minimum duration tone increases from just a few milliseconds to 50 msec. This result is in agreement with previous findings, that very brief tones produce larger or faster estimates of integration rate. In terms of the exponential model, then, the

---

* Because the frequency distribution across species was extremely skewed toward lower frequencies, correlations were computed using the logarithm of frequency. Also, because the distribution of time constants was also very skewed toward small values, the log of the time constant was used.

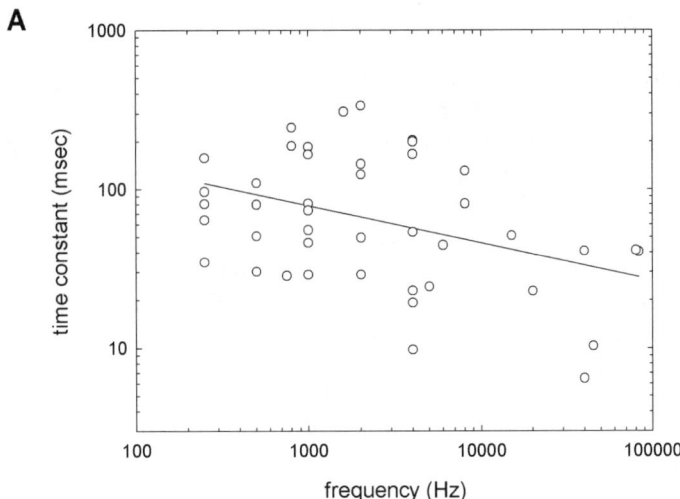

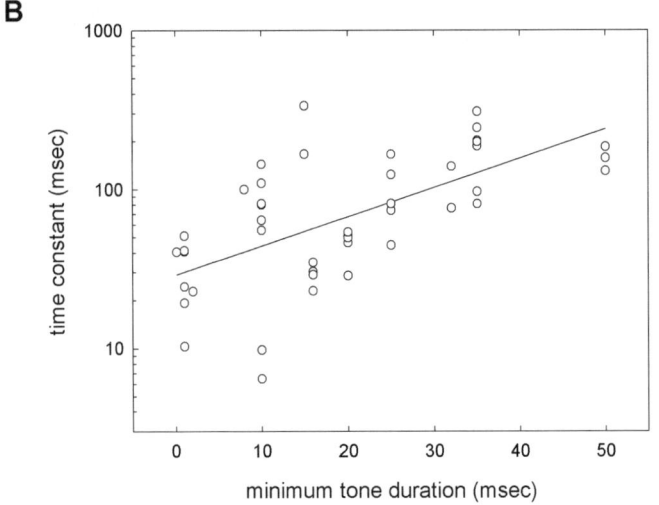

FIGURE 3.2 (A) Scatterplot showing the relationship between the rate of integration as expressed by the time constant in the exponential model and frequency. Each point represents data collected from one species at a particular frequency. (B) Scatterplot showing the relationship between the time constant and shortest duration tone selected for presentation in each study. Each point represents data collected from one species at one frequency.

nonlinear component of integration rate is sensitive to the minimum duration selected for study, and this fact must be taken into consideration when comparing results across species.

With the caveats from the above findings in mind, we can proceed with a comparison of temporal integration across species. Distributions of the estimated rate and scaling parameter values for the exponential model are displayed in

Auditory Temporal Integration in Primates: A Comparative Analysis

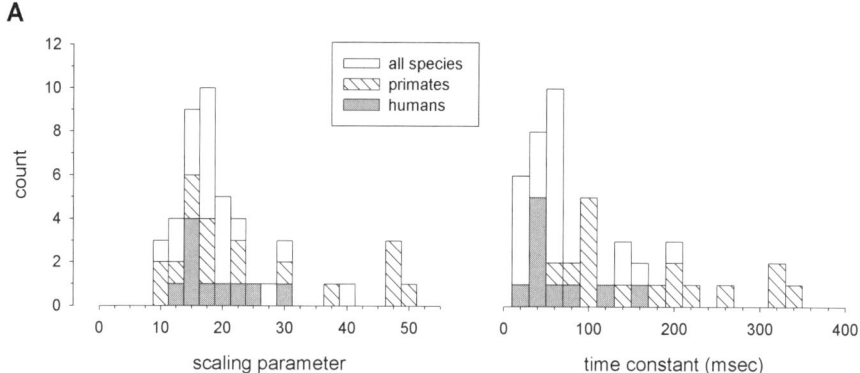

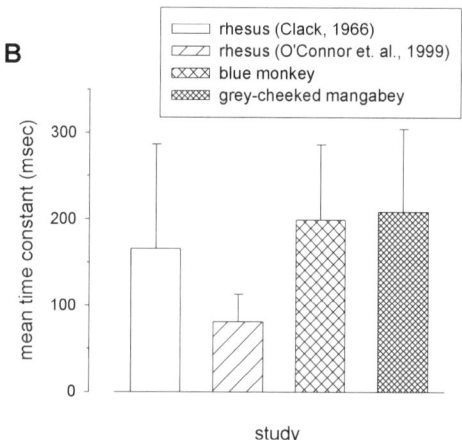

FIGURE 3.3 (A) Distributions are shown for the scaling parameter k (left panel) and time constant τ (right panel), estimated from fitting Eq. (3.4) to the threshold-duration functions of each species, at each tone frequency tested. Distributions are shown across all species (open bars), for primates only (hatched bars), and for humans only (shaded bars). (B) Time constants averaged over frequency for three primate species (rhesus monkeys, blue monkeys, and grey-cheeked mangabeys) are shown. Error bars represent one standard deviation.

Figure 3.3A. The distribution for each parameter is broken down according to three categories: all species (open bars), primates only (hatched bars), and humans (shaded bars). One notable characteristic of all the distributions is their high variability. Both the scaling parameter and time constant have skewed distributions covering large ranges, and the primate values (though not the human values) extend fully across these ranges.

A difference does appear to exist between the scaling and rate parameters in how they are distributed as a function of species category. In the case of the scaling parameter, all the distributions peak over roughly the same value (~15–17). For the time constant, however, the human distribution appears shifted toward smaller values, having a peak over ~40 msec, with no other primates exhibiting time constants less than 50 msec. In fact, if humans are excluded, the time-constant distribution for the remaining primates appears to occupy primarily the tail of the total (across species) distribution, with only a couple of values extending into the peak portion of the distribution, 0 to 80 msec.

We used the Mann–Whitney rank sum test to examine the statistical significance of these observed differences. Primates exhibit longer time constants than nonprimates, even when humans are included [$T(19,26) = 322.0$; $p < 0.01$], with primates having a median time constant of 81.2 msec, and nonprimates a median τ of 41.3 msec. Humans exhibit smaller time constants than other primates [$T(10,16) = 72$; $p < 0.0001$], with humans displaying a median $\tau = 32.5$ msec and non-human primates a median $\tau = 144.9$ msec. No significant difference, however, was observed between humans and all other species.

The finding that primates are more likely to have longer time constants than other species seems a surprising result, and it is important to consider whether this is consistent across the order or a consequence of one species or study. The primate time constants are derived from three species: the rhesus monkey (*Macaca mulatta*),[22,25] blue monkey (*Cercopithecus mitis*),[24] and grey-cheeked mangabey (*Cercocebus albigena*).[24] Figure 3.3B shows that three of these studies report relatively long time constants. The only study reporting a mean time constant less than the median is the rhesus monkey study conducted by O'Connor et al.,[22] in which the mean value of 80.7 msec is about half the next lowest mean value of 166.0 msec reported by Clack.[25] For each study, however, these mean values are averaged over four frequencies and there is considerable variation over these measurements. That the group differences were small relative to their variance was confirmed by a Kruskal–Wallis ANOVA on ranks [$F(3) = 1.67$; $p > 0.05$]. The longer time constants among primates, then, cannot be attributed to a single study or species, though the small number of these studies and species should make us cautious about accepting this result unequivocally.

We may consider the small human time constants to be related to speech perception where the demands of rapid auditory processing might necessitate fast integration rates, so the differences between humans and non-human primates are not necessarily surprising. But Figure 3.3A shows that the human time-constant data make up fewer than half of the values falling below the median (81.2 msec), prompting the question: What other species show fast time constants? These turn out to be mice (range, 24–51 msec), chinchillas (26–54 msec), dolphins (10–23 msec), bats (6–40 msec), parakeets (9–18 msec), and rhesus monkeys (45–336 msec, most recent study[22]). The species displaying time constants longer than the median are cats (130–185 msec), field sparrows (139 msec), rhesus monkeys (45–336 msec, earlier study[25]), blue monkeys (97–309 msec), and grey-cheeked mangabeys (82–309 msec). The human range (9–144 msec) also displayed some values greater than the median, although these were in the minority. This list demonstrates the considerable variation in the range of time constants existing both within and across species.

What might distinguish between species having short or long time constants? As discussed above, two candidate factors are tone frequency and minimum stimulus duration. Species displaying short time constants are more likely to have been tested on high-frequency tones, and this is true of mice, dolphins, and bats, though not of chinchillas, parakeets, rhesus monkeys, or humans. Recall, also, that long time constants are more likely to be found with long minimum stimulus durations. Four of the five species displaying predominately long time constants — cats, blue monkeys, grey-cheeked mangabeys, and field sparrows — were tested with stimulus sets having relatively long minimum durations (>30 msec).

It would seem, then, that a great deal, though not all, of the variation in the rate of integration, as measured by the time constant in the exponential model, can be accounted for by the values of two physical experimental variables: tone frequency and minimum stimulus duration. Differences in integration rate based on species factors cannot be ruled out; we do not know, for example, whether the small time constants related to high-frequency testing are caused by the high-frequency tones or instead result from the high-frequency sensitivity of the tested species. For two of these species, dolphins and mice, testing on relatively low-frequency tones (4 and 5 kHz, respectively) yielded small time constants (20 and 24 msec, respectively), which supports the notion that high-frequency sensitivity may play a role in temporal integration. The finding that humans have smaller time constants than other primates is also intriguing and cannot be explained fully by the long minimum-duration tones used in the case of the blue monkey and grey-cheeked mangabey, as rhesus monkeys were tested on short minimum-duration stimuli. The effects of the experimental factors, however, make it difficult at this point to unequivocally identify any species-related variables associated with differences in integration rate.

IV. SUMMARY AND CONCLUSIONS

In this work, we examined auditory temporal integration in a variety of primate and nonprimate species to determine whether any systematic phylogenetic differences could be found. To do this, we applied an exponential, leaky-integrator model to previously obtained threshold data to determine the linear scaling and nonlinear rate components of temporal integration.

The rate of integration, as defined by the time constant, was found to be related to two independent experimental variables across species and studies: tone frequency was negatively related to the size of the time constant (higher tone frequencies were associated with smaller time constants), while the minimum stimulus duration selected for study was positively related to the time constant (long minimum-duration tones were associated with large time constants).

Time-constant values were also found to depend on species category: primates were found to have longer time constants than other species, and humans were found to have shorter time constants than other primates. These trends are not easily separable, however, from the relationship between the time constant and the experimental variables noted above. Several of the species showing small time constants were tested predominantly on high frequencies, and most of the species exhibiting long time constants were tested with relatively long minimum-duration stimuli.

The systematic variation we find in the rate of integration, then, seems to be more closely associated with particular physical experimental variables, than with any obvious species factors. Therefore, although we found evidence that non-human primates exhibit slower rates of integration than most other species, this result must be offered with some skepticism. The finding that humans show faster rates of integration than other primates may be more solid, but it is difficult to ascertain the significance of this result, as several other species also exhibited time constants within or below the human range. Our work does clearly show the importance of designing experiments on temporal integration with care, particularly with regard to the choice of experimental parameters if cross-species comparisons are intended.

Though the evidence for species differences in auditory temporal integration seems equivocal, strong evidence exists for species differences in one aspect of auditory processing, namely spectral information; numerous studies, for example, show that humans are better at discriminating tone differences than most other species, including non-human primates.[23,30] Furthermore, comparative work with humans and macaques suggests that humans are better at performing more complex spectral discriminations. Sinnott and colleagues[31-34] have performed much of this work and have shown that macaques are worse than humans at discriminating vowel sounds, a result that might imply a species-specific difference in perceiving human speech or, alternatively, a more general species difference in processing spectral information. A recent study by O'Connor et al.[35] supports the latter conclusion. We compared the abilities of humans and rhesus macaques to discriminate stimuli having sine-modulated power spectra, also known as ripple stimuli. Like natural sounds such as human speech and other vocalizations, these stimuli contain peaks and troughs in their power spectra and may vary in the frequency separation between these peaks and troughs (sine-profile frequency or ripple density). We determined sensitivity for detecting the depth of modulation (spectral contrast) of the sine-profile stimuli and found humans to be more sensitive across the range of sine-profile frequencies tested. These results cannot be explained in terms of spectral resolution (peripheral narrow-band filtering) differences between the species, suggesting instead that some difference in a central, frequency-integration process is responsible.

The results showing spectral processing differences between humans and macaques may point to a more satisfactory way of doing comparative work on auditory temporal integration, as well as other types of auditory processing. Natural sounds such as vocalizations are dynamic and complexly structured, varying in frequency as a function of time, while most work on temporal integration has used static, simple sounds such as pure tones and noise. These simple stimuli are easily generated and manipulated in the laboratory but may place insufficient informational demands on auditory systems (which have evolved to process more complex sounds under noisy conditions) to reveal interesting species differences. A better strategy may be to use stimuli structured to resemble or approximate biologically relevant natural sounds, in either the time or frequency domain, or both. It seems likely that stimuli of this kind would be more effective in probing the processing mechanisms evolution has built into nervous systems and revealing their operation.

ACKNOWLEDGMENTS

This work was supported by NIH Grant MH56649-01, NIDCD Grant DC02514-01A1, and the Sloan Foundation. We thank Asif Ghazanfar for helpful comments on a previous version of this manuscript.

REFERENCES

1. Shannon, C.E. and Weaver, W., *The Mathematical Theory of Communication*, The University of Illinois Press, Urbana, 1949.
2. Gabor, D., Theory of communication, *J. Inst. Elec. Eng. (London)*, 93(part III), 29, 1947.
3. Gabor, D., Acoustical quanta and the theory of hearing, *Nature*, 159, 591, 1947.
4. Graham, C.H. and Margaria, R., Area and the intensity-time relation in the peripheral retina, *Am J. Physiol.*, 113, 299, 1935.
5. Stevens, S.S., Duration, luminance, and the brightness exponent, *Percept. Psychophysiol.*, 1, 96, 1966.
6. Hughes, J.W., The threshold of audition for short periods of stimulation, *Proc. Roy. Soc. London B*, 133, 486, 1946.
7. Garner, W.R. and Miller, G.A., The masked threshold of pure tones as a function of duration, *J. Exp Psychol.*, 37, 293, 1947.
8. Munson, W.A., The growth of auditory sensation, *J. Acoust. Soc. Am.*, 55, 584–591, 1947.
9. Plomp, R. and Bouman, M.A., Relation between hearing threshold and duration for tone pulses, *J. Acoust. Soc. Am.*, 31, 749, 1959.
10. Zwislocki, J., Theory of temporal auditory summation, *J. Acoust. Soc. Am.*, 22, 1046, 1960.
11. Eddins, D.A. and Green, D.M., Temporal integration and temporal resolution, in *Hearing*, B.C. J. Moore, Ed., Academic Press, San Diego, CA, 1995, p. 206.
12. Garner, W.R., The effect of frequency spectrum on temporal integration in the ear, *J. Acoust. Soc. Am.*, 19, 808, 1947.
13. Gerken, G.M., Gunnarson, A.D., and Allen, C.M., Three models of temporal summation evaluated using normal hearing and hearing-impaired subjects, *J. Speech Hear. Res.*, 26, 256, 1983.
14. Gerken, G.M., Bhat, V.K.H., and Hutchison-Clutter, M., Auditory temporal integration and the power function model, *J. Acoust. Soc. Am.*, 88, 767, 1990.
15. Watson, C.S. and Gengel, R.W., Signal duration and frequency in relation to auditory sensitivity, *J. Acoust. Soc. Am.*, 46, 989, 1969.
16. Johnson, C.S., Relation between absolute threshold and duration of tone pulses in the bottlenosed porpoise, *J. Acoust. Soc. Am.*, 43, 757, 1966.
17. Henderson, D., Temporal summation of acoustic signals by chinchilla, *J. Acoust. Soc. Am.*, 46, 474, 1969.
18. Ehret, G., Temporal auditory summation for pure tones and white noise in the house mouse (*Mus musculus*), *J. Acoust. Soc. Am.*, 59, 1421, 1976.
19. Fay, R.R. and Coombs, S., Neural mechanisms in sound detection and temporal summation, *Hearing Res.*, 10, 69, 1983.
20. Dooling, R.J. and Searcy, M.H., Temporal integration of acoustic signals by the budgerigar (*Melopsittacus undulatus*), *J. Acoust. Soc. Am.*, 77, 1917, 1985.

21. Saunders, S.S. and Salvi, R.J., Psychoacoustics of normal adult chickens: thresholds and temporal integration, *J. Acoust. Soc. Am.*, 94, 83, 1993.
22. O'Connor, K.N., Barruel, P., Hajalilou, R., and Sutter, M.L., Auditory temporal integration in the rhesus macaque (*Macaca mulatta*), *J. Acoust. Soc Am.*, 106, 654, 1999.
23. Fay, R.R., *Hearing in Vertebrates: A Psychophysics Databook*, Hill-Fay, Winnetka, IL, 1988.
24. Brown, C.H. and Maloney, C.G., Temporal integration in two species of Old World monkeys: blue monkeys (*Cercopithecus mitis*) and grey-cheeked mangabey (*Cercocebus albigena*), *J. Acoust. Soc. Am.*, 79, 1058, 1986.
25. Clack, T.D., Effect of signal duration on the auditory sensitivity of humans and monkeys (*Macaca mulatta*), *J. Acoust. Soc. Am.*, 40, 1140, 1966.
26. Costalupes, J.A., Temporal integration of pure tones in the cat, *Hear. Res.*, 9, 43, 1983.
27. Ayrapet'yants, E. Sh. and Konstantinov, A.I., *Echolocation in Nature*, An English translation of the National Technical Information service, JPRS 63328-1 and −2, Springfield, VA, 1974.
28. Wall, L.G., Ferraro, J.A., and Dunn, D.E., Temporal integration in the chinchilla, *J. Aud. Res.*, 21, 29, 1981.
29. Dooling, R.J., Temporal summation of pure tones in birds, *J. Acoust. Soc. Am.*, 77, 1917, 1985.
30. Long, G.R., Psychoacoustics, in *Comparative Hearing: Mammals*, R.R. Fay and A.N. Popper, Eds.), Springer–Verlag, New York, 1994, pp. 18–56.
31. Sinnott, J.M., Beecher, M.D., Moody, D.B., and Stebbins, W.C., Speech sound discrimination by monkeys and humans, *J. Acoust. Soc. Am.*, 60, 687, 1976.
32. Sinnott, J. M. and Kreiter, N.A., Differential sensitivity to vowel continua in Old World monkeys (*Macaca*) and humans, *J. Acoust. Soc. Am.*, 89, 2421, 1991.
33. Sinnott, J., Peterson M., and Hopp, S., Frequency and intensity discrimination in humans and monkeys, *J. Acoust. Soc. Am.*, 78, 1977, 1985.
34. Sinnott, J.M., Owren, M.J., and Petersen, M.R., Auditory frequency discrimination in primates: species differences (*Cercopithecus, Macaca, Homo*), *J. Comp. Psychol.*, 101, 126, 1987.
35. O'Connor, K.N., Barruel, P., and Sutter, M.L., Global processing of spectrally complex sounds in macaques (*Macaca mullata*) and humans, *J. Comp. Physiol. A*, 186, 903, 2000.

4 Mechanisms of Acoustic Perception in the Cotton-Top Tamarin

Cory T. Miller, Daniel J. Weiss, and Marc D. Hauser

CONTENTS

I. Introduction..43
    A. Subject Species..45
II. Acoustic Perception in Cotton-Top Tamarins...46
    A. Natural Signals 46
        1. Units of Perception 46
        2. Functionally Relevant Acoustic Features......................................49
    B. Speech Signals: Units of Perception..53
III. Discussion 56
References..57

I. INTRODUCTION

Communication involves transmitting a signal encoded with information that can be interpreted by a receiver and used to mediate behavioral responses and decisions. In order for communication to function properly, some level of co-evolution between signal producer and receiver must have occurred; otherwise, signals would either be ignored or misinterpreted. Although this relationship is evident in every communication system, specialized systems offer a unique opportunity to observe how the specific features of a system interact to facilitate communication. For example, in vocal communication systems, individuals communicate by emitting vocalizations that are then interpreted by the auditory system of conspecifics. The information content and structure of the signals are manifested in a suite of acoustic variables that conform to species-typical boundaries. In a specialized system, particular acoustic features are encoded in the vocal signal and transmitted to the receiver who, in turn, has evolved a perceptual system to interpret particular features of the signal. Subtle differences in acoustic structure can be interpreted to indicate vastly different pieces of information. To decipher these signals, researchers must understand the interaction between acoustic features within the call and the behaviors that are elicited by such features. Specialized systems of vocal communication offer us an important opportunity to investigate this relationship most effectively.

For example, the Tungara frog (*Physalaemus pustolosus*) has evolved a specialized vocal communication system to mediate mate attraction. In this species, male frogs congregate around ponds and produce advertisement calls to attract sexually receptive females.[1] Acoustic analyses show that males produce calls containing two acoustically distinct syllable types, most often a single-frequency, modulated whistle, dubbed the *whine*, followed by one to five short *chucks*.[2] To test the functional significance of the two syllable types, Ryan and colleagues conducted a series of phonotaxis experiments in which competing stimuli were presented to females from two equidistant speakers. Results showed that, although the whine was sufficient to elicit selective phonotaxis, females preferentially approached calls with chucks over calls without chucks and preferentially approached calls with more chucks over those with fewer chucks.[2,3] Additional tests showed that closely related species shared this bias for chucks, despite the fact that in none of these species do males produce advertisement calls with chucks.[4] These results suggest that sexual selection and female perceptual biases have influenced the acoustic structure of the advertisement call. This elegant series of studies illustrates the importance of examining the interplay between production acoustics and perceptual response in a specialized system and can serve as a model for many studies of auditory perception in animals.

Here we review the mechanisms underlying acoustic perception in non-human primates (hereafter *primates*), focusing on the cotton-top tamarin as a potential model system. Throughout, we emphasize three aspects of our approach that we feel are crucial. The first of these is the importance of focusing on one highly specialized signaling system with a clearly defined context and function. Second, we describe the significance of detailed acoustic analyses of a particular signal, both for understanding production mechanisms and for providing the foundation for our third approach, perceptual analyses. Finally, we extract the meaningful units of this vocal signal (i.e., acoustic units that elicit specific behavioral responses) using a variety of experimental tasks. We argue that a detailed understanding of the perceptual world of an animal can only come from using different techniques that explore both similar and different processes. Our discussion is organized into two main sections: (1) the perception of natural signals, and (2) the perception of speech signals. In both sections, we focus on the units of perception problem;[5] in Section II.A.2, we additionally explore the functionally relevant acoustic features, or the acoustic features that are pertinent for classification of the signal according to a specific parameter (i.e., call type, species, sex, etc.) of natural vocal signals. Both natural vocal communication and speech perception raise interesting but different problems with respect to the evolution of auditory mechanisms. Specifically, studies investigating the perception of species-typical signals provide insights into how different features of these signals are specialized to solve species-typical acoustic problems. Studies of speech perception, on the other hand, illustrate which aspects of the perceptual system are sensitive to more general acoustic problems, ones that may have been critical to the evolution of our own system of communication.

A. Subject Species

The cotton-top tamarin, a small New World monkey endemic only to the northern forests of Colombia, is one of several tamarin species that inhabit much of the rainforest areas of Central America and the Amazon basin.[6] All studies described here were conducted on members of a captive colony that are housed at the Primate Cognitive Neuroscience Laboratory, Harvard University.

Cotton-top tamarins produce a range of acoustically distinct vocal signals.[7] Our research, however, focuses on a single vocalization, the combination long call (CLC; see Figure 4.1).[7-9] The CLC consists of a concatenation of two distinct syllable types: chirps and whistles. Typically, a call contains one to three chirps followed by two to five whistles. We focus on this call type for five reasons. First, historically, our most sophisticated understanding of auditory mechanisms has come from in-depth analyses of specialized systems, such as bat echolocation, electric fish signaling, birdsong, and barn owl sound localization (see review chapters in Hauser and Konishi[10]). Tamarin long calling behavior is a specialized system ripe for mechanistic analyses. Second, tamarins produce CLCs more than any other vocalization, thus we can record CLCs from a large number of individuals. Third, tamarins emit CLCs in the context of physical isolation or separation from other group members and appear to use them both to re-establish contact and possibly for mate choice.[11] Fourth, subjects not only orient to this call when they hear it but also respond with antiphonal calls. An antiphonal call occurs when an individual responds to hearing a CLC by producing its own CLC. Fifth, acoustic analyses suggest that the CLCs encode information about caller identity, sex, and group membership.[9]

To study auditory perception in cotton-top tamarins, we adopted an approach that is relatively rare in the field but that we consider essential. In particular, we have looked at the problem of acoustic perception by using a wide variety of techniques with the aim of triangulating on robust, method-independent findings. In cases where results diverge, our goal is to better understand the sources of such variation, including the relationship between processing mechanisms and behavioral responses. The research described here is based on four different experimental methods: *habituation discrimination*,[9,12] *isolated antiphonal calling*,[8,13,14] *social antiphonal calling*,[15] and *selective phonotaxis*.[11]

In habituation-discrimination experiments,[16,17] subjects are presented with a series of exemplars from one stimulus class (e.g., CLCs produced by one individual) until the subject no longer responds by orienting toward the speaker location. Following habituation, subjects receive a test trial involving an exemplar from a putatively different stimulus class (e.g., a CLC produced by a different individual). The logic underlying this procedure is that an orienting response in the test trial can be taken as evidence that subjects perceptually discriminate the two classes of stimuli, while a failure to respond suggests that subjects perceptually cluster the two stimuli into one class.

The antiphonal calling paradigm takes advantage of the fact that when tamarins hear a CLC, they often respond, antiphonally, with a CLC. As tamarins do not antiphonally call with CLCs to other sounds, we can probe the causes of this response by manipulating particular acoustic features of the CLC and assess which features cause increases or decreases in antiphonal calling rates. If antiphonal calling rates decline in response to a particular manipulation, we can conclude that this particular feature plays an important role in eliciting the antiphonal response. In the isolated antiphonal calling assay, we present normal and manipulated CLCs through a hidden loudspeaker to subjects removed from their home cages and isolated in an acoustic chamber. The social antiphonal calling paradigm, in contrast, is conducted with subjects that are still in the colony room and in contact with one or more members of the colony.

The fourth assay, selective phonotaxis,[1,18] involves placing a subject equidistant between two hidden speakers. Stimuli are sequentially broadcast from each speaker and then subjects are permitted to approach one or both of the speaker locations. If a meaningful difference between the stimuli is perceived, subjects will selectively approach one speaker location over the other. This technique has been used with great success in studies of insects and frogs to investigate the acoustic features driving mate choice and species recognition.[1,18] It has the potential to be an important tool for studies of auditory perception in primates.[11]

II. ACOUSTIC PERCEPTION IN COTTON-TOP TAMARINS

A. Natural Signals

1. Units of Perception

Like anurans and songbirds, many primates produce vocalizations that consist of several temporally distinct acoustic units (e.g., lemurs, spider monkeys, gelada baboons, gibbons, and chimpanzees; for review, see Miller and Ghazanfar[19]). In some cases, the signal consists of units that are acoustically similar, while in other cases the units are made up of a variety of acoustically different units. It is often unclear whether each individual unit serves a unique function (frogs[20]), whether functionality is derived from the concatenation of all syllables (songbirds[21]), or both. If we are to understand how vocal signals influence behavior, it is critical to address questions about the functional organization of acoustic units within vocalizations (i.e., the units of perception) by exploring how animals perceive such sounds.

In contrast to the numerous studies of frog and bird vocalizations addressing the units of perception problem,[2,22,23] relatively few primate studies have addressed the question of perceptual units.[19] Mitani and Marler,[24] for example, analyzed the songs of male agile gibbons (*Hylobates agilis*) and observed that syllable order was significantly nonrandom; a system of rules appeared to constrain both syllable order and the frequency with which certain syllables occurred together. Based on these observations, they used a playback experiment to investigate whether gibbons were perceptually sensitive to the acoustic organization of syllables within their species-typical song. In this experiment, subjects were played either normal songs or songs with the syllables rearranged in a species-atypical order. Using several behaviors as dependent measures, Mitani and Marler reported that subjects responded similarly across most behaviors to both normal and manipulated song. However, gibbons produced significantly fewer territorial squeak calls in response to the rearranged songs, suggesting that syllable order is perceptually salient and that the entire song represents a meaningful unit of perception.

Because the tamarin CLC has a multisyllabic structure, it provides a particularly useful signal for exploring the unit of perception question. Our original study was designed to investigate what aspects of a CLC mediate subjects' antiphonal responses. Do tamarins antiphonally call in response to a specific syllable type (i.e., chirps or whistles), or do they respond to some sequential combination of these two syllables?[8] To answer these questions, we used the isolated antiphonal calling paradigm described above. In each trial, we presented subjects with one of the following stimulus types: an isolated chirp, an isolated whistle, a complete CLC, or a series of chirps arranged to mimic the temporal pattern of the CLC. We found that subjects antiphonally called significantly less to individual syllables than to a CLC but responded equally to the chirp series and the CLC. These results suggest (1) that the entire CLC serves as at least one meaningful unit of perception for cotton-top tamarins, (2) that individual components of the CLC are insufficient to mediate the strongest level of a species-typical antiphonal response, and (3) that the temporal properties of the CLC may be more important than the spectral information for eliciting an antiphonal response. In other words, tamarins may significantly weight temporal features such as the shape of the amplitude waveform and interpulse interval when classifying a signal as a species-typical CLC. This experiment does not, however, rule out other meaningful units of perception such as individual syllables or combinations of syllables that may elicit either different degrees of antiphonal calling or other behavioral responses.

To address this last point, we conducted a follow-up study in which we explored whether individual syllables within the CLC represented meaningful units of perception. In particular, we wanted to test whether only cohesive, continuous whistles would be considered meaningful perceptual units. For this investigation, we conducted an auditory continuity experiment[25] using isolated whistle syllables that we extracted from naturally produced CLCs.[14] In the first condition, we presented subjects with four stimulus types: two baseline stimuli (complete CLCs and unmanipulated whistles) and two test stimuli (whistles with white noise inserted into the middle [middle-noise] and whistles with a gap of silence inserted into the middle [middle-silence]; see Figure 4.1). Results indicated that subjects antiphonally called at equal levels to unmanipulated whistles and middle-noise whistles, but called significantly less to middle-silence whistles.

The second condition was designed to test whether our subjects' responses in the first condition were mediated by the mere presence of any continuous acoustic signal. We again broadcast the same two baseline stimuli, but the manipulated whistles were different. The first manipulation consisted of an initial 150 msec of whistle followed immediately by a segment of white noise (end-noise; Figure 4.1); the second manipulation was the temporal opposite, with the initial segment consisting of white noise and the last 150 msec consisting of a normal whistle (begin-noise; Figure 4.1). Subjects antiphonally called significantly less to both manipulated whistles than to normal whistles. These data suggest that only cohesive whistle units bound by species-typical parameters at the onset and offset are meaningful perceptual units.

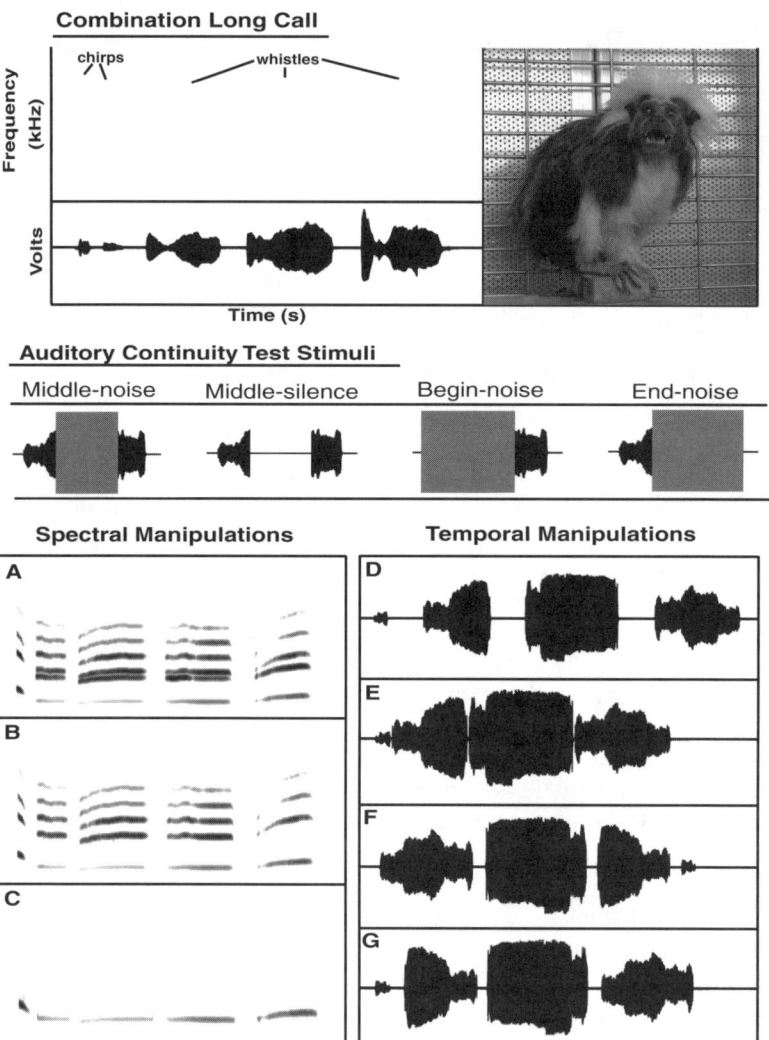

FIGURE 4.1 Stimuli used in acoustic perception experiments with cotton-top tamarins. (Top) A spectrogram and amplitude waveform of a combination long call are shown on the left. The two syllable types (chirps and whistles) are noted. A picture of an adult tamarin during production of a CLC is shown to the right. (Middle) The four whistle manipulations used in experiments on auditory continuity by Miller et al.[14] are shown. (Bottom) Spectrograms of some spectral manipulations used by Weiss et al.[15] are shown to the left: (A) mistuned harmonic structure, (B) missing the second harmonic, and (C) fundamental frequency only. The amplitude waveforms of some temporal manipulations used by Ghazanfar et al.[13] are depicted on the right: (D) expanded interpulse interval, (E) no interpulse interval, (F) global reversal, and (G) local reversal.

Page 48 — Below is the correct version of Figure 4.1.

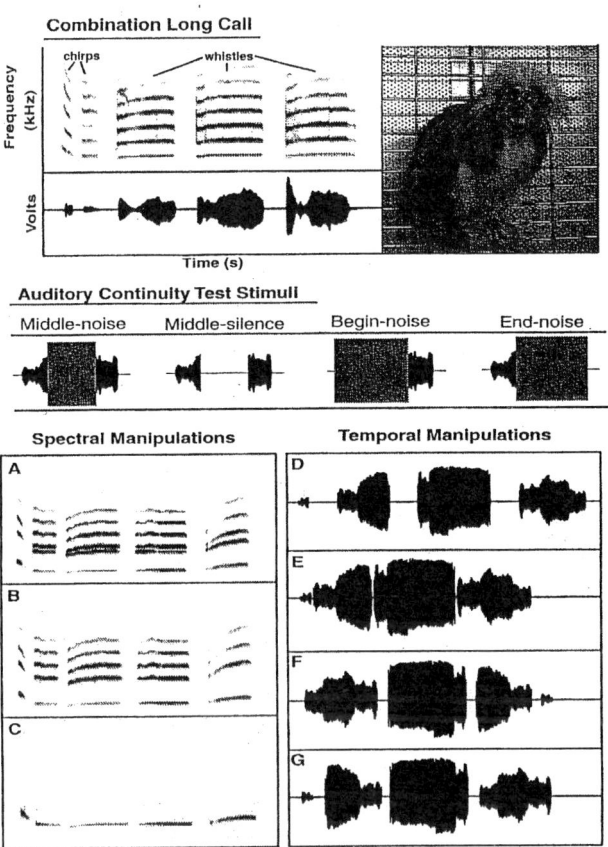

FIGURE 4.1 Stimuli used in acoustic perception experiments with cotton-top tamarins. (Top) A spectrogram and amplitude waveform of a combination long call are shown on the left. The two syllable types (chirps and whistles) are noted. A picture of an adult tamarin during production of a CLC is shown to the right. (Middle) The four whistle manipulations used in experiments on auditory continuity by Miller et al.[14] are shown. (Bottom) Spectrograms of some spectral manipulations used by Weiss et al.[15] are shown to the left: (A) mistuned harmonic structure, (B) missing the second harmonic, and (C) fundamental frequency only. The amplitude waveforms of some temporal manipulations used by Ghazanfar et al.[13] are depicted on the right: (D) expanded interpulse interval, (E) no interpulse interval, (F) global reversal, and (G) local reversal.

September 6, 2002

Dear Customer:

Thank you for your purchase of *Primate Audition: Ethology and Neurobiology*, edited by Asif A. Ghazanfar.

On the reverse is a corrected version of Figure 4.1, which appears on page 48 of the book. A spectrogram was inadvertently omitted from the original figure.

We sincerely regret any inconvenience this may have caused you. Please let us know if we can be of any assistance regarding this title or any other titles that CRC Press publishes.

Best regards,

CRC Press LLC

#0956/0-8493-0956-5

Few studies have investigated the units of perception in primate communication. These data, however, are ultimately necessary to provide a true comparative look at auditory perceptual mechanisms. Our work on tamarins is a first step in trying to bridge this gap. Overall, we found that despite the multisyllabic structure of CLCs, tamarins perceive the entire call as a meaningful unit of perception. Neither syllable type alone was sufficient to elicit the antiphonal response from isolated individuals; rather, this behavior was released only after the tamarins perceived a series of syllables. Further, results indicated that the individual syllables within the CLC are represented as cohesive units of perception that cannot be divided or manipulated outside the species-typical parameters. Although these data provide important initial insights into how tamarins represent acoustic units, many questions remain. If both syllable types elicit comparable levels of antiphonal calling, what is the functional significance of two syllable types in the CLC? Do tamarins distinguish the chirp sequence as a perceptual unit distinct from the whistle sequence? Future work will explore these questions in an attempt to better understand how tamarins perceive this species-typical vocal signal.

2. Functionally Relevant Acoustic Features

To truly understand a communication system, one must ascertain which perceptually salient features mediate a receiver's behavioral responses. In the absence of such information, it is impossible to make true comparisons across different species' communication systems. Thus, one of the main goals of research on acoustic perception is uncovering the necessary and sufficient features required for animals to recognize and discriminate different vocalizations.[5,26] To achieve this goal, we must understand the interaction between call production and perception. A complete study requires the researcher to determine how an animal responds to each vocalization within its repertoire, then record such vocalizations and extract a suite of call-specific acoustic features. Next, one must determine which of the main information-bearing parameters of the vocalization are actually functionally relevant for perceivers. One of the most effective ways of accomplishing this is to manipulate individual features of the call while holding all other parameters constant. Surprisingly, such comprehensive research programs are rare in studies of non-human primates, especially when contrasted with the number of such studies with avian and anuran species.[1,18,27]

An excellent example of an analysis of functionally relevant features comes from psychophysical experiments conducted with two distinctive variants of the "coo" call of the Japanese macaque (*Macaca fuscata*). These vocalizations — the smooth early (SEH) and smooth late (SLH) high coos — differ in the relative temporal position of the peak frequency, as well as the social context in which they are produced.[28] Psychophysical studies have demonstrated that macaques can distinguish between these two coo variants.[29] Subsequent studies, using synthetic exemplars, revealed that discrimination of these call types depends on frequency but not amplitude modulation,[30] and that the distinction between SEH and SLH may be categorical[31] (for a failure to replicate, with a slightly different procedure, see Hopp et al.[32]). In addition, Le Prell and Moody[33] manipulated the spectral features of synthetic coo calls and showed that Japanese macaques rely on the relative amplitude of the harmonics for call categorization. Taken together, these results suggest that Japanese macaques can use a suite of different acoustic features to guide classification of these calls.

A number of other studies have explored the functionally relevant features of primate vocalizations using methods different from those carried out with Japanese macaques. These include psychophysical studies contrasting vervet monkey and human perception[34] and field playback studies with rhesus macaques designed to explore the underlying perception of conspecific and temporally manipulated conspecific vocalizations,[35] as well as a number of field and laboratory experiments on syllable order[24,36] and categorical perception.[37-39] Together, these studies have helped elucidate how perceivers use different acoustic features to classify vocalizations into functionally significant categories.

Our examination of the functionally relevant acoustic features of tamarin calls has taken a two-pronged approach. The first prong consists of detailed acoustic analyses aimed at uncovering the overall anatomy of signals within the repertoire. The second prong consists of playback experiments designed to assess whether the features extracted from acoustic analyses match up with the features used by perceivers during classification and the initiation of species-typical and context-appropriate responses.

Using multiple discriminant analyses, Weiss and colleagues[9] found that CLC exemplars could be reliably assigned to three functional classes based on a suite of spectrotemporal features: individual identity, sex, and cage-group membership. These findings do not, however, address whether tamarin perceivers actually extract information at each of these levels nor does it tell us which features are important; rather, our analyses suggest that such information is potentially available as a source of recognition. To address whether tamarins actually perceive identity differences embedded in the CLC, we conducted a series of perceptual experiments.

First, Weiss and colleagues[9] ran a series of habituation-discrimination playback experiments designed to assess whether the calls of two different individuals could be discriminated on the basis of the acoustic properties alone. Isolated subjects were habituated to a series of of CLCs produced by one individual and then a test CLC from a novel individual was played; both opposite- and same-sex pairings were tested. Results showed that subjects responded when caller identity changed, but failed to respond (i.e., transferred habituation) when caller identity was held constant and a novel exemplar from the same individual was played instead (control condition). Follow-up experiments revealed an asymmetry between the acoustic analyses of individual identity and the tamarin's capacity to discriminate among vocal signatures: whereas all colony members have distinctive vocal signatures, not all individuals are equally discriminable based on the habituation-discrimination paradigm. Namely, the acoustic variation in call structure between CLCs produced by two individuals is not always sufficiently different to permit discrimination in this paradigm. Additional evidence supporting this conclusion comes from social antiphonal calling playback experiments in the colony room. In these experiments, we removed two individuals residing in different cages from the colony room and then played back the CLC of one of these individuals. We found that the caller's cagemate antiphonally responded significantly more often than the cagemate of the other animal removed from the

Mechanisms of Acoustic Perception in the Cotton-Top Tamarin

homeroom.[15] This shows that the antiphonal calling response is not simply mediated by the absence of individuals from the colony room, but rather to the selective removal and calling behavior of a group mate. Overall, these studies demonstrate that individual identity is not only encoded in the CLCs of cotton-top tamarins but is also perceptually salient for receivers in two different experimental contexts.

In a series of selective phonotaxis experiments that manipulated the caller's familiarity, Miller and colleagues[11] presented tamarins with CLCs produced by three classes of individual: their cagemate, a familiar non-cagemate (i.e., animal from the same homeroom but not the same cage), and foreign tamarins (Figure 4.2). Results

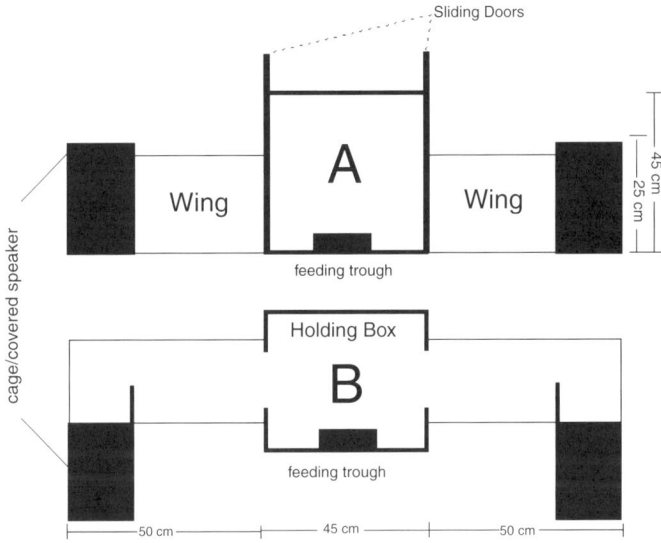

	Experimental Conditions		
	Cagemate	vs	Familiar NonCagemate
Male:	4		1
Female:	4		1
	Familiar NonCagemate	vs	Foreign
Male:	2		3
Female:	5		0
	Cagemate	vs.	Foreign
Male:	1		4
Female:	4		1

FIGURE 4.2 Basic experimental setup and results of Miller et al.[11] (Top) Schematic drawing of the phonotaxis apparatus: A, front view; B, top view. (Bottom) The table depicts the number of first approaches made by males and females toward each stimulus class for each experimental condition.

indicated that subjects demonstrated selective phonotaxis across all experimental conditions, suggesting that subjects perceived meaningful differences between all three stimulus classes. In addition, Miller et al. found that adult males showed a preference for first approaching foreign adult females over familiar ones, whereas adult females first approached cagemate males and homeroom adult males over foreign males (Figure 4.2). These data suggest that tamarins can extract information about the caller's identity and relative familiarity from the acoustic structure of a CLC and use this information to mediate behavioral decisions.

Having established that tamarins can extract information about individual identity from the acoustic structure of the CLC, using three different procedures, Weiss and colleagues[40] used habituation-discrimination playback experiments to determine which acoustic parameters were functionally relevant to the animals in call classification. Specifically, the study used synthetic replicas of naturally produced CLCs to test the contribution of harmonic structure to the perception of CLCs. In the first condition, subjects were habituated to a series of calls from one individual and then were played a synthetic call from the same individual. If subjects transferred habituation by failing to show an orienting response or to produce an antiphonal call (i.e., treated the natural and synthetic exemplars as functionally similar), then a synthetic call with the second harmonic deleted was played (Figure 4.1B). Because acoustic analyses had shown that the second harmonic was one of the most prominent harmonics with respect to amplitude, we predicted that changing this feature would elicit a strong response. Surprisingly, a significant proportion of subjects transferred habituation to the manipulated call, showing that this manipulation failed to elicit a perceptually meaningful change. In the next conditions, the test calls (both natural and synthetic) consisted of a CLC with the fundamental frequency deleted. These conditions presented an interesting test, because in humans the fundamental frequency plays a significant role in voice recognition.[41] In parallel with the first condition, the tamarins showed no evidence of perceiving this change as functionally meaningful, transferring habituation to the test stimulus.

In the third condition, we eliminated all of the upper harmonics, leaving only the fundamental frequency (Figure 4.1C). In contrast to the first two conditions, a significant proportion of our test subjects responded to this manipulation, suggesting that although individual harmonics may not be particularly salient, overall harmonic structure may be significant in individual recognition. The fourth condition of this study explored whether mistuning the second harmonic constitutes a perceptually meaningful change (Figure 4.1A). In parallel with condition three, a significant proportion of subjects responded to this manipulation, suggesting that tamarins are sensitive to harmonic relationships. Overall, these experiments begin to address questions pertaining to which acoustic features are used by tamarins to classify the CLC as a species-typical vocal signal.

In a series of playback experiments, Ghazanfar and colleagues[8,13] investigated the acoustic cues underlying the tamarin's antiphonal calling response. They first played normal CLCs alternating with calls that were locally reversed (played in normal sequence, but individual syllables reversed; see Figure 4.1G) or globally reversed (entire call was temporally reversed; see Figure 4.1F). Across all these conditions, subjects responded with similar rates of antiphonal calling. Likewise, when subjects heard calls in which the fundamental frequency (and corresponding harmonics) was either reduced by one half (lowered in pitch) or doubled (raised in pitch), they responded at an equal rate to the manipulated stimuli and normal CLCs. No change in antiphonal calling rate was evident, even when the intersyllable interval was expanded to twice the maximum (Figure 4.1D) observed in the colony. However, when the intersyllable interval was deleted (Figure 4.1E), leaving a continuous but amplitude-modulated signal, subjects responded at a significantly higher rate. Interestingly, subjects were more likely to antiphonally call during the playback (i.e., interrupt the putative caller) of an expanded CLC than during a normal CLC. The vast majority of these interruptions occurred right after, or in the middle of, the third syllable. This pattern suggests that either a certain number of syllables or a certain amount of energy must be present in order to elicit an antiphonal response.

Two final conditions explored the importance of temporal patterns of signal intensity: amplitude-modulated white noise (species-typical amplitude envelope applied to white noise) and gated white noise (syllables replaced with blocks of white noise of equal duration). Here, subjects showed no difference in antiphonal calling rates to amplitude-modulated calls and normal CLCs, but a significant decrease in calling rates to the gated white noise stimuli. From these results, Ghazanfar and colleagues concluded that antiphonal calling in socially isolated tamarins is largely mediated by two temporal cues: specifically, the amplitude envelope and intersyllable interval. In addition, based on results from this experiment as well as a previous study,[8] the authors postulate a syllable threshold (at least three syllables) necessary to elicit an antiphonal response.

In summary, our two-pronged research program on cotton-top tamarins has begun to elucidate the functionally relevant features underlying tamarins' perceptions of CLCs. With respect to production, the CLC encodes different levels of information concerning caller identity, including individual, sex, and group.[9] With respect to perception, we have begun an extensive project, guided by the production data, which explores how the perceptually salient features of the CLC mediate both an orienting and antiphonal response. Future experiments will focus on additional feature manipulations, as well as investigating changes in perception across different contexts (e.g., in the presence of multiple individuals) and different methods.

B. Speech Signals: Units of Perception

Spoken language appears to activate different circuits of the human brain than sounds associated with either human emotion (e.g., laughter, crying) or music. The fact that specialized, and even dedicated neural circuitry is recruited for speech perception is certainly not surprising, especially when one considers the evolution of other species-specific communication systems. Exploration of this comparative database reveals that the rule in nature is one of special design, whereby natural selection builds (blindly, of course) adaptations suited to past and current environmental pressures. Thus, by looking at the communicative problems that each organism faces, we find signs of special design, including the dance of the honeybee, electric signaling of mormyrid fishes, the song of Passerine birds, and the foot drumming of kangaroo rats. The question of interest in any comparative analysis then becomes which aspects of the communicative system are uniquely designed for the species of interest and which are conserved. In the case of speech perception, we know that many peripheral mechanisms (ear, cochlea, and brainstem) have been largely conserved in mammals.[5,42] The focus of this section is to investigate which components of speech perception are mediated by a specialized phonetic mode and which by a more general mammalian auditory mode. Evidence that nonhuman animals parse speech signals in the same way as humans provides evidence against the claim that such capacities evolved for speech perception, arguing instead that they evolved for more general auditory functions and were subsequently co-opted by the speech system.

Human speech perception involves segregating an acoustic continuum into a suite of perceptually meaningful categories. As one particular example, when we listen to an artificially created acoustic continuum running from /ba/ to /pa/, we are able to discriminate and label between-category exemplars, but not within-category exemplars. When this phenomenon, termed *categorical perception*, was first discovered, it was thought to be specifically tailored to the speech signal. Liberman and colleagues[43] posited that categorical perception was uniquely human and special to speech. In response to this claim, the phenomenon of categorical perception was soon explored in (1) adult humans, using non-speech acoustic signals as well as visual signals; (2) human infants, using a habituation procedure with the presentation of speech stimuli; and (3) animals, using operant techniques.[44] Results showed that categorical perception could be demonstrated for non-speech stimuli in adults and for speech stimuli in both human infants and non-human animals,[45] suggesting that the mechanism underlying categorical perception in humans is shared with other animals and may have evolved at least as far back as our divergence point with birds. Although this finding does not rule out the importance of categorical perception in speech processing, it strongly suggests that the underlying mechanism is unlikely to have evolved for speech. In other words, the capacity to treat an acoustic continuum as being composed of discrete acoustic categories is a general auditory mechanism that evolved before humans began producing and perceiving the sounds of speech.

The comparative speech perception research program described above could be criticized for the following two reasons. First, speech scientists might argue that all of the studies conducted to date are based on extensive training regimes and thus fail to show what animals spontaneously perceive or, more appropriately, *how* they actually perceive the stimuli. Second, because nearly all studies have focused on aspects of categorical perception, one might also argue that the range of phenomena explored is narrow and thus fails to capture the essential design features of language.[46] Building on the elegant studies described above, we have been pushing the development of methodological tools that involve no training and can be used with animals or human infants, thereby providing a more direct route to understanding which mechanisms are spontaneously available to animals for processing speech and which are uniquely human.

The first habituation-discrimination experiment on speech perception in animals[12] explored whether the capacity of human infants to discriminate between, and subsequently acquire, two natural languages is based on a mechanism that is uniquely human or shared with other species. More generally, this experiment sought evidence that, like humans, non-human primates are capable of extracting the rhythmic properties of language and use this level of analysis to distinguish one language from another. To explore this problem, we asked whether French-born human neonates and cotton-top tamarins can discriminate sentences of Dutch from sentences of Japanese, and whether the capacity to discriminate these two languages depends on whether they are played in a forward (i.e., normal) or backward direction. Given the fact that adult humans process backward speech quite differently from forward speech, we expected to find some differences, though not necessarily in both species. Neonates failed to discriminate the two languages played forward. Rather than run the backward condition with natural speech, we decided to synthesize the sentences and run the experiment again, with new subjects. One explanation for the failure with natural speech was that discrimination was impaired by the significant acoustic variability imposed by the different speakers. Consequently, synthetic speech provides a tool for looking at language discrimination, while eliminating speaker variability. When synthetic speech was used, neonates showed discrimination of the two languages, but only if the sentences were played in the normal, forward direction. In contrast to the neonates, tamarins showed discrimination of the two languages played in a forward direction, for both natural and synthetic sentences. Like neonates, they also failed to discriminate Dutch from English when the sentences were played backward. This suggests that in the absence of training, both human neonates and tamarins are able to use speech-specific acoustic cues to discriminate between two human languages.

A real-world problem facing the human infant is how to segment the continuous acoustic stream of speech into functional units, such as words and phrases. How, more specifically, does the infant know where one word ends and another begins? Saffran et al.[47] designed an experiment to test whether human infants are equipped with mechanisms that enable them to extract statistical regularities from a sequence of syllables. Eight-month-old infants were first familiarized for two minutes with a continuous string of synthetically created syllables. Within this continuous acoustic stream, some three-syllable sequences had a 1.0 probability of occurring together (words), some had a transitional probability of 1.0 for the first two syllables to occur together but only 0.33 for the third syllable (part-words), while others were never associated (non-words). Results showed that infants were sensitive to the transitional probabilities between words, being more responsive to non-words and part-words than to words. Although this kind of computation might help children on their path to acquiring a language, these results do not inform the relevant comparative questions. Specifically, are the computations that human infants use to segment a continuous acoustic stream uniquely human and, equally important, special to language?

To address this question, Hauser et al.[48] used the original Saffran et al.[47] stimuli and a similar experimental assay in order to attempt a replication with cotton-top tamarins of the statistical learning effects observed with human infants. Like the infants, several outcomes are possible. If tamarins simply respond to novelty as opposed to familiarity, then they might show a significantly higher level of response (i.e., orienting to the concealed speaker) in the word-vs.-non-word condition but fail to show a difference between the word and part-word conditions. In other words, because the first two syllables are familiar in the part-word comparison, while the third is novel, this difference may be insufficient to differentiate the two test items. In contrast, if tamarins compute the transitional probabilities, then non-words and part-words are both novel and should elicit a greater number of responses when contrasted with words. Like human infants, tamarins oriented to playbacks of non-words and part-words more often than to words. This result is powerful, not only because tamarins show the same kind of response as do human infants, but because the methods and stimuli were largely the same and involved no training.

Building on many elegant earlier studies, our research program on speech perception in tamarins has attempted to broaden the field of inquiry to experiments other than categorical perception and employ methods other than operant conditioning. We show that areas of speech other than categorical perception can be studied in non-human animals and that such data are key to elucidating questions about which mechanisms of speech perception are unique to humans and which are shared by other animals. Further, we argue that employing experimental assays that use subjects' spontaneous behavior as the dependent measure is an effective way to study speech perception in non-human animals. By employing comparable experimental assays in humans and non-humans, we will better inform our claims about the uniqueness of speech perception in the evolution of communication.

III. DISCUSSION

This chapter provides a review of some of the mechanisms underlying acoustic perception in non-human primates, focusing in particular on our studies of cotton-top tamarins. Although our understanding of such mechanisms is nowhere near as precise as that of vocal perception in other taxa, several aspects of the system are ideally suited for more in-depth analyses of how individuals process species-specific and heterospecific vocal signals. Moreover, because human and non-human primates share such close proximity, comparative studies of primate vocal perception can shed light on the extent to which different mechanisms are homologous as opposed to analogous. These data will not only impact our understanding of auditory perception at the behavioral and perceptual levels, but will also generate insights into the neural mechanisms underlying primate communication.[49-51] For example, when tamarins produce antiphonal responses, what circuitry is recruited to coordinate both the rapid recognition of a species-typical call followed by the initiation of an appropriate species-typical response? Do tamarins, like other species,[52,53] possess combination-sensitive neurons that respond specifically to the multisyllabic acoustic structure of species-typical signals? When an individual habituates to a class of stimuli, where in the auditory system does such habituation arise? Would it be possible to run the habituation-discrimination experiments at the single-unit level, testing neuronal populations from different levels of the auditory stream? Given the rapid developments of tools for examining neural function, it should be possible to answer these questions in the near future. If the spectacular discoveries from insects, bats, songbirds, and anurans are any indication, we are in for some fascinating findings from primates as well.

REFERENCES

1. Ryan, M.J., *The Tungara Frog: A Study in Sexual Selection and Communication*, University of Chicago Press, 1985.
2. Ryan, M.J., Female mate choice in a Neotropical frog, *Science*, 209, 523–525, 1980.
3. Rand, A.S. and Ryan, M.J., The adaptive significance of a complex vocal repertoire in a Neotropical frog, *Zeitschrift fur Tierpsychologie*, 57, 209–214, 1981.
4. Ryan, M.J., Fox, J.H., Wilczynski, W., and Rand, A.S., Sexual selection for sensory exploitation in the frog, *Physalaemus pustulosus*, *Nature*, 343, 66–67, 1990.
5. Hauser, M.D., *The Evolution of Communication*, MIT Press, Cambridge, MA, 1996.
6. Rylands, A.B., *Marmosets and Tamarins: Systematics, Behaviour, and Ecology*, Oxford University Press, London, 1993.
7. Cleveland, J. and Snowdon, C.T., The complex vocal repertoire of the adult cotton-top tamarin, *Saguinus oedipus oedipus*, *Zeitschrift fur Tierpsychologie*, 58, 231–270, 1981.
8. Ghazanfar, A.A., Flombaum, J.I., Miller, C.T., and Hauser, M.D., The units of peception in cotton-top tamarin (*Saguinus oedipus*) long calls, *J. Comp. Physiol. A*, 187, 27–35, 2001.
9. Weiss, D.J., Garibaldi, B.T., and Hauser, M.D., The production and perception of long calls in cotton-top tamarins (*Saguinus oedipus*), *J. Comp. Psychol.*, 11, 258–271, 2001.
10. Hauser, M.D. and Konishi, M., *The Design of Animal Communication*, MIT Press, Cambridge, MA, 1999.
11. Miller, C.T., Miller, J., Costa, R.G.D., and Hauser, M.D., Selective phonotaxis by cotton-top tamarins (*Saguinus oedipus*), *Behaviour*, 138, 811–826, 2001.
12. Ramus, F., Hauser, M.D., Miller, C.T., Morris, D., and Mehler, J., Language discrimination by human newborns and cotton-top tamarin monkeys, *Science*, 288, 349–351, 2000.
13. Ghazanfar, A.A., Smith-Rohrberg, D., Pollen, A., and Hauser, M.D., Temporal cues in the antiphonal calling behaviour of cotton-top tamarins, *Animal Behav.*, in press.
14. Miller, C.T., Dibble, E., and Hauser, M.D., Amodal completion of acoustic signals by a nonhuman primate, *Nature Neurosci.*, 4, 783–784, 2001.
15. Jordan, K., Weiss, D.J., and Hauser, M.D., Individual recognition and antiphonal responses to loud contact calls produced by cotton-top tamarins, *Am. J. Primatol.*, in press.
16. Cheney, D.L. and Seyfarth, R.M., Assessment of meaning and the detection of unreliable signals by vervet monkeys, *Animal Behav.*, 36, 477–486, 1988.
17. Hauser, M.D., Functional referents and acoustic similarity: field playback experiments with rhesus monkeys, *Animal Behav.*, 55, 1647–1658, 1998.
18. Gerhardt, H.C., Acoustic communication in two groups of closely related treefrogs, *Adv. Study Behav.*, 30, 99–167, 2001.
19. Miller, C.T. and Ghazanfar, A.A., Meaningful acoustic units in nonhuman primate vocal behavior, in *The Cognitive Animal*, M. Bekoff, C. Allen, and G. Burghardt, Eds., MIT Press, Cambridge, MA, 2002.
20. Narins, P.M. and Capranica, R.R., Communicative significance of the two-note call of the treefrog *Eleutherodactylus coqui*, *J. Comp. Physiol. A*, 127, 1–9, 1978.
21. Searcy, W.A. and Nowicki, S., Functions of song variation in song sparrows, in *The Design of Animal Communication*, M.D. Hauser and M. Konishi, Eds., MIT Press, Cambridge, MA, 1999, pp. 577–596.

22. Nelson, D.A. and Marler, P., Categorical perception of a natural stimulus continuum: birdsong, *Science*, 244, 976–978, 1989.
23. Searcy, W.A., Podos, J., Peters, S., and Nowicki, S., Discrimination of song types and variants in song sparrows, *Animal Behav.*, 49, 1219–1226, 1995.
24. Mitani, J. and Marler, P., A phonological analysis of male gibbon singing behavior, *Behaviour*, 109, 20–45, 1989.
25. Bregman, A.S., *Auditory Scene Analysis: The Perceptual Organization of Sound*, MIT Press, Cambridge, MA, 1990.
26. Bradbury, J.W. and Vehrencamp, S.L., *Principles of Animal Communicatoin*, Sinauer Associates, Sunderland, MA, 1998.
27. Emlen, S.T., An experimental analysis of the parameters of bird song eliciting species recognition, *Behaviour*, 41, 130–171, 1972.
28. Green, S., Variation of vocal pattern with social situation in the Japanese macaque (*Macaca fuscata*): a field study, in *Primate Behavior*, Vol. 4, L.A. Rosenblum, Ed., Academic Press, New York, 1975, pp. 1–102.
29. Petersen, M.R., Beecher, M.D., Zoloth, S.R., Moody, D.B., and Stebbins, W.C., Neural lateralization of species-specific vocalizations by Japanese macaques, *Science*, 202, 324–326, 1978.
30. May, B.J., Moody, D.B., and Stebbins, W.C., The significant features of Japanese monkey coo sounds: a psychophysical study, *Animal Behav.*, 36, 1432–1444, 1988.
31. May, B., Moody, D.B., and Stebbins, W.C., Categorical perception of conspecific communication sounds by Japanese macaques, *Macaca fuscata*, *J. Acoust. Soc. Am.*, 85, 837–847, 1989.
32. Hopp, S.L., Sinnott, J.M., Owren, M.J., and Petersen, M.R., Differential sensitivity of Japanese macaques (*Macaca fuscata*) and humans (*Homo sapiens*) to peak position along a synthetic coo call continuum, *J. Comp. Psychol.*, 106, 128–136, 1992.
33. Le Prell, C.G. and Moody, D.B., Perceptual salience of acoustic features of Japanese monkey coo calls, *J. Comp. Psychol.*, 111, 261–274, 1997.
34. Owren, M.J., Acoustic classification of alarm calls by vervet monkeys (*Cercopithecus aethiops*) and humans. II. Synthetic calls, *J. Comp. Psychol.*, 104, 29–40, 1990.
35. Hauser, M.D., Orienting asymmetries in rhesus monkeys: the effect of time-domain changes on acoustic perception, *Animal Behav.*, 56, 41–47, 1998.
36. Robinson, J. G., An analysis of the organization of vocal communication in the titi monkey *Callicebus moloch*, *Zeitschrift fur Tierpsychologie*, 49, 381–405, 1979.
37. Masataka, N., Categorical responses to natural and synthesized alarm calls in Goeldi's monkeys (*Callimico goeldii*), *Primates*, 24, 40–51, 1983.
38. Masataka, N., Psycholinguistic analyses of alarm calls of Japanese monkeys (*Macaca fuscata fuscata*), *Am. J. Primatol.*, 5, 111–125, 1983.
39. Snowdon, C.T. and Pola, Y.V., Interspecific and intraspecific responses to synthesized pygmy marmoset vocalizations, *Animal Behav.*, 26, 192–206, 1978.
40. Weiss, D.W. and Hauser, M.D., Perception of harmonics in the combination long call of cotton-top (*Saguinus oedipus*), *Animal Behav.*, in press.
41. Handel, S., Timbre perception and auditory object identification, in *Hearing*, B.C.J. Moore, Ed., Academic Press, San Diego, CA, 1995.
42. Stebbins, W.C., *The Acoustic Sense of Animals*, Harvard University Press, Cambridge, MA, 1983.
43. Liberman, A.M., Cooper, F.S., Shankweiler, D.P., and Studdert-Kennedy, M., Perception of the speech code, *Psychol. Rev.*, 74, 431–461, 1967.
44. Harnad, S., *Categorical Perception: The Groundwork of Cognition*, Cambridge University Press, Cambridge, U.K., 1987.

45. Kuhl, P.K., On babies, birds, modules and mechanisms: a comparitive approach to the acquisition of vocal communication, in *The Comparative Psychology of Audition*, R.J. Dooling and S.H. Hulse, Eds., Lawrence Erlbaum, Hillsdale, NJ, 1989, pp. 379–422.
46. Trout, J.D., The biological basis of speech: what to infer from talking to animals, *Psychol. Rev.*, 108, 523–549, 2001.
47. Saffran, J.R., Aslin, R.N., and Newport, E.L., Statistical learning by 8-month-old infants, *Science*, 274, 1926–1928, 1996.
48. Hauser, M.D., Newport, E., and Aslin, R., Segmentation of the speech stream in a non-human primate: statistical learning in cotton-top tamarins, *Cognition*, 78, B53–B64, 2001.
49. Ghazanfar, A.A. and Hauser, M.D., Neuroethology of primate communication: substrates for the evolution of speech, *Trends Cognitive Sci.*, 10, 377–384, 1999.
50. Ghazanfar, A.A and Hauser, M.D., The auditory behaviour of primates: a neuroethological perspective, *Curr. Opin. Neurobiol.*, 11, 712–720, 2001.
51. Wang, X., On cortical encoding of vocal communication sounds in primates, *Proc. Natl. Acad. Sci.*, 97, 11843–11849, 2000.
52. O'Neill, W.E. and Suga, N., Target range-sensitive neurons in the auditory cortex of the mustache bat, *Science*, 251, 565–568, 1979.
53. Margoliash, D. and Fortune, E.S., Temporal and harmonic combination-sensitive neurons in the zebra finch's HVc, *J. Neurosci.*, 12, 4309–4326, 1992.

5 Psychophysical and Perceptual Studies of Primate Communication Calls

Colleen G. Le Prell and David B. Moody

CONTENTS

I. Introduction 62
   A. Methods for Studying Primate Vocalization Perception
      in Laboratory Settings 63
      1. Standard Threshold Assessment 63
      2. Category Identification 63
      3. Multidimensional Scaling 64
      4. Acoustic Processing vs. Communicative Processing 65
   B. Acoustic Features of Primate Vocalizations 66
      1. Frequency-Based Acoustic Cues in Communication Calls 66
      2. Amplitude-Based Acoustic Cues in Communication Calls 66
      3. Phase-Based Acoustic Cues in Communication Calls 67
II. Results from Studies of Macaque Vocalizations 67
   A. Macaque Monkeys: An Overview 67
   B. Coo Calls 67
      1. Field Observations of Coo Call Vocalizations 67
      2. Laboratory Experiments on Macaque Vocal Discrimination 68
      3. Do Discrete SEH/SLH Category Boundaries Exist? 72
      4. Frequency Cues vs. Amplitude Cues 73
      5. Summary 76
   C. Scream Vocalizations 76
      1. Field Observations 77
      2. Laboratory Experiments on Vocal Discrimination 77
      3. Do Discrete Category Boundaries Exist? 78
      4. Summary 78
III. Conclusions and Future Directions 78
Acknowledgments 79
References 80

I. INTRODUCTION

Studies of animal communication systems have been largely composed of comparisons with human language. Indeed, one of the most obvious reasons to study primate communication is to identify parallels between human and non-human primate signal production and use. With the exception of grunts,[1] the signals used by human and non-human primate species tend to be acoustically different. These acoustic differences result from differences in vocal tract morphology and articulatory apparatuses (for reviews, see Snowden[2] and Fitch and Hauser[3]). Investigators have thus taken two distinctly different approaches to the study of primate communication skills. In one approach, primate communication abilities have been assessed by the extent to which they learn human communication systems, including speech,[4,5] sign language,[6-9] and arbitrary symbols.[10,11] This search for parallels with human language is a bias introduced by our familiarity with the syntax of human language and by the methods developed for analyzing human speech sounds. In this review, we will instead evaluate the production and perception of species-typical vocal signals by monkeys. We will focus on macaques because of the wealth of field and laboratory studies examining their vocalizations.

The combination of field and laboratory studies is critical for a complete understanding of an animal's communicative repertoire. Initially, correlation of vocalizations to socioecological context and/or a change in an animal's behavior is presumed to suggest meaningful signal classes.[12-17] Playback studies typically follow observational work. Demonstration that playback of various signals reliably elicits different responses further supports meaningful signal class.[12,18-22] In laboratory-based experiments, animals can be isolated and perception of vocal signals further examined.[23-30] This approach can be used to define the stimulus parameters to which animals are (or are not) sensitive; that is, we can assess the minimum change necessary for discriminable differences.

Presumably, test animals will be relatively more sensitive to those features that are communicatively relevant and for which specialized processing mechanisms may exist (for discussion, see Beecher et al.,[23] Zoloth et al.,[30] Marler,[31] and Petersen et al.[32]). Acoustic features of vocalizations relevant to communication can be suggested based on increased sensitivity to changes in those features. Alternatively, one can train subjects to classify natural stimuli into different categories and then examine the rules subjects use to define those categories.[33] A third strategy recently applied to the study of primate vocal behavior uses the statistical technique of multidimensional scaling (MDS).[34-36]

In this review, we will integrate field and laboratory studies examining macaque communication. First, we describe the benefits and caveats of various laboratory tasks used to test different hypotheses. Second, we present the current evidence for the primary acoustic features used by Japanese and rhesus macaques to discriminate their vocalizations.

A. Methods for Studying Primate Vocalization Perception in Laboratory Settings

1. Standard Threshold Assessment

Sensitivity of non-human primates to various acoustic features has been examined using techniques based on those used to determine human sensitivity. To assess absolute sensitivity, monkeys are trained to perform an operant response upon detection of a suprathreshold tone. Tone level and tone frequency can be varied and thresholds estimated using a variety of psychophysical methods (for review, see Niemiec and Moody[37]).

This task has two basic variations (for review, see Dooling and Okanoya[38]). In the *go/no-go* test procedure, subjects listen to a repeating standard. After a random number of standard stimulus presentations, a test stimulus that differs along some dimension, such as frequency or intensity, is presented. In contrast, in the *same/different* procedure, subjects are presented with two stimuli, which can be either the same or different. Subjects are required to perform one response when the stimuli are the same or an alternative response when they detect a difference between the two stimuli.

Animals ranging from rodents to primates can be readily trained to perform this simple task, and data showing the salience of single acoustic features can be obtained within relatively brief time periods. This task is quite useful for determining sensitivity to simple acoustic parameters, and such data are often used to suggest that a given acoustic feature could be important within more complex vocalization stimuli. The most significant limitation on this task is that an animal is highly trained for attending to a single acoustic parameter. Thus, what we are in fact measuring is how well subjects can learn to attend to a single acoustic feature, rather than the extent to which the acoustic parameter is typically attended.[39]

2. Category Identification

The goal of category identification testing in the laboratory is to determine if animals can sort vocalizations presumed to have different meanings into different categories based on vocalization acoustics. Owren used category identification procedures to determine if vervet monkeys discretely categorize snake and eagle alarm call classes.[28,29] Free-ranging vervet monkeys produce predator-specific alarm calls that elicit distinct and adaptive responses from conspecifics.[15–17] Specifically, leopard alarms result in tree-climbing, eagle alarms result in looking to the sky and retreating into dense brush, and snake alarms result in looking to the ground. To examine alarm call categorization, Owren used a two-choice operant task. Subjects were required to make contact with a response tube during a period of silence. Following a variable-duration silent interval, subjects were presented with either an eagle or a snake alarm call. The subject was then required to push the tube either to the right, or to the left, depending on the perceived call type.

Identification of perceptual categories has also been probed for Japanese macaque subjects tested with coo call vocalizations. These experiments were based on field observations suggesting that different types of coo calls were used in specific social situations.[13] Perceptual testing for natural categories of Japanese macaque coo calls was conducted using a discrimination-based, go/no-go paradigm (for review, see Moody[33] and Stebbins[39]). The go/no-go test paradigm is similar to the two-choice task used by Owren;[28,29] however, in the go/no-go paradigm, subjects are trained to treat vocalizations from only one category as target stimuli that require an active response. The subject must perform an operant holding response during presentation of non-target stimuli and, upon detecting a vocalization from the target category, perform a release response in order to earn a food pellet. These techniques have been used extensively by Moody, Stebbins, and their colleagues to examine coo call classification by Japanese macaques.[23,26,27,30,32,40] In particular, these investigations have focused on the smooth-early-high (SEH) and smooth-late-high (SLH) coo calls described by Green.[13]

One advantage of categorical testing is that perceptual classification of species-typical vocalizations can be examined. In addition, these tests can be used to examine the phenomenon known as categorical perception. Categorical perception is evident when graded stimuli are perceived not as falling along a continuum but rather as belonging to discrete categories sharply separated by some natural category boundary. In addition, using the appropriate task, it is more difficult to discriminate among stimuli that belong to the same category than stimuli belonging to different categories. Numerous examples of categorical perception of speech phonemes by non-human species are available (for review, see Kuhl[41]). The ability of Japanese macaques to perceive SEH and SLH stimuli categorically has been examined by multiple groups.[25,27] These results will be discussed in greater detail later in this review.

A primary criticism of category identification tests is that subjects are trained by the investigator to use different categories of stimuli.[42] Thus, categorization schemes revealed during perceptual tests are only biologically relevant to the extent that the trained categories represent natural, meaningful, signal classes. This point is highlighted by categorical perception tests in which speech sounds have been categorically perceived by animals. Speech sounds clearly are not biologically relevant for animal subjects; therefore, in this example, the categorical perception is a function of general auditory system function, rather than a language-specific processing specialization.

3. Multidimensional Scaling

Multidimensional scaling (MDS) allows investigators to examine perceptual processes in the absence of training to attend to specific call features. MDS uses statistical procedures to translate the psychological concept of similarity into the geometrical concept of distance.[43] Non-human subjects are trained to perform an operant response following detection of a target stimulus. Multiple stimuli are presented in different pairings of standard and test stimulus. If perception of speech phoneme similarity was under investigation, a subject might be presented with

/da/–/ta/ stimulus pairs on some trials, and /ba/–/da/ or /ta/–/ba/ stimulus pairs on other trials. Based on latency of response to stimulus change, estimates of stimulus similarity would then be determined. The assumption underlying this model is that two stimuli perceived to be very different are very easy to discriminate, thus the operant reporting response is performed rapidly. Stimuli not easily discriminated require a longer discrimination decision process and thus a longer response period.[44]

The stimulus presentation procedure and the required operant response have two variants. One set of procedures requires that subjects perform an observing response during presentation of a repeating standard, followed by a reporting response signaling the detection of a test stimulus that varies from the repeating standard (i.e., a go/no-go task).[45] The alternative procedure requires a subject to perform an observing response to initiate presentation of two calls, which can be either the same or different. If, and only if, the two stimuli differ, then the subject must perform a reporting response (i.e., a same/different task).[44,46–48]

Multidimensional scaling has been used to examine perception of vowels by humans.[49–53] Results from MDS investigations paralleled those obtained using other signal detection strategies.[50] Given this success, MDS was adapted for use in an animal model to examine the perception of simple acoustic stimuli,[47] speech,[54,55] and species-specific calls.[34–36,44–46,48]

Multidimensional scaling procedures present a seemingly ideal solution to the problems encountered in previous laboratory analyses that examined natural perceptual categories of non-human subjects. The MDS procedure does not require that subjects undergo training to attend to an investigator-selected criterion. Instead, subjects are allowed to respond based on any acoustic feature. Placement of vocalizations within the perceptual space is then correlated with acoustic features of the calls. MDS tasks, however, are subject to the criticism that subjects are performing repetitive discrimination tasks. Memory demands imposed by the task may not accurately reflect the demands imposed during communicative interactions with vocalizing animals.

4. Acoustic Processing vs. Communicative Processing

Because different psychophysical tasks have sometimes yielded different results, it seems that some tasks assess speech-like phenomena and others do not (for discussion, see Moody et al.[56]). Effects such as these are often described using a variety of terms. For example, Kuhl[41] distinguishes between "auditory" and "phonetic" processing. Sinnott et al.[57] make a similar distinction, classifying processing as "sensory" or "memory." Finally, Terbeek[58] distinguishes "psychophysical mode" from "language mode." The fundamental distinction for each of these categories appears to be a separation of acoustic cues, which can be resolved by general auditory processes, from communication cues, for which specialized processing mechanisms presumably exist. The identification of perceptual predispositions to attend to meaningful dimensions is a critical element in distinguishing between acoustic cues and communication cues.[31]

An ideal demonstration separating acoustic and communicative processing mechanisms has yet to be designed; however, several types of evidence could indicate communicative processing is used. First, species differences in absolute sensitivity to variation along some acoustic dimension could reflect heightened sensitivity to specific acoustic features that are communicatively important. Second, the demonstration that graded stimuli are consistently perceived in specific categories may reflect that such categories are communicatively meaningful. Evidence that stimuli are used in specific behavioral contexts or elicit specific responses from conspecific group members enhances such an interpretation. Finally, identifying the acoustic features that are most salient to members of a species, in the absence of specific training to attend to those features, could be taken to indicate that members of that species preferentially attend to such features. Each of these approaches has been used to study the perception of coo calls and, more recently, scream vocalizations by macaque monkeys. Results from these studies will be discussed in detail following a brief review of acoustic cues within a vocal signal.

B. Acoustic Features of Primate Vocalizations

Simple acoustic stimuli can be defined in terms of frequency, amplitude, and phase. Although the perceptual correlates of frequency (pitch) and amplitude (loudness) have been clearly identified, the perceptual consequences of phase manipulations remain less well understood. Complex acoustic stimuli, such as primate vocalizations, can be defined as having multiple frequency components that independently vary along the amplitude and phase dimensions. Descriptions of each of these physical characteristics of sounds are readily available.[59,60]

1. Frequency-Based Acoustic Cues in Communication Calls

Many primate vocalizations are composed of a fundamental frequency of phonation (F_0) and several harmonic multiples of the fundamental (H2, H3, H4, and, sometimes, H5). It is standard to refer to the fundamental as F_0[42,61] or F0,[62,63] although it has also been referred to as H1,[64] or the first harmonic.[65] Common frequency-based differences in communication signals are differences in the maximum or minimum value of F_0 or differences in the rate (or magnitude) of change across F_0. The temporal pattern of changes in the F_0 contour may also be described.

2. Amplitude-Based Acoustic Cues in Communication Calls

Modulation of the amplitude contours of harmonic components in vocal signals is determined primarily by properties of the vocal tract of the vocalizing animal. Because the vocal tract shape reinforces energy at particular frequencies, the amplitude of certain harmonics that accompany F_0 may be modified as a function of the supralaryngeal vocal tract (for reviews, see Fitch and Hauser[3] and Owren and Linker[63]). That is, some frequencies are amplified, whereas others are damped; these frequencies define the transfer function of the vocal tract. More thorough descriptions of vocal tract anatomy and function in primates are available elsewhere.[3,62,63,66]

The relationship between vocal tract transfer functions and vocalization acoustics has been described in Sykes's monkeys.[62] In that investigation, one source of acoustic variation was changing articulation. It seems likely that rhesus monkeys also have sufficient control over their articulatory gestures to effectively change vocal tract configuration and thus vocalization acoustics.[67–69] For example, dropping of the mandible during coo production is accompanied by an increase in dominant frequency of the coo vocalization.[68] The hypothesis that non-human primates may control some articulation gestures does not necessarily imply purposeful control of the supralaryngeal vocal tract. Instead, vocal tract configuration may be determined as a consequence of facial configurations that depend on emotional state (for a brief review, see Hauser et al.[68]). This relationship is complicated, however, as unique facial configurations may have become associated with specific social situations as information-bearing dimensions of vocal signals were maximized (for review, see Fitch[66]). For example, vocal tract length determines formant dispersion and thus provides an acoustic cue indicating body size of the vocalizer. To acoustically maximize estimates of vocalizer body size by conspecifics, a vocalizing animal might purse its lips with a nearly closed mouth. This facial configuration is, in fact, used during aggressive interactions across a wide range of species (for review, see Fitch[66]).

3. Phase-Based Acoustic Cues in Communication Calls

Variation in temporal envelope of a vocal signal is largely due to variation in stimulus amplitude; however, phase differences could also contribute. Moody et al. demonstrated that macaque subjects are able to resolve acoustic differences produced as a function of phase changes.[70] Because these experiments were conducted using three-tone complexes, however, the data do not provide direct evidence that macaque monkeys attend to phase cues in vocal signals. Primate vocalizations contain a great deal of variation in the frequency and amplitude domains, suggesting vocalizations should be more readily distinguished on those bases. It remains possible, however, that temporal envelope shape, which varies as a function of both phase and signal amplitude, provides an important acoustic signal in the perception of primate vocalizations.

III. RESULTS FROM STUDIES OF MACAQUE VOCALIZATIONS

A. Macaque Monkeys: An Overview

Macaques are Old World monkeys belonging to the Cercopithecidae family and the Cercopithecinae subfamily.[71] Geographical ranges and primary ecological niches for numerous species are well known.[72] In brief, *Macaca* species are found primarily in Northern Africa and Asia, and, although native to South Asia, at least one species (*M. mulatta*) has been introduced to Cayo Santiago in the West Indies. The vocal repertoires of various macaque species, especially Japanese and rhesus macaques, have been particularly well investigated.

B. Coo Calls

1. Field Observations of Coo Call Vocalizations

Coo calls are produced by free-ranging and captive macaques of many species, particularly during feeding and troop movement.[13,14,42,61,64,67,73–85] These are tonal vocalizations, with energy concentrated into harmonically related bands. A number of categorization schemes for their different coo calls have been proposed for the Japanese macaque. Itani[80] described 15 variants of "sounds generally emitted in peaceful states of emotion." These variants were differentiated based on the type of sound (e.g., light, calm, nasal, soft, shrill, deep, flat, low) as well as subtle differences, such as intonation. Differences in gender, age, and situational specificity were noted across classes.

While Itani[80] does not specifically refer to the observed calls as coo calls, it appears likely that coos are the general class of vocalizations, given the overlap in acoustic characteristics and the situational usage of these calls and calls later termed "coo calls." In the categorization scheme proposed by Green, seven fundamental frequency patterns within coo vocalizations were noted.[13] Green argued that these coo subtypes were roughly correlated with ten specific social interaction patterns. Inoue[79] has since suggested that the coo classes described by Green were not related to context or internal motivational state, but rather were a consequence of age-related variation. In addition, Inoue[79] reported greater variability in the number of fundamental frequency patterns, describing 55 possible patterns of coo modulation.

Examining rhesus macaque coos does not clarify the basis for modulation of coo calls. Rowell and Hinde[84] describe a class of "clear" calls, which appear to be coo (or coo-like) calls. These calls are composed of harmonic components, they have no marked rise or fall in pitch, and they tend to be used during movement or separation and when food is presented. Precise classifications were not attempted, and large individual differences were noted. Hauser[61] more closely examined variation in rhesus monkey coos by observing four prototypical contours for calls produced by a single animal. When discriminant function analysis was applied to the coos produced by this animal, the context in which the coo was produced could be predicted; however, coo context could not be reliably predicted when the data set included coos produced by multiple individuals. In contrast, coo calls could be correctly assigned to individual animals within this larger data set; that is, vocalizer identity could be determined.

2. Laboratory Experiments on Macaque Vocal Discrimination

A number of laboratory tests have examined macaque discrimination of coo calls. The earliest laboratory experiments tested the hypothesis that two coo subtypes, the smooth-early-high (SEH) and smooth-late-high (SLH) coo calls, could be reliably discriminated by Japanese macaques. Green[13] proposed that Japanese macaque SEH and SLH coo calls were defined by a boundary two thirds of the way through the call duration. Calls in which the transition from rising to falling frequency contour occurred during the first two thirds of the call were termed SEH, and calls during which this transition occurred in the final one third of the call were termed SLH. In the laboratory, Japanese macaques learned to discriminate SEH and SLH calls.[23,30] More importantly, they learned this particular discrimination task more quickly than control species.[23,30] Because the Japanese macaque subjects experienced greater difficulty in learning to classify coo calls based on pitch, general differences in task acquisition could not be used to explain the species difference in SEH/SLH discrimination ability (Figure 5.1). These results are particularly compelling, in that the test and control species share similar audiograms but show striking differences in their ability to detect differences in specific frequency-based acoustic features of communication stimuli.

Japanese macaque subjects were later shown to perform the SEH/SLH discrimination task more proficiently when coo calls were presented to the right ear rather than the left ear.[32,40] This processing asymmetry was not evident in control species. Cortical lesion studies support the hypothesis that hemisphere-specific specialized processing

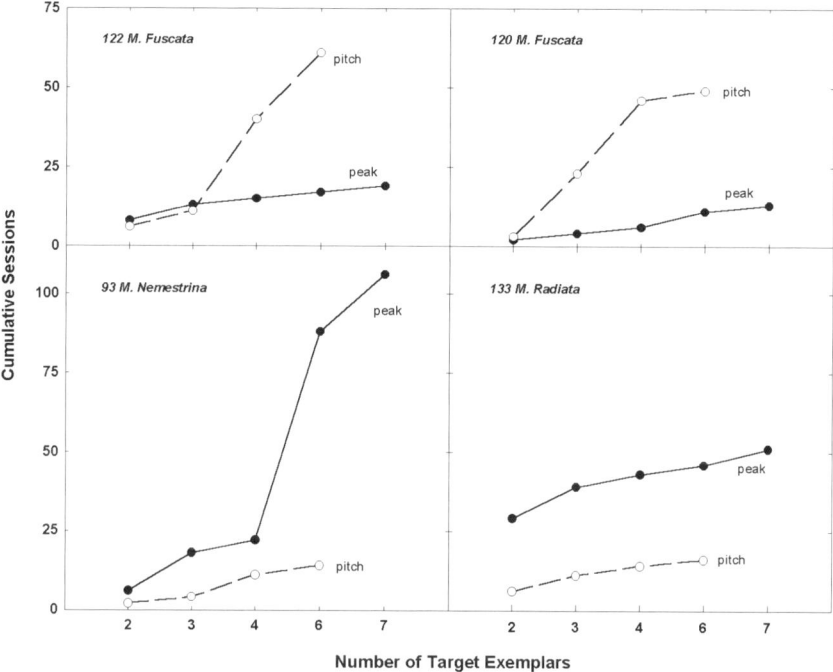

FIGURE 5.1 Cumulative learning curves for two discrimination tasks are shown for four subjects drawn from different macaque species. In one task (the "pitch" discrimination task), stimulus classification was based on the starting pitch of a coo vocalization (less than or greater than 600 Hz). In the other task (the "peak" discrimination task), stimulus classification was based on the relative placement of a frequency peak, early in the coo call or late in the coo call. *Macaca fuscata* subjects, for whom the peak-sorting task was presumed to be communicatively relevant, learned the peak-sorting task most quickly regardless of the order in which they were trained in the tasks. In contrast, the closely related *Macaca nemestrina* and *Macaca radiata* subjects learned the pitch-sorting task most quickly. In these diagrams, each point represents the total (cumulative) number of sessions necessary to meet an arbitrary performance criterion. The total number of target exemplars in each stage of training is shown along the *x*-axis. The targets were discriminated from an equal number of background stimuli. The subjects on the left were tested on the pitch task first; those on the right were first tested on the peak task. (Adapted with permission from Zoloth et al.[30]. Copyright 1979, American Association for the Advancement of Science.)

centers are used by Japanese macaque subjects discriminating SEH and SLH coo calls.[86,87] This ear advantage would not be expected based on the general auditory capabilities of the species tested and could not have been as readily probed in a field setting. Thus, laboratory-based studies were critical to determining that Japanese macaques have a processing asymmetry in the discrimination of coo call vocalizations.

The perceptual salience of the acoustic feature differentiating SEH and SLH coo calls was carefully probed by May and colleagues.[26] They tested the hypothesis that placement of the frequency inflection was a salient acoustic feature, as shown by the monkey's ability to classify different vocalizations based on this feature. They then demonstrated that manipulation of the frequency inflection altered classification of coos as SEH or SLH.[26,27] The ability of Japanese monkeys to discriminate SEH and SLH stimuli has since been demonstrated by other groups.[25,86] However, we point out that in the task used by Hopp et al.,[25] subjects were trained to discriminate any detectable difference in the temporal placement of a frequency inflection. Thus, while subjects in all of the above investigations were trained to attend to the same dimension, subjects used by Hopp et al. were not specifically taught to use the SEH and SLH categories.

While this evidence supports the hypothesis that SEH and SLH coo calls are communicatively distinct signals, these studies had a fundamental limiting factor. Specifically, the subjects in most of the above investigations were trained to attend to the investigator-defined SEH/SLH categories.[42] To address this criticism, we tested Japanese macaque subjects using MDS procedures to identify perceptual similarity among coo stimuli.[35,36] Our data did not support the hypothesis that SEH and SLH coo calls form discrete perceptual categories in the absence of specific training to use these categories. Instead, overlap in the perception of SEH and SLH stimuli was evident, suggesting much more graded perception of the SEH and SLH coo call stimuli. This result perhaps helps to explain the controversy surrounding the question of discrete category boundaries for the discrimination of SEH and SLH coo call stimuli by Japanese macaque subjects.

Perceptual data from rhesus monkeys have focused on identity-related cues rather than context-specific variation. Discriminant function analyses have been used to demonstrate that rhesus monkey coos contain significant information about vocalizer identity and matriline relationships.[61,88] Although the ability of rhesus monkeys to resolve specific identity-based acoustic cues in laboratory settings has not been examined, field-based evidence shows that rhesus macaques respond differently to coos produced by kin and non-kin.[89] Although coos are produced by rhesus monkeys in different social contexts, including food-related contexts as well as grooming and troop movement, systematic differences in coos produced in different social contexts have not been reported.[14,61]

We recently completed an investigation demonstrating that rhesus monkeys can learn to classify their calls on the basis of the Japanese macaque SEH/SLH coo classification scheme or a high/low-frequency modulation scheme.[90] Each of the rhesus subjects learned these two tasks at approximately equivalent rates (Figure 5.2). We believe the best interpretation of these data is that rhesus macaques can learn a sorting task based on either temporal or frequency-based characteristics of communication stimuli. Although the ability to sort communication stimuli based

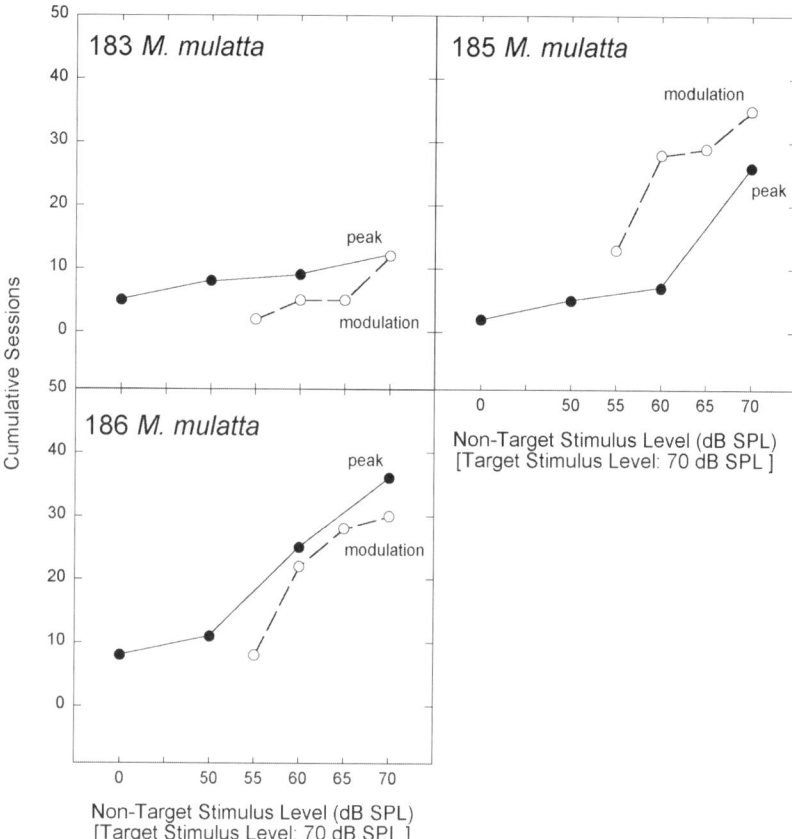

FIGURE 5.2 Cumulative learning curves for two discrimination tasks are shown for three rhesus macaque (*Macaca mulatta*) subjects. In one task (the "modulation" discrimination task), stimulus classification was based on the extent of frequency modulation within a coo vocalization. The investigator-defined classes included high-frequency modulation (500–750 Hz) and low-frequency modulation (122–269 Hz). Peak placement was systematically varied within these two categories, occurring early in the call (0.2–0.4), across the middle of the call (0.4–0.6), or toward the end of the call (0.6–0.8). In the other task (the "peak" discrimination task), stimulus classification was based on the relative placement of a frequency peak, early in the coo call or late in the coo call. Rhesus subjects learned the two tasks at approximately equal rates. In these diagrams, each point represents the total (cumulative) number of sessions necessary to meet an arbitrary performance criterion. During the training for each task, level differences were used to distinguish the target stimuli (to which the subjects responded by releasing a metal response cylinder) from non-target stimuli. During presentation of non-target stimuli, subjects maintained contact with the response cylinder. The magnitude of the level difference cue is depicted along the x-axis. For each task, the target stimuli were presented at approximately 70-dB SPL, and the non-target stimuli were attenuated by decreasing increments until target and non-target stimuli were at equivalent sound levels. The subjects were all tested on the peak task first.

on different acoustic features suggests that those features could be communicatively meaningful, the fact that these two tasks were learned at equivalent rates suggests that neither of these tasks is communicatively relevant. There is currently no evidence suggesting that rhesus actually use either temporal cues or frequency modulation cues in coo calls for communication with conspecifics.

3. Do Discrete SEH/SLH Category Boundaries Exist?

To test the discreteness of the SEH/SLH category boundary, May and colleagues[27] trained Japanese macaques to categorize SEH and SLH coo calls and then conducted generalization tests. Synthetically generated coo stimuli were designed to systematically probe the category boundary. Using this approach, the SEH/SLH category boundary was not as discrete as proposed by Green.[13] Japanese macaque subjects trained to respond to either SEH or SLH category exemplars all demonstrated high response rates to unfamiliar coo calls containing peaks in the mid-call region (from 40 to 60% of the call duration).[27] Although May et al.[27] proposed that calls with mid-vocalization peak placements may be atypical, Owren and Casale[42] have since shown that, at least within their captive colony, such calls are produced quite regularly.

Although coos with mid-coo peak placements were not classified discretely, appropriate classification of coo calls in which the frequency peak was located near the call endpoint was considered evidence of categorical learning.[26,27] In addition, by using test procedures designed to specifically probe the ability to detect a change in temporal peak position, May et al.[27] found that discrimination was enhanced at the category boundary, as expected in categorical perception; however, observations of enhanced discriminability were highly dependent on the task the subject performed. It was only under conditions of high uncertainty (i.e., a task in which memory plays a prominent role and arguably is akin to a phonetic task) that the enhanced discrimination at category boundaries was readily apparent. Low-uncertainty testing, using tests in which monkey subjects could directly compare the test stimulus to a repeating standard in a presumably acoustic (non-phonetic) task, revealed limited enhancement at the category boundaries (for additional discussion, see Moody et al.[56]).

Hopp and colleagues[25] also examined categorical perception of coo calls by Japanese macaques. Subjects in that investigation performed a same/different discrimination task. Because this task is much like the low-uncertainty task used by May et al.,[27] one could predict that discrimination enhancement at the category boundary should be limited. In fact, no enhancement of discrimination at the hypothesized category boundary was observed. As noted by Hopp et al.,[25] the subject populations differed in that the subjects used by May et al.[27] had received previous training defining specific coo call categories, whereas the subjects used by Hopp et al. had not received such training. Thus, Hopp et al.[25] concluded that the monkeys used by May et al.[26,27] may have performed discriminations based on spectral characteristics of the frequency sweep rather than on the temporal position of the frequency peak. In contrast, they argued that their monkeys discriminated stimuli based on temporal position of the peak given a steady-state interval that may have directed their attention to temporal features of the vocal signals.[25]

Taken together, the evidence for categorical perception of Japanese macaque coo calls based on variation in the temporal placement of a frequency inflection is not definitive. When discrimination behavior that appears to reflect categorical discrimination has been observed, highly trained subjects in very specific tasks have been used. However, selection of the task, based on memory demands, is critical to the demonstration of categorical perception in human as well as animal subjects.[27,56,57,91] In addition to being task specific, these results can be criticized on the basis that the tests used a relatively small sample of coos, perhaps failing to reflect the full variety of coo call acoustics. Also, age, gender, and identity characteristics were not systematically probed or correlated with the observed classification behaviors in the experiments conducted by May et al.[26,27] We have assessed the impact of at least one of these factors, subject training. Using the same coo stimuli as May et al.,[26,27] but subjects without specific training to attend to SEH and SLH categories, we found that discrimination strategies were not based simply on frequency contours.[35,36] Instead, at least one investigation suggested that harmonic amplitude may influence perceptual similarity of coo calls.[35] In that investigation, perceptual similarity of coos was determined using MDS analysis of discrimination latencies. Coos with dissimilar frequency peak placements (early or late), but sharing the unique characteristic of a temporary shift in the relative levels of the harmonics (i.e., the second harmonic transiently louder than the first), were perceived as similar (Figures 5.3 and 5.4).

4. Frequency Cues vs. Amplitude Cues

Because coos sharing the unique characteristic of a temporary shift in the relative levels of the harmonics were perceived as similar, we also examined the perceptual consequences of altering the relative levels of coo harmonics.[35] Manipulating the relative levels of the harmonics resulted in shifts in macaque subject estimates of coo call perceptual similarity.[35]

Given that changes in relative harmonic level appear to influence macaque perception of coo call similarity, we explicitly assessed threshold sensitivity for detection of a change in the relative energy distribution in coo call harmonics.[92] We did so by incrementally altering the amplitude of a single harmonic of a synthetic coo call during a restricted time segment. Coo call stimuli were modeled after those used in our earlier experiments.[35,36] The results confirmed that the perceptual changes revealed in our first MDS experiments were a consequence of suprathreshold stimulus manipulations,[35] and the lack of effect of amplitude manipulations in a later experiment was a consequence of subthreshold stimulus manipulations.[36] In addition to detecting amplitude changes in coo calls, Japanese macaques readily detect amplitude changes in vowel formants as well as non-formant components in vowel-like stimuli.[93,94]

Variation in harmonic energy distribution is presumably a function of vocal tract configuration. Because changes in articulation posture can alter vocal tract transfer functions,[62,67–69] variation in harmonic energy distribution may be correlated with changes in vocal tract configuration. Thus, relative harmonic level pattern might provide a communicative cue in the absence of visual contact.[65,95]

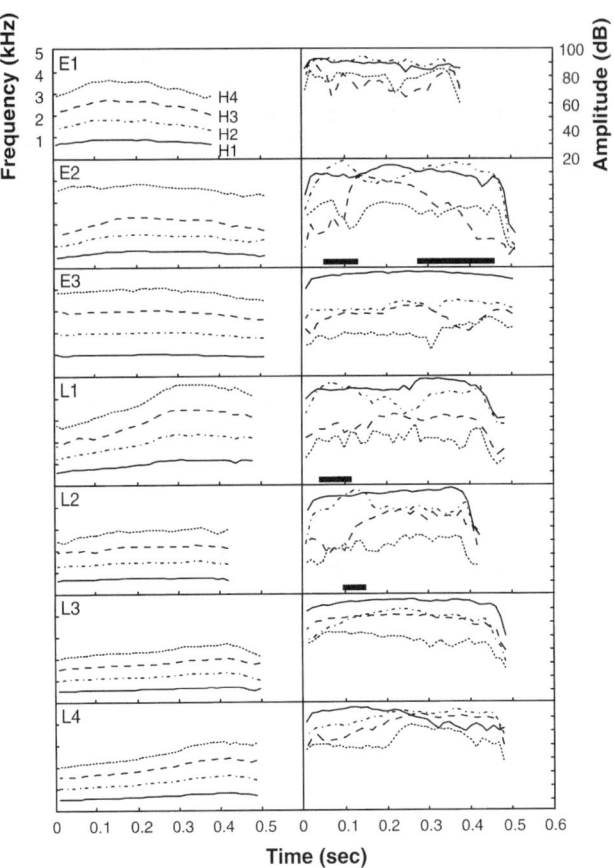

FIGURE 5.3 Summary of frequency (left; H1, H2, H3, H4) and amplitude (right) of field-recorded Japanese macaque coo calls. Frequency and amplitude of the four highest level harmonic components of coo calls were extracted using the Signal software package (Engineering Design, Belmont, MA) to conduct multiple fast Fourier transforms (FFTs) of the digitized waveforms. A Hanning window was used for the FFTs, with a 512-point transform size. Successive transforms were at 5-point intervals. E1, E2, and E3 are smooth-early-high (SEH) coo calls; L1, L2, L3, and L4 are smooth-late-high (SLH) coo calls. (Steven Green provided all field-recorded coo calls.) Calls E2, L1, and L3 each contain a segment where the level of H2 becomes greater than that of H1 (see bars). (Adapted with permission from Le Prell and Moody.[35] Copyright 1997 by the American Psychological Association.)

Alternatively, it may be that amplitude cues such as these provide a cue linking the vocalization to the identity of the vocalizing animal. We note that this amplitude-based cue has yet to be definitively associated with social or environmental context or animal identity.

The specific importance of amplitude-based cues can only be addressed through additional examination of coo calls recorded from identified animals during known situations. Libraries of coo call vocalizations that could be analyzed exist.[13,42,96] Only by looking at a large number of vocalizations from different

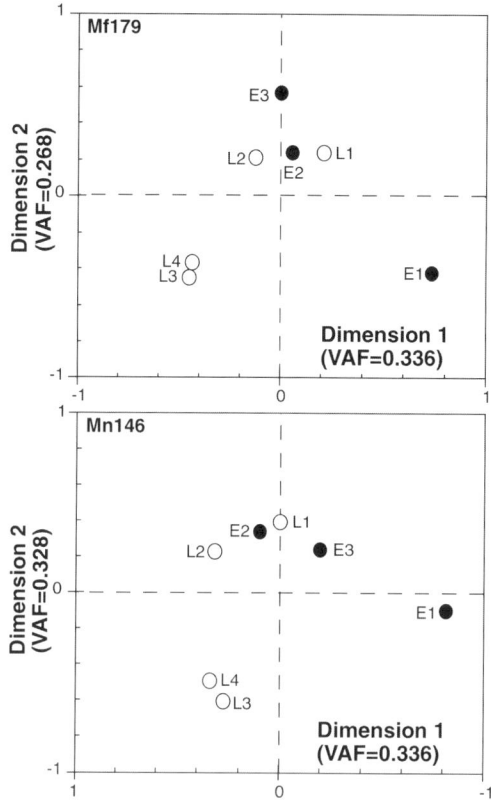

FIGURE 5.4 Multidimensional scaling stimulus space based on response latency data obtained from two macaque subjects (top, Mf179; bottom, Mn146) after discriminations of synthetic coo stimuli closely modeled after field-recorded coos (see Figure 5.3). Macaque subjects maintained contact with a response bar during repeated presentation of a standard coo stimulus and released the bar upon detection of a target coo that differed from the repeating standard. All coo stimuli were presented as standard and target in combination with each of the other coo stimuli. Because the spatial representations indicate perceptual similarity through relative distance between stimuli, for Mn146 we have flipped the axes (i.e., multiplied the values by −1) to maintain the spatial location of stimulus clusters across subjects. Spatial clustering revealed that E1 was perceptually distinct. L3 and L4 were perceived as similar to each other but different from the other coos in the test. Finally, E2, L1, and L2 were perceptually similar; E3 was perceived as more similar to these stimuli than to the other coos in the test set. The single characteristic shared by E2, L1, and L2 was a brief period when the level of H2 increased above the level of H1. (From Le Prell and Moody.[35] Copyright 1997 by the American Psychological Association.)

contexts and correlating the contexts of vocalizations with the patterns of relative harmonic level will any relationship between context, identity, and amplitude-based cues clearly emerge.

5. Summary

Despite the large number of macaque species that produce coo calls, situation-specific use of coo variants is limited to the reports provided by Green[13] and, to a lesser extent, Lillehei and Snowdon.[82] In the latter report, individual differences among call acoustics were emphasized for stumptail macaques. Specific calls could not always be attributed to variations in social context, but situation-specific use of SEH (young alone) and SLH calls (young to mother) was described.[82] We note here, however, that subtle differences in situational specificity are evident. While Green[13] had also reported that SEH coos were produced by young alone, he noted that SLH calls were produced by estrous females, not by young animals vocalizing at their mother.

Taken together, the situation-specific coo call usage described by Green[13] may better reflect individual differences in the acoustics of calls produced by different animals, as suggested by Inoue.[79] We note, though, that stumptail macaques do produce similar coos in some corresponding situations, even after accounting for individual differences.[82] Although this observation supports the conclusion of Green that at least some coo calls are associated with relatively specific social situations,[13] individual differences also appear to be important in Japanese macaque coo call production and perception. Sound spectrum analysis of wild Japanese macaque coo calls resulted in correct identification for two thirds of the adult animals tested.[83] In addition, Japanese macaque mothers seem to be able to identify the vocalizations of their offspring.[97]

Although we consider it unlikely that all of the available data on coo call production and perception reflect only individual identity cues (i.e., no context-specific communicative value), we cannot exclude this possibility. While an interpretation emphasizing individual differences initially appears contrary to the laboratory data available to date, these data are not fundamentally difficult to reconcile. If production of SEH coos rather than SLH coos (or vice versa) reflects an individual animal bias, the ability of an animal to learn to classify calls using these categories nonetheless represents the important communicative task of assigning identity to a vocalizer. In summary, we believe that multiple acoustic features define coo calls. These features no doubt include frequency-based characteristics such as maximum or minimum frequency and extent of frequency modulation. In addition, it now seems certain that vocal tract morphology and articulation result in identity-specific acoustic information. It is critical that the salience of identity-related acoustic cues are specifically probed in a controlled laboratory environment to carefully isolate the effects of familiarity with a known vocalizer from a general ability to discriminate among the vocalizations of different (and unknown) conspecifics.

C. Scream Vocalizations

Scream vocalizations are a second class of vocalizations that macaque monkeys produce. Screams have been particularly well characterized for rhesus monkeys.[12,84,98,100] Pigtailed macaque screams have also been subject to fairly detailed analyses, and a limited amount of information is available with respect to screams produced by stumptail macaques and Sulawesi crested black macaques.[99,101,102]

1. Field Observations

Free-ranging rhesus monkeys on Cayo Santiago, Puerto Rico, produce scream vocalizations within the context of submission.[12,14,88] The presumed purpose of these vocalizations is recruitment of aid during attack by another monkey.[12,99,102,103] Five subclasses of scream vocalizations (arched, tonal, noisy, pulsed, undulated) have been identified,[12] although not all screams are readily assigned to a single category.[104–106] Functionally, the acoustic morphology of rhesus screams co-varies with social ranking of the aggressor and the severity of aggression directed at the caller.[12,99] While multiple macaque monkey species produce similar screams, species differences in situation-specific use have been observed.[99] The observed differences may be a function of species-specific variation in intensity of aggression.[99]

When analyses are restricted to rhesus monkeys, the association of call classes with opponent attributes deviates from that predicted by chance alone, but the calls do not sort perfectly by context.[12] Gouzoules et al.[98] later attributed this finding to "fuzzy logic." Fuzzy logic allows monkeys to categorize opponents not just as "higher ranking" but as "much higher ranking" or "just a little bit higher ranking." Thus, variation in responses to higher ranking opponents does not indicate random variation but, rather, distinctions among higher ranking opponents.

2. Laboratory Experiments on Vocal Discrimination

Given that rhesus monkeys appear to produce scream vocalization subtypes as a function of specific aggressive interactions, we recently tested the hypothesis that arched and tonal scream signals are perceived as belonging to discrete categories.[105] Because there is some overlap in the acoustic characteristics of these calls,[12] one of the main goals of our experiment was to determine if rhesus monkeys can be trained to sort scream vocalizations into discrete categories. Although Gouzoules et al.[12] reported that approximately 90% of the scream calls they recorded could be classified into relatively discrete categories, other observations reveal more gradation and heterogeneity of call bouts, suggesting a graded system.[104–106]

Rhesus macaques were trained to respond to either arched or tonal scream targets using a go/no-go procedure.[105] Various screams belonging to the non-target category were presented prior to target stimulus presentation. Although subjects readily learned to respond appropriately to the training stimuli, ambiguous treatment of unfamiliar stimuli was frequently observed during generalization testing. We therefore concluded that rhesus screams are produced and perceived along a graded continuum. In addition, we proposed that the rhesus subjects attended to a number of acoustic features. Since then, we have tested rhesus monkey subjects with additional digitally synthesized scream exemplars to isolate the acoustic features that determine the gradation across tonal and arched classes. We found that manipulations of the maximum value of F_0 were highly effective in altering stimulus classification responses. These effects were enhanced by changes in the rate of frequency change and frequency bandwidth. Manipulations of harmonic structure were effective in some, but not all, stimulus manipulations.[107]

3. Do Discrete Category Boundaries Exist?

The variety of acoustic features that seem to influence rhesus monkey scream classifications in a laboratory setting is consistent with a significant grading of scream categories.[105,107] The observation of screams containing both tonal and arched features[105–107] also supports the graded vocalization scheme proposed by Rowell.[84,100] Thus, although we only tested two scream call types, it seems unlikely that discrete category boundaries exist.

4. Summary

Multiple types of scream vocalizations are produced by several different macaque species. Levels of intertroop aggression vary across these species, and the use of different scream types varies across situations when the multiple species are compared. Within a given species (i.e., rhesus macaques), production of particular scream types is correlated with specific social altercations but some variability is associated with this observation. When the perception of two different screams was studied in a laboratory setting, the scream classes were perceptually graded. Taken together, the data suggest that scream vocalizations are used to convey information about the nature of aggressive attacks, but the information conveyed is not highly specific.

IV. CONCLUSIONS AND FUTURE DIRECTIONS

Macaque monkeys produce a variety of vocal signals. In this chapter, we have reviewed two such signals: coo calls, which are considered affiliative signals, and screams, which are produced during agonistic encounters. Early work with Japanese macaque coo calls suggested that SEH and SLH coos could be discretely categorized communication signals. In all of the laboratory investigations, however, the subjects were specifically trained to attend to the SEH/SLH frequency-based acoustic cue.[23,26,27,30,32,40] In the few more recent investigations in which macaque subjects were *not* trained to attend to this acoustic dimension, cues in addition to the SEH/SLH frequency-based cue also appeared to influence perception.[35,36] Moreover, other investigations examining context-specific variation in coo production by Japanese and rhesus macaques failed to identify such a relationship.[14,61,79,80] Indeed, cues related to individual animal identity or matrilineal relatedness have now been widely described for macaque species.[61,67,79,88,96] Taken together, the above data suggest that the wealth of acoustic information available to and used by macaques discriminating coo calls is greater than previously thought; however, we note that this suggestion is based on discriminant function analyses and on field observations of animals that respond more quickly to vocalizations produced by kin, relative to non-kin. Demonstration that macaques can discriminate calls produced by different individuals, either kin or non-kin, within a controlled laboratory environment is critical to the argument that acoustic cues that code identity are perceptually salient and useful to macaque monkeys.

The pattern of amplitude modulation within a vocal signal is related to individual-specific characteristics of the vocal tract. Presumably, amplitude cues in vocal signals have not previously been the subject of detailed study because both distance and environment can affect amplitude-based cues.[108-110] However, if subjects attend to the relative amplitudes of the harmonics, rather than some measure of overall amplitude, variation as a function of distance will be reduced because the subjects are attending to differences in harmonic levels. That is, they may judge changes in the level of one harmonic relative to the other harmonics of the same call. Because of the ability of the auditory system to perceptually segregate multiple acoustic sources,[111,112] such comparisons are possible, at least in theory. Perceptual segregation of amplitude-based cues has not been specifically demonstrated in a behavioral task, however.

The current data set indicates that more fine-grained analyses of vocalizations produced in various settings are needed if we are to ever understand the "language" systems used by primate species. The type of task used to probe vocalization systems in future studies is critical. Studies in which animals are trained to attend to specific dimensions clearly identify that subjects belonging to closely related species can process specific acoustic features differently. For example, comparisons across macaque species showed that subjects with similar auditory abilities, as assessed with simple stimuli, dealt differently with various acoustic features of conspecific communication signals (e.g., Figure 5.1). Such results are quite compelling indeed. We have very sensitive measures of general auditory ability for these macaque species; however, there are no known differences in their ability to detect acoustic stimuli or resolve frequency or intensity-based differences.[113-119] The original reports concerning macaque discrimination of SEH and SLH stimuli were quite likely influenced by the categorical training provided to the subjects. Nonetheless, species differences in the ability to learn to categorize SEH and SLH coos were obvious. The demonstration by more recent MDS-based studies (for which no explicit categorical training was provided) that additional acoustic cues may be important in the perception of SEH and SLH calls does not invalidate the previous reports of species differences in sensitivity to this acoustic feature. Instead, these data suggest that the perception of coo calls depends on multiple acoustic features and is thus much richer than originally proposed.

To summarize, it is true that an acoustic feature must be detectable and discriminable to be useful for communication, and we have argued that increased sensitivity to a specific acoustic feature could suggest that the feature is communicatively relevant. However, the fact that an acoustic feature is detectable and discriminable, even if highly discriminable, cannot be taken as proof that the feature is actually used for communication. Thus, future analyses of vocalizations in field and laboratory settings are critical.

ACKNOWLEDGMENTS

Portions of this manuscript are based on a manuscript submitted in partial fulfillment of the requirements for the Doctor of Philosophy degree in the Horace H. Rackham School of Graduate Studies at the University of Michigan.[95] Manuscript preparation was supported by NIH-NIDCD F32DC00367 (CGL) and NIH-NIDCD P01-DC00078 (DBM).

REFERENCES

1. McCune, L. et al., Grunt communication in human infants (*Homo sapiens*), *J. Comp. Psychol.*, 110, 27, 1996.
2. Snowdon, C.T., Language capacities of nonhuman animals, *Yrbk. Phys. Anthropol.*, 33, 215, 1990.
3. Fitch, W.T. and Hauser, M.D., Vocal production in nonhuman primates: acoustics, physiology, and functional constraints on "honest" advertisement, *Am. J. Primatol.*, 37, 191, 1995.
4. Kellogg, W.N. and Kellogg, L.A., *The Ape and The Child*, Hafner Publishing, New York, 1933.
5. Hayes, K.J. and Hayes, C., The intellectual development of a home-raised chimpanzee, *Proc. Am. Philos. Soc.*, 95, 105, 1951.
6. Gardner, R.A. and Gardner, B.T., Teaching sign language to a chimpanzee, *Science*, 165, 1969.
7. Gardner, B.T. and Gardner, R.A., Two-way communication with an infant chimpanzee, in *Behavior of Nonhuman Primates*, Vol. 4, A. Schrier and F. Stollnitz, Eds., Academic Press, New York, 1971, p. 117.
8. Patterson, F.G., The gestures of a gorilla: language acquisition in another pongid, *Brain Lang.*, 5, 72, 1978.
9. Miles, C.L., Apes and language: the search for communicative competence, in *Language in Primates: Perspectives and Implications*, A. deLuce and H.T. Wilder, Eds., Springer–Verlag, New York, 1983, p. 43.
10. Premack, D., Language in chimpanzee?, *Science*, 172, 808, 1972.
11. Rumbaugh, D.M., Gill, T.V., and Von Glaserfeld, E.C., Reading and sentence completion by a chimpanzee, *Science*, 198, 731, 1973.
12. Gouzoules, S., Gouzoules, H., and Marler, P., Rhesus monkey (*Macaca mulatta*) screams: representational signalling in the recruitment of agonistic aid, *Animal Behav.*, 32, 182, 1984.
13. Green, S., Variation of vocal pattern with social situation in the Japanese monkey (*Macaca fuscata*): a field study, in *Primate Behavior*, Vol. 4, L. Rosenbloom, Ed., Academic Press, New York, 1975, p. 1.
14. Hauser, M.D. and Marler, P., Food-associated calls in rhesus macaques (*Macaca mulatta*). I. Socioecological factors, *Behav. Ecol.*, 4, 194, 1993.
15. Seyfarth, R.M., Cheney, D.L., and Marler, P., Monkey responses to three different alarm calls: evidence of predator classification and semantic communication, *Science*, 210, 801, 1980.
16. Seyfarth, R.M., Cheney, D.L., and Marler, P., Vervet monkey alarm calls: semantic communication in a free-ranging primate, *Animal Behav.*, 28, 1070, 1980.
17. Struhsaker, T.T., Auditory communication among vervet monkeys (*Cercopithecus aethiops*), in *Social Communication Among Primates*, S.A. Altmann, Ed., University of Chicago Press, 1967, p. 281.
18. Cheney, D.L. and Seyfarth, R.M., Recognition of other individuals' social relationships by female baboons, *Animal Behav.*, 58, 67, 1999.
19. Cheney, D.L., Seyfarth, R.M., and Silk, J.M., The role of grunts in reconciling opponents and facilitating interactions among adult female baboons, *Animal Behav.*, 50, 249, 1995.
20. Fischer, J., Hammerschmidt, K., and Todt, D., Local variation in Barbary macaque shrill barks, *Animal Behav.*, 56, 623, 1998.

21. Rendall, D. et al., The meaning and function of grunt variants in baboons, *Animal Behav.*, 57, 583, 1999.
22. Zuberbühler, K., Noë, R., and Seyfarth, R.M., Diana monkey long-distance calls: messages for conspecifics and predators, *Animal Behav.*, 53, 589, 1997.
23. Beecher, M. et al., Perception of conspecific vocalizations by Japanese macaques: evidence for selective attention and neural lateralization, *Brain Behav. Evol.*, 16, 443, 1979.
24. Brown, C.H., Beecher, M.D., Moody, D.B., and Stebbins, W.C., Locatability of vocal signals in Old World monkeys: design features for the communication of position, *J. Comp. Physiol. Psychol.*, 93, 806, 1979.
25 Hopp, S.L., Sinnott, J. M., Owren, M.J., and Petersen, M.R., Differential sensitivity of Japanese macaques (*Macaca fuscata*) and humans (*Homo sapiens*) to peak position along a synthetic coo call continuum, *J. Comp. Psychol.*, 106, 128, 1992.
26 May, B., Moody, D., and Stebbins, W., The significant features of Japanese macaque coo sounds: a psychophysical study, *Animal Behav.*, 36, 1432, 1988.
27 May, B., Moody, D., and Stebbins, W., Categorical perception of conspecific communication sounds by Japanese macaques, *Macaca fuscata*, *J. Acoust. Soc. Am.*, 85, 837, 1989.
28. Owren, M., Acoustic classification of alarm calls by vervet monkeys (*Cercopithecus aethiops*) and humans (*Homo sapiens*). I. Natural calls, *J. Comp. Psychol.*, 104, 20, 1990.
29. Owren, M., Acoustic classification of alarm calls by vervet monkeys (*Cercopithecus aethiops*) and humans (*Homo sapiens*). II. Synthetic calls, *J. Comp. Psychol.*, 104, 29, 1990.
30. Zoloth, et al., Species-specific perceptual processing of vocal sounds by monkeys, *Science*, 204, 870, 1979.
31. Marler, P.R., Avian and primate communication: the problem of natural categories, *Neurosci. Biobehav. Rev.*, 6, 87, 1982.
32. Petersen, M.R. et al., Neural lateralization of species-specific vocalizations by Japanese macaques (*Macaca fuscata*), *Science*, 202, 324, 1978.
33. Moody, D.B., Classification and categorization procedures, in *BioMethods*, Vol. 6, *Methods in Comparative Acoustics*, G.M. Klump, R.J. Dooling, R.R. Fay, and W.C. Stebbins, Eds., Birkhauser Verlag, Basel, Switzerland, 1995, p. 293.
34. Brown, C.H., Sinnott, J.M., and Kressley, R.A., Perception of chirps by Sykes's monkeys (*Cercopithecus albogularis*) and humans (*Homo sapiens*), *J. Comp. Psychol.*, 108, 243, 1994.
35. Le Prell, C.G. and Moody, D.B., Perceptual salience of acoustic features of Japanese monkey coo calls, *J. Comp. Psychol.*, 111, 261, 1997.
36. Le Prell, C.G. and Moody, D.B., Factors influencing the salience of temporal cues in the discrimination of Japanese monkey coo calls, *J. Exp. Psychol. Animal Behav. Proc.*, 26, 261, 2000.
37. Niemiec, A.J. and Moody, D.B., Constant stimulus and tracking procedures for measuring sensitivity, in *BioMethods*, Vol. 6, *Methods in Comparative Acoustics*, Klump, G.M., Dooling, R.J., Fay, R.R., and Stebbins, W.C., Eds., Birkhauser Verlag, Basel, Switzerland, 1995, 65.
38. Dooling, R.J. and Okanoya, K., Psychophysical methods for assessing perceptual categories, in *BioMethods, Vol. 6, Methods in Comparative Acoustics*, G.M. Klump, R.J. Dooling, R.R. Fay, and W.C. Stebbins, Eds., Birkhauser Verlag, Basel, Switzerland, 1995, p. 307.

39. Stebbins, W.S., Uncertainty in the study of comparative perception: a methodological challenge, in *BioMethods*, Vol. 6, *Methods in Comparative Acoustics*, G.M. Klump, R.J. Dooling, R.R. Fay, and W.C. Stebbins, Eds., Birkhauser Verlag, Basel, Switzerland, 1995, p. 331.
40. Petersen, M.R. et al., Neural lateralization of vocalizations by Japanese macaques: communicative significance is more important than acoustic structure, *Behav. Neurosci.*, 98, 779, 1984.
41. Kuhl, P.K., Theoretical contributions of tests on animals to the special-mechanisms debate, *Exp. Biol.*, 45, 233, 1986.
42. Owren, M. and Casale, T., Variations in fundamental frequency peak position in Japanese macaque (*Macaca fuscata*) coo calls, *J. Comp. Psychol.*, 108, 291, 1994.
43. Davison, M.L., *Multidimensional Scaling*, Krieger, Malabar, FL, 1992.
44. Dooling, R.J., Perception of complex, species-specific vocalizations by birds and humans, in *The Comparative Psychology of Audition: Perceiving Complex Sounds*, R.J. Dooling and S.H. Hulse, Eds., Lawrence Erlbaum, Hillsdale, NJ, 1989, p. 423.
45. Okanoya, K. and Kimura, T., Acoustical and perceptual structures of sexually dimorphic distance calls in Bengalese finches (*Lonchura striata domestica*), *J. Comp. Psychol.*, 107, 386, 1993.
46. Dooling, R.J., Perception of vocal signals by budgerigars (*Melopsittacus undulatus*), *Exp. Biol.*, 45, 195, 1986.
47. Dooling, R.J. et al., Perceptual organization of acoustic stimuli by budgerigars (*Melopsittacus undulatus*). I. Pure Tones, *J. Comp. Psychol.*, 101, 139, 1987.
48. Dooling, R.J. et al., Perceptual organization of acoustic stimuli by budgerigars (*Melopsittacus undulatus*). II. Vocal signals, *J. Comp. Psychol.*, 101, 367, 1987.
49. Fox, R.A., Auditory contrast and speaker quality variation in vowel perception, *J. Acoust. Soc. Am.*, 77, 1552, 1985.
50. Iverson, P. and Kuhl, P.K., Mapping the perceptual magnet effect for speech using signal detection theory and multidimensional scaling, *J. Acoust. Soc. Am.*, 97, 553, 1995.
51. Kewley-Port, D. and Atal, B.S., Perceptual differences between vowels located in a limited phonetic space, *J. Acoust. Soc. Am.*, 85, 1726, 1989.
52. Pols, L.C.W., van der Kamp, L.J.T., and Plomp, R., Perceptual and physical space of vowel sounds, *J. Acoust. Soc. Am.*, 46, 458, 1969.
53. Singh, S. and Woods, D.R., Perceptual structure of 12 American English vowels, *J. Acoust. Soc. Am.*, 49, 1861, 1971.
54. Okanoya, K. and Dooling, R.J., Obtaining acoustic similarity measures from animals: a method for species comparisons, *J. Acoust. Soc. Am.*, 83, 1690, 1988.
55. Dooling, R.J., Okanoya, K., and Brown, S.D., Speech perception by budgerigars (*Melopsittacus undulatus*): the voiced–voiceless distinction, *Percept. Psychophysiol.*, 46, 65, 1989.
56. Moody, D.B., Stebbins, W.C., and May, B.J., Auditory perception of communication signals by Japanese monkeys, in *Comparative Perception*, Vol. II, *Complex Signals*, W.C. Stebbins and M.A. Berkley, Eds., John Wiley & Sons, New York, 1990, p. 311.
57. Sinnott, J.M. et al., Speech sound discrimination by monkeys and humans, *J. Acoust. Soc. Am.*, 60, 687, 1976.
58. Terbeek, D., A cross-language multidimensional scaling study of vowel perception, in *Working Papers in Phonetics*, Vol. 37, Department of Linguistics, University of California, Los Angeles, 1977, p. 1.
59. Stevens, S.S. and Davis, H.D., *Hearing: Its Psychology and Physiology*, American Institute of Physics, New York, 1983.

60. Yost, W.A. and Nielsen, D.W., *Fundamentals of Hearing: An Introduction*, 2nd ed., Holt, Rhinehart & Winston, New York, 1985.
61. Hauser, M.D., Sources of acoustic variation in rhesus macaque (*Macaca mulatta*) vocalizations, *Ethology*, 89, 29, 1991.
62. Brown, C.H. and Cannito, M.P., Modes of vocal variation in Sykes's monkey (*Cercopithecus albogularis*) squeals, *J. Comp. Psychol.*, 109, 398, 1995.
63. Owren, M.J. and Linker, C.D., Some analysis methods that may be useful to acoustic primatologists, in *Current Topics in Primate Vocal Communication*, E. Zimmermann, J.D. Newman, and U. Jurgens, Eds., Plenum Press, New York, 1995, p. 1.
64. Bauers, K.A. and de Waal, F.B.M., "Coo" vocalizations in stumptailed macaques: a controlled functional analysis, *Behaviour*, 119, 143, 1991.
65. Moody, D.B., Garbe, C.M., and Niemiec, A.J., Auditory communication in Japanese macaques: salience of acoustic stimulus features, in *Advances in Hearing Research*, G.A. Manley, G.M. Klump, C. Koppl, H. Fastl, and H. Oeckinghaus, Eds., World Scientific Publishers, Singapore, 1995, p. 512.
66. Fitch, W.T., Vocal tract length and formant frequency dispersion correlate with body size in rhesus macaques, *J. Acoust. Soc. Am.*, 102, 1213, 1997.
67. Hauser, M., Articulatory and social factors influence the acoustic structure of rhesus monkey vocalizations: a learned mode of production?, *J. Acoust. Soc. Am.*, 91, 2175, 1992.
68. Hauser, M.D., Evans, C.S., and Marler, P., The role of articulation in the production of rhesus monkey (*Macaca mulatta*) vocalizations, *Animal Behav.*, 45, 423, 1993.
69. Hauser, M. and Schön-Ybarra, M., The role of lip configuration in monkey vocalizations: experiments using xylocaine as a nerve block, *Brain Lang.*, 46, 232, 1994.
70. Moody, D.B., Le Prell, C.G., and Niemiec, A.J., Monaural phase discrimination by macaque monkeys: use of multiple cues, *J. Acoust. Soc. Am.*, 103, 2618, 1998.
71. Simons, E.L., *Primate Evolution: An Introduction to Man's Place in Nature*, Macmillan, New York, 1972.
72. Napier, J.R. and Napier, P.H., *A Handbook of Living Primates*, Academic Press, New York, 1967.
73. Bertrand, M., The behavioral repertoire of the stumptail macaque, *Bibl. Primatol.*, 11, 1, 1969.
74. Chevalier-Skolnikoff, S., The ontogeny of communication in the stumptail macaque (*Macaca arctoides*), in *Contributions to Primatology*, Vol. 2, S. Karger, New York, 1974.
75. Dittus, W.P.J., Toque macaque food calls: semantic communication concerning food distribution in the environment, *Animal Behav.*, 32, 470, 1984.
76. Grimm, R.J., Catalogue of sounds of the pig-tailed macaque (*Macaca nemestrina*), *J. Zool.*, 152, 361, 1967.
77. Hohmann, G., Vocal communication of wild bonnet macaques (*Macaca radiata*), *Primates*, 30, 325, 1989.
78. Hohmann, G. and Herzog, M.O., Vocal communication in lion-tailed macaques (*Macaca silenus*), *Folia Primatol.*, 45, 148, 1985.
79. Inoue, M., Age gradations in vocalization and body weight in Japanese monkeys (*Macaca fuscata*), *Folia Primatol.*, 51, 76, 1988.
80. Itani, J., Vocal communication of the wild Japanese monkey, *Primates*, 4, 11, 1963.
81. Kaufman, I.C. and Rosenblum, L.A., A behavioral taxonomy for *Macaca nemestrina* and *Macaca radiata* based on longitudinal observation of family groups in the laboratory, *Primates*, 7, 205, 1966.

82. Lillehei, R.A. and Snowdon, C.T., Individual and situational differences in the vocalizations of young stumptail macaques (*Macaca arctoides*), *Behaviour*, 65, 270, 1978.
83. Mitani, M., Voiceprint identification and its application to sociological studies of wild Japanese monkeys (*Macaca fuscata yakui*), *Primates*, 27, 397, 1986.
84. Rowell, T.E. and Hinde R.A., Vocal communication by the rhesus monkey (*Macaca mulatta*), *Proc. Zool. Soc. Lond.*, 138, 279, 1962.
85. Simons, R.C. and Bielfert, C.F., An experimental study of vocal communication between mother and infant monkeys (*Macaca nemestrina*). *Am. J. Phys. Anthropol.*, 38, 455, 1973.
86. Heffner, H.E. and Heffner, R.S., Temporal lobe lesions and perception of species-specific vocalizations by macaques, *Science*, 226, 75, 1984.
87. Heffner, H.E. and Heffner, R.S., Effect of restricted cortical lesions on absolute thresholds and aphasia-like deficits in Japanese macaques, *Behav. Neurosci.*, 103, 158, 1989.
88. Rendall, D., Owren, M.J., and Rodman, P.S., The role of vocal tract filtering in identity cueing in rhesus monkey (*Macaca mulatta*) vocalizations, *J. Acoust. Soc. Am.*, 103, 602, 1998.
89. Rendall, D., Rodman, P.S., and Emond, R.E., Vocal recognition of individuals and kin in free-ranging rhesus monkeys, *Anim. Behav.*, 51, 1007, 1996.
90. Le Prell, C.G., unpublished data, 2001.
91. Kuhl, P.K. and Miller, J. D., Speech perception by the chinchilla: identification functions for synthetic VOT stimuli, *J. Acoust. Soc. Am.*, 63, 905, 1975.
92. Le Prell, C.G. and Moody, D.B., Detection thresholds for intensity increments in a single harmonic of synthetic Japanese monkey coo calls, *J. Comp. Psychol*, in press.
93. Sinnott, J.M. et al., A multidimensional scaling analysis of vowel discrimination in humans and monkeys, *Percept. Psychophysiol.*, 8, 1214, 1997.
94. Le Prell, C.G., Niemiec, A.J., and Moody, D.B., Macaque thresholds for detecting increases in intensity: effects of formant structure, *Hear. Res.*, 162, 29, 2001.
95. Le Prell, C.G., Significance of Frequency- and Amplitude-Based Temporal Cues in Japanese Macaque Coo Vocalizations, UMI Dissertation Services, Ann Arbor, MI, 1998.
96. Owren, M.J. et al., "Food" calls produced by adult female rhesus (*Macaca mulatta*) and Japanese (*M. fuscata*) macaques, their normally raised offspring, and offspring cross-fostered between species, *Behaviour*, 120, 218, 1992.
97. Pereira, M.E., Maternal recognition of juvenile offspring coo vocalizations in Japanese macaques, *Animal Behav.*, 34, 935, 1986.
98. Gouzoules, H., Gouzoules, S., and Tomaszycki, M., Agonistic screams and the classification of dominance relationships: are monkeys fuzzy logicians?, *Animal Behav.*, 55, 51, 1998.
99. Gouzoules, H. and Gouzoules, S., Agonistic screams differ among four species of macaques: the significance of motivation-structural rules, *Anim. Behav.*, 59, 501, 2000.
100. Rowell, T.E., Agonistic noises of the rhesus monkey (*Macaca mulatta*), *Symp. Zool. Soc. Lond.*, 8, 91, 1962.
101. Gouzoules, H. and Gouzoules, S., Sex differences in the acquisition of communicative competence by pigtail macaques (*Macaca nemestrina*), *Am. J. Primatol.*, 19, 163, 1989.
102. Gouzoules, S. and Gouzoules, H., Design features and developmental modification of pigtail macaque, (*Macaca nemestrina*), agonistic screams, *Animal Behav.*, 37, 383, 1989.

103. Gouzoules, H., Gouzoules, S., and Marler, P., Vocal communication: a vehicle for the study of social relationships, in *The Cayo Santiago Macaques*, R.G. Rawlins and M.J. Kessler, Eds., State University of New York Press, Albany, 1986, p. 111.
104. Kitko, R., Gesser, D., and Owren, M.J., Noisy screams of macaques may function to annoy conspecifics, *J. Acoust. Soc. Am.*, 106, 2221, 1999.
105. Le Prell, C.G., Hauser, M.D., and Moody, D.B., Discrete or graded variation within rhesus monkey screams? Psychophysical experiments on classification, *Animal Behav.*, 63, 47, 2002.
106. Owren, M.J. and Nederhouser, M., Functional implications of rhesus macaque scream acoustics, paper presented at the Animal Behavior Society Meeting, Atlanta, GA, 2000.
107. Le Prell, C.G. and Moody, D.B., Critical features of arched and tonal rhesus monkey scream vocalizations, unpublished data.
108. Brown, C.H. and Waser, P.M., Environmental influences on the structure of primate vocalizations, in *Primate Vocal Communication*, D. Todt, P. Goedeking, and D. Symmes, Eds., Springer–Verlag, Berlin, 1988, p. 51.
109. Brown, C.H., The acoustic ecology of East African primates and the perception of vocal signals by grey-cheeked mangabeys and blue monkeys, in *The Comparative Psychology of Audition: Perceiving Complex Sounds*, R.J. Dooling and S.H. Hulse, Eds., Lawrence Erlbaum, Hillsdale, NJ, 1989, p. 201.
110. Brown, C.H. and Sinnott, J.M., The perception of complex acoustic patterns in noise by blue monkey (*Cercopithecus mitis*) and human listeners, *Int. J. Comp. Psychol.*, 4, 79, 1990.
111. Hulse, S.H., MacDougall-Shackleton, S.A., and Wisniewski, A.B., Auditory scene analysis by songbirds: stream segregation of birdsong by European starlings (*Sturnus vulgaris*), *J. Comp. Psychol.*, 111, 3, 1997.
112. Wisniewski, A.B. and Hulse, S.H., Auditory scene analysis in European starlings (*Sturnus vulgaris*): discrimination of song segments, their segregation from multiple and reversed conspecific songs, and evidence for conspecific song categorization, *J. Comp. Psychol.*, 111, 337, 1997.
113. Gourevitch, G., Detectability of tones in quiet and in noise by rats and monkeys, in *Animal Psychophysics: The Design and Conduct of Sensory Experiments*, W.C. Stebbins, Ed., Appleton-Century-Crofts, New York, 1970, p. 67.
114. Jackson, L.L., Heffner, R.S., and Heffner, H.E., Free-field audiogram of the Japanese macaque (*Macaca fuscata*), *J. Acoust. Soc. Am.*, 106, 3017, 1999.
115. Lasky, R.E., Soto, A.A., Luck, M.L., and Laughlin, N.K., Otoacoustic emission, evoked potential, and behavioral auditory thresholds in the rhesus monkey (*Macaca mulatta*), *Hear. Res.*, 136, 35, 1999.
116. Pfingst, B.E. et al., Pure tone thresholds for the rhesus monkey, *Hear. Res.*, 1, 43, 1978.
117. Stebbins, W.C., Hearing of Old World monkeys (*Cercopithecinae*), *Am. J. Phys. Anthrop.*, 38, 357, 1973.
118. Sinnott, J.M., Owren, M.J., and Petersen, M.R., Auditory frequency discrimination in primates: species differences (*Cercopithecus, Macaca, Homo*), *J. Comp. Psychol.*, 101, 126, 1987.
119. Sinnott, J. M., Petersen, M.R., and Hopp, S.L., Frequency and intensity discrimination in humans and monkeys, *J. Acoust. Soc. Am.*, 78, 1977, 1985.

6 Primate Vocal Production and Its Implications for Auditory Research

W. Tecumseh S. Fitch

CONTENTS

I. Introduction 87
II. Principles of Primate Vocal Production 89
    A. The Myoelastic–Aerodynamic Theory of Phonation 89
    B. The Source–Filter Theory of Vocal Production 90
    C. Nonlinear Dynamics in Vocal Production 93
    D. Summary of Vocal Production 94
III. Primate Vocal Adaptations 94
    A. Laryngeal Air Sacs 95
    B. Vocal Membranes 96
    C. Vocal Tract Elongation 96
IV. Putting It to Work: Using the Production Perspective
for Analysis and Synthesis 97
    A. Why Synthesis? 97
    B. Linear Prediction in Bioacoustics 98
    C. Analysis and Synthesis Using Linear Prediction 100
V. Conclusion 102
References 103

I. INTRODUCTION

Vocal production forms an important nexus in acoustic communication, lying at the intersection of physics, physiology, neurobiology, and evolution. The foundation of vocal production is provided by physical acoustics, which specifies the basic principles that underlie any sound production system, including those of nonlinear dynamics (for understanding phonation) and linear acoustics (sound propagation through the vocal tract and into the environment). An equally important foundational field is comparative anatomy and physiology. There is considerable variety in the structure and function of primate vocal anatomy that must be grasped before the diversity of primate vocal repertoires can be fully understood. Acoustics and physiology provide the necessary foundation for understanding the function of the vocal mechanism that the primate nervous system has evolved to control (in terms of motor function) and perceive (in terms of auditory function). Finally, the functioning of the vocal apparatus influences the evolution of primate vocal communication systems. This has negative and positive aspects, because the details of the vocal production system both constrain the signals that can be produced and provide adaptive avenues within which new signals evolve. Thus, an understanding of the basic mechanisms of vocal production is an important component of the ethological approach to auditory perception and the evolution of communication.[1-3]

The intent of this chapter is to demonstrate the relevance of vocal production to scientists mainly interested in primate acoustic communication and auditory neuroscience. Two points will be emphasized, one theoretical and the other practical. First, it is now well understood that different species have different needs and have thus evolved anatomical and neural specializations that are fine-tuned to the physical and social environments in which the species normally functions.[4] An obvious example is provided by the ultrasonic vocalizations of mouse lemurs[5,6] which allow them to communicate without being overheard by predators. The 233 species of primates, divided into 13 different families,[7] live in a very wide variety of habitats from mangrove forest to dry savanna and have an equally impressive diversity of vocal anatomy and vocal communication systems. An auditory neuroscience that wishes to move beyond the study of "the monkey" will need to appreciate the rich variability among the many different non-human primate species and will use these many experiments of nature to achieve a deeper understanding of the evolution of primate communication and the neural mechanisms that subserve it.

Second, and more practically, an understanding of vocal production is necessary for the application and development of new analysis and synthesis techniques for studying vocal communication and auditory perception. Major progress in understanding speech perception was made only after a deep understanding of the physics and physiology of speech production had been attained, particularly the codification of the source–filter theory of speech production.[8,9] This understanding allowed researchers to create synthetic speech signals, systematically varying particular acoustic parameters and observing their perceptual effects. This approach was fundamental to our current understanding of speech perception,[10] and it seems likely that the same will be true in the (still nascent) field of primate vocal production. Fortunately, much of our hard-won understanding of speech production appears to apply to non-human primate vocal production. This means that the tools and concepts of speech science can be adapted, with some modifications, to the study of non-human primate communication.

The following condensed overview of primate vocal production focuses on aspects of vocal production that seem likely to be important to auditory neuroscientists. To conserve space, we will assume some understanding of speech acoustics, as there are many easily accessible introductions to this topic.[1,11,12] A survey of the considerable variability observed in the primate larynx and vocal tract follows the brief introduction to vocal acoustics. Owing to their strong reliance on acoustic communication, primates exhibit a wide variety of interesting vocal modifications, which, although poorly understood at present, provide a morphological basis for the wide variation in vocal repertoires. This vocal diversity is probably correlated with variations in the perceptual mechanisms associated with species-specific vocal perception. Finally, a description of the implications of vocal production for analysis and synthesis of primate vocalizations focuses on the utility of linear prediction for analysis of vocal tract function and for the synthesis of natural-sounding acoustic stimuli and addresses both the power and potential pitfalls of this technique. When properly used, such techniques hold substantial promise for furthering the empirical study of primate vocal communication.

II. PRINCIPLES OF PRIMATE VOCAL PRODUCTION

The vocal tract of a terrestrial vertebrate is made up of three components: the lungs, which provide an air flow and the source of energy for phonation; the source, which converts this flow into sound; and a vocal tract, which filters this sound before releasing it into the environment. In most vertebrates and all terrestrial mammals, the source is made up of a pair of vocal folds contained within the larynx. (For a review of vertebrate vocal production in general, see Fitch and Hauser;[13] reviews of mammalian laryngeal anatomy are available in Negus[14] and Harrison.[15])

A. The Myoelastic–Aerodynamic Theory of Phonation

Phonation is the process whereby the steady flow of air out of (or in some cases into) the lungs is converted into sound — that is, rapid pressure fluctuations in air. In primates, this process is accomplished by the vocal folds, which are set into self-sustained oscillations by the air flow. The vocal folds move rapidly in and out of the air stream, gating it into a series of pressure pulses or "puffs" of air and thus creating sound waves with a repetition frequency corresponding to the rate of vocal fold vibrations. This rate is termed the fundamental frequency (abbreviated F_0) and is the primary determinant of the pitch of the emitted sound (thus the term *pitch* should never be used to refer to other frequency components of a vocalization, such as the spectral peaks associated with formants). Typically, the spectrum of the sound created at the larynx includes a prominent component at the fundamental frequency, along with higher spectral components, termed harmonics, at integral multiples of F_0.

Some early researchers believed that each pulse emitted from the larynx was accompanied by an active tensing and relaxation of the vocal fold musculature, the so-called "active" theory of phonation. This theory was proven false for humans by the demonstration of phonation in larynges excised from cadavers and thus deprived of nervous stimulation.[16,17] In such an excised larynx preparation, the vocal folds of a "dead" larynx can be set into normal vibration by simply adjusting their anatomical position and passing warm humid air between them at the proper flow rate and pressure. Such experiments led to the explicit formulation of the myoelastic–aerodynamic theory by van den Berg,[17] which has subsequently been verified with minor modifications by many researchers.[18–20] This theory holds that phonation results from the interaction between myoelastic restoring forces within the vocal folds and aerodynamic driving forces exerted by air flow on the exposed surfaces of the vocal folds. These forces together establish a self-oscillating system, meaning that no active firing of neurons at the rate of vocal fold vibrations is necessary for phonation. Instead, neural signals to the laryngeal musculature (carried by branches of the vagus nerve) control fundamental frequency by varying the tension of the folds and their configuration.

Although the myoelastic–aerodynamic theory is widely accepted for human phonation and probably applies to most other mammals as well,[21,22] at least one example of active phonation is known in mammals: purring by domestic cats. Purring is accomplished via active contractions of the laryngeal musculature at a rate (20–30 Hz) duplicating the purr F_0.[23,24] Because of the limitations on the possible rates of muscular contraction, such active phonation can only be expected to rates below about 40 Hz.[25] Purring is seen in most other felids,[26] and purr-like vocalizations are seen in some non-human primates (e.g., squirrel monkeys, *Saimiri*[27]). Therefore, the possibility that very low-frequency primate vocalizations might be accomplished via an active process should not be ignored. However, the vast majority of primate vocalizations have F_0s considerably exceeding the limits of muscular contraction and are almost certainly produced using the more passive, myoelastic–aerodynamic mechanism.

Although the most common mode of vocal production in primates is phonation, other sound-generating sources are also available. The most primitive source in vertebrates is exemplified by hissing, a sound produced by members of all tetrapod classes. In hissing, air flows within a narrow constriction, generating turbulence. Turbulent noise has a very broad and aperiodic frequency spectrum. Human whispering is produced in this way, with the turbulence formed at the larynx between the non-vibrating vocal folds, and various fricative consonants (e.g., /s/, /sh/, /f/) are produced via constrictions formed elsewhere in the vocal tract (at the teeth or lips).

B. The Source–Filter Theory of Vocal Production

Sound produced in the larynx must travel through the air-filled chambers of the throat (pharynx) and the nasal and oral cavities before being emitted into the environment (Figure 6.1). These air-filled chambers collectively comprise the vocal tract. Like any volume of air, the air contained within the vocal tract has multiple natural modes of vibration or resonance frequencies at which it prefers to vibrate. These vocal tract resonances are referred to as formants (from the Latin *formare*, "to shape"). Formants act as filters that shape the spectrum of the sound created by the source, preferentially passing energy at the formant frequencies. Human speech depends strongly on formants; vowels in all human languages are distinguished by their formant frequencies, and highly accurate speech perception is possible using synthetic signals that discard all other acoustic information besides formants.[28] Formants are manifested acoustically as broad peaks in the spectrum, often encompassing many harmonics. Formants are often particularly prominent in aperiodic (harsh or "noisy") sounds, where the sound source has a very broad frequency spectrum and no harmonics that could potentially be confused with the formants.

The source–filter theory of speech production[8,9] is based on the observation that the contributions to speech of the laryngeal source and the vocal tract filter are clearly separable and additive. The central tenet of this theory is that source and

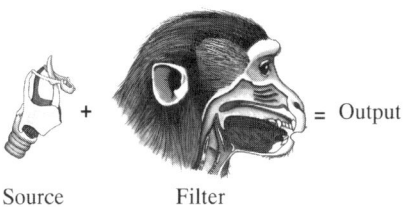

FIGURE 6.1 Schematic of the source–filter theory of vocal production. Sound production starts with the source, which turns the flow of air from the lungs into acoustic energy. In mammals, the source is typically the larynx, which contains the vibrating vocal folds, or vocal cords. These slap together to create a source signal composed of a fundamental frequency and many integer-multiple harmonics of that frequency. This sound then passes through the vocal tract, composed of the pharyngeal, oral, and nasal cavities. The air in these cavities possesses numerous vocal tract resonances, or formants, which filter the source signal to produce an output signal that is the combination of source and filter.

filter are independent. In humans, this is readily demonstrated by the fact that one can sing many different pitches (indicating different rates of vocal fold vibration) with a single vowel or produce many different vowels (indicating different formant frequencies) on the same pitch. The independence of source and filter in vocal production comes as a surprise to physicists familiar with the acoustics of wind instruments, because virtually all wind instruments rely on a strong coupling between source and "filter." Thus, the vibration frequency of the lips on the trombone (which are the source of acoustic energy) is largely determined by the frequencies of the resonances of the air column in the instrument, which are in turn determined by its length. Thus, in wind instruments (trumpets, flutes, clarinets, etc.) source–filter theory does not apply. This fundamental difference is often overlooked, providing a perennial source of confusion in the bioacoustics literature (see Fitch and Hauser[1] for more detail).

The existence of a vocal tract between the larynx and the environment is an anatomical fact for all primates, and the laws of physics dictate that the vocal tract will have resonant frequencies. Thus, it is somewhat surprising that the investigation and appreciation of formants in primate vocal production, despite work by early researchers (e.g., Lieberman[29] and Andrew[30]), are fairly recent phenomena that have blossomed only in the last decade. The reasons for this neglect are obscure. Researchers may have been influenced by the notion that speech is special and, given the central importance of formants to speech, that formants are also special (uniquely human). More pragmatically, it is difficult to demonstrate conclusively that a particular spectral peak represents a formant (i.e., results from a vocal tract resonance) rather than a concentration of spectral energy originating from the source. The most conclusive demonstration of formants comes from experimental studies of vocalizations produced in light gases such as helium. Because the speed of sound is faster in a helium/oxygen atmosphere, formant frequencies are increased, and spectral analysis will show an upward shift of formants in this situation. Experiments with light gases have demonstrated the importance of vocal tract filtering in bird song[31,32] and cetacean clicks.[33] Similar work demonstrates that formant-like spectral peaks in frog vocalizations are *not* the product of vocal tract filtering.[34] To the author's knowledge, no research on non-human primates vocalizing in helium has been published, and other methods have been used to demonstrate the existence of formants in non-human primate vocalizations.

Early workers observed spectral peaks in the vocalizations of apes, noted that the peaks were in roughly the appropriate frequency ranges given the animal's vocal tract length, and hypothesized that these features represented formants.[29] Later workers documented correlations between spectral peaks and facial movements, providing further evidence that such peaks represented formants in macaques and baboons,[30,35,36] and researchers began to apply signal processing techniques appropriate for formants to non-human primates.[37] Perhaps the strongest evidence for formants in non-human primate calls came from the demonstration of strong correlations between vocal tract length, as measured from radiographs of anesthetized macaques, and formant frequencies, as measured from the same individuals' calls.[38] Because resonance frequencies are determined primarily by the length of the vocal tract, a tight correlation between multiple spectral peaks and vocal tract length can only be explained if the peaks indeed represent formants. The calls studied were aperiodic pant-threat calls, thus avoiding any potential confusion between formants and harmonics (Figure 6.2). This technique has since been applied successfully to other mammals,[39,40] as well as to other primates (Fitch, in preparation; J. Fischer, personal communication). It is now widely accepted that the calls of many non-human primates possess formants, and given the clear ability of macaques to perceive formants,[41] recent research has begun to focus on the possible roles of formants in primate communication systems.[42–45] This research endeavor has been significantly aided by an understanding of the production mechanism (see below).

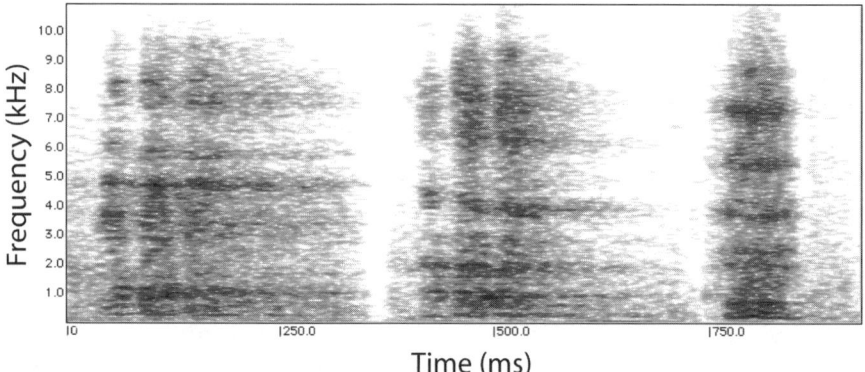

FIGURE 6.2 Spectrogram of a pant-threat vocalization from three different rhesus macaques. These calls are not clearly voiced and thus do not have a fundamental frequency and harmonics and show formant frequencies clearly. The clearly visible horizontal areas indicate the formant frequencies or vocal tract resonances. The calls come from females of increasing weight: 2.6 kg, 5.3 kg, and 9.2 kg, from left to right. The *x*-axis is time in msec; the *y*-axis is frequency in kHz.

C. Nonlinear Dynamics in Vocal Production

The most recent advance in our understanding of non-human vocal production has come from the application of the principles and tools of nonlinear dynamics to animal vocalizations[46,47] (see Fitch et al.[45] for a more detailed introduction). Because the phonation mechanism is irreducibly nonlinear (as shown, for example, by the presence of harmonics in the source spectrum), it is surprising that the first applications of nonlinear dynamical theory to vocal production are quite recent, appearing first for human baby cries[48] and soon after for adult human pathological voices.[49] These seminal insights were quickly appreciated and augmented by leading speech researchers;[50] however, the significance of this research for normal speech is limited because healthy adults typically avoid the bifurcations, subharmonics, and chaos that are signatures of nonlinear dynamics. This limitation does not appear to apply in the case of non-human vocalizations, where such irregular phenomena are quite common (Figure 6.3).[45–47] Thus, nonlinear dynamical theory is probably of greater significance in non-human primate vocalizations than in speech.

The fundamental insight of nonlinear dynamics is that simple deterministic systems can produce very complex and unpredictable output. Although initially understood for weather patterns by Lorenz,[51] the importance of this insight for biological systems was recognized by biologists[52,53] and has since played an important role in many branches of biology, including physiology[54] and neuroscience.[55,56] In vocal production, the two vocal folds provide the simple system of two coupled oscillators, and all of the classic nonlinear phenomena (bifurcations, subharmonics, biphonation, and deterministic chaos) have been observed in this system. A hallmark of nonlinear dynamical systems is deterministic chaos, a very broad-spectrum, unpredictable behavior that nonetheless bears traces of order. It is now clear that the

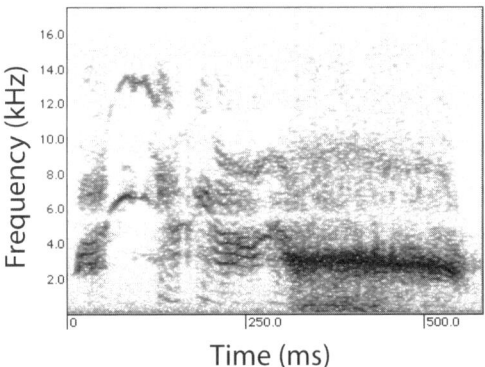

FIGURE 6.3 Spectrogram of a scream vocalization from a juvenile rhesus macaque, illustrating a series of nonlinear transitions. The call starts with a relatively low fundamental frequency, with clear harmonics, visible as horizontal stripe, but quickly transitions to a very high-frequency (around 7 kHz) tonal chirp. At about 100 msec, it returns to the lower frequency regimes until around 300 msec, when it transitions to an irregular chaotic regime, with a clear concentration of energy around 3 kHz, which continues for the rest of the call. The x-axis is time in msec; the y-axis is frequency in kHz.

scream vocalizations that are produced under highly aversive circumstances by most mammals (including humans) represent deterministic chaos. A second signature of nonlinear systems is provided by bifurcations: abrupt transitions from one oscillatory regime to another (Figure 6.3). Such bifurcations can easily be observed in spectrograms of primate vocalizations, particularly in screams, but also in other vocalization types.[45] They may appear as "windows" of tonality in otherwise aperiodic vocalizations or as windows of chaos in otherwise tonal calls. Thus, nonlinear phenomena may provide a neurally "cheap" way to generate a variety of vocal signals and thus to enrich the vocal repertoire.

Although we still have little detailed understanding of the role of nonlinear phenomena in perception, these new insights into vocal production may have interesting implications for auditory neuroscience. First, seemingly disparate phenomena (such as subharmonics or biphonation, which look tonal, and chaos, which looks noisy) often sound surprisingly similar to the human ear. From the perspective of motor theories of perception,[57] this is not surprising. Because the underlying motor-control signals for these superficially different call types might be identical, they would be perceived as similar. If this perceptual equivalence also applies for non-human primates, it would be intriguing to investigate at what level of the nervous system this perceptual equivalence is manifested. Second, nonlinear phenomena may provide distinctive vocal signatures to highlight identity or group membership.[58] Finally, nonlinear phenomena such as screams are often produced in aversive situations and are perceived as unpleasant. It has been hypothesized that their unpleasantness derives from their unpredictability in much the same way that prey species elude capture via unpredictable escape routines.[59] By this hypothesis, it is impossible to habituate to, and thus to ignore, an ever-changing "protean" sound.[44,45]

D. Summary of Vocal Production

Air flow from the lungs is converted into sound by the source, typically the vibrating vocal folds housed in the larynx. The folds can vibrate in several modes. In the vibratory regime most resembling speech, the folds produce a periodic output consisting of a fundamental frequency and its integer-multiple harmonics. Such sounds have a clear pitch determined by the fundamental frequency. Other vibratory regimes of the larynx include deterministic chaos, which underlies screams and other irregular vocalizations. The source-generated sound then passes through the vocal tract, which filters out some frequencies while selectively allowing the passage of others frequencies. The resulting spectral peaks correspond to vocal tract resonances or formants, which are present in many non-human vocalizations, and can be discriminated by macaques. Given the central importance of formant frequencies in human speech perception, it will be very interesting to investigate the degree to which non-human primates use formant frequencies in their communication systems.

III. PRIMATE VOCAL ADAPTATIONS

The principles described above apply to all primates (and indeed to most mammals). Within this unifying framework, however, members of the primate order exhibit an impressive diversity of vocal adaptations: morphological peculiarities that play a role in vocal production. A well-known example is the huge larynx and hyoid apparatus of howler monkeys (genus *Alouatta*) which, adjusted for body size, is the largest larynx among primates. The howler hyolaryngeal source fills the entire space between the jaw and sternum and allows these relatively small monkeys to produce extremely loud, low-pitched vocalization.[60,61] Such differences in the relative size of the larynx can be observed even in closely related species.[62] Following is a brief review of three additional types of adaptations: air sacs, vocal membranes, and vocal tract elongation. The introduction of new imaging tools[63] to the study of primate vocal anatomy should lead to rapid progress in understanding such adaptations.

A. Laryngeal Air Sacs

Many non-human primates possess out-pouchings of the epithelium lining the larynx or vocal tract into diverticula, termed *air sacs*.[14,64] These can be small (the size of a pea in macaques) or large (6 liters in orangutans). In most of these, the opening of the sac is within millimeters of the vocal folds, which strongly suggests an acoustic function. The most common type of air sac is a subhyoid sac, a thin-walled diverticulum lining the hollowed-out body of the hyoid bone and opening into the larynx directly above the vocal folds. This type of sac is seen in almost all Old World monkeys (including vervets, baboons, and macaques) with the exception of a few colobine species.[62] It is also seen in some New World monkeys (*Aotus*, *Pithecia*, *Ateles*, *Lagothrix*, and *Alouatta*), but not in such common laboratory species as *Saimiri* or *Cebus*.[65] Although humans do not normally have air sacs, they do sometimes appear congenitally or with disease, a condition known as laryngocele, which is observed in some 2% of humans.[66] Laryngoceles duplicate precisely the location and structure of great ape air sacs.

Unfortunately, except for the work of Gautier[67] demonstrating a decrease in amplitude of calls from a DeBrazza's monkey (*Cercopithecus neglectus*) with a deflated sac, few empirical data are available to distinguish among the many hypotheses that have been offered for air sac function.[1] These hypotheses include the possibility that air sacs serve as impedance-matching systems, allowing low-frequency sounds to be more efficiently emitted into the environment (as proposed in frogs[68,69]); that they serve as accessory lungs to increase the amount of air in an expiration and thus to increase pressure[70] and/or prolong vocalization;[1] or that they serve as coupled resonators to support phonation at particular frequencies. Less plausible hypotheses include the notion that they serve as oxygen storage mechanisms[14] — implausible because the sacs must be filled with expired air that will be poor in oxygen. Inflated air sacs could even serve as inflatable life-preservers during swimming! (Walrus air sacs appear to serve this function, along with their use in sound production.[71]) Because air sacs may support different functions in different species (or even in the same individual), it is currently difficult to determine which of these hypotheses, if any, applies in any particular species or call type.

An important reason that air sac function deserves scientific attention is that all great apes possess large inflatable air sacs,[70,72,73] but humans lack them. Parsimony leads to the conclusion that the common ancestor of all these species, which would also be the common ancestor of apes and humans, possessed an air sac (other hypotheses require its independent evolution in separate ape species). Thus, the common ancestor of chimpanzees and humans had an air sac, which was lost in humans after our divergence from chimps. This loss of a vocal adaptation is as striking as the descent of the human larynx, but has received much less attention.[74] Understanding why humans lost these air sacs, of course, requires an understanding of the role or air sacs in extant primates.

B. Vocal Membranes

Vocal membranes (often referred to as *vocal lips* after the German "Stimmlippe") are another common feature of the non-human primate vocal apparatus. Vocal membranes are thin, rostrally directed extensions of the vocal folds. They are present in a great diversity of non-human primates; however, it is not always clear from the anatomical literature which species have them. Many of the Old World monkeys have vocal membranes (e.g., *Papio*, *Macaca*, and *Cercopithecus*), as do chimpanzees[75] and gibbons.[76] Most New World monkeys also appear to have vocal membranes (*Cebus*, *Callithrix*, *Aotus*, *Saimiri*, and *Leontopithecus*).[77,78]

Experimental investigations of the function of primate vocal membranes are lacking, but they are widely presumed to enable the production of high-frequency vocalization due to their extremely low mass.[64] This idea is supported by the fact that most microchiropteran bats also have vocal membranes, which are clearly implicated in the production of their ultrasonic echolocation calls.[79-81] Computer models of the vocal membranes suggest that, in addition to supporting high-frequency vocalization, they could improve efficiency of phonation and thus increase amplitude or decrease energy expenditure for loud calls.[82] Vocal membranes may also result in increased coupling between the larynx and vocal tract, potentially leading to an increase in nonlinear phenomena.

C. Vocal Tract Elongation

A final interesting morphological oddity, vocal tract elongation, is seen in our own species and is also known in various other mammals. The oral vocal tract stretches from the larynx to the lips, while the nasal tract spans the larynx and nostrils. Thus, theoretically, the vocal tract can be elongated in three ways: (1) by extending the lips (which many primates do facultatively during calling); (2) by extending the nostrils (by elongating the nose, as in male proboscis monkeys[83]); or (3) by lowering the larynx (as seen in humans, as well as in some deer[40] and large cats, such as lions and leopards[84]). All of these changes have the same effect: by elongating the vocal tract, they lower formant frequencies. It has been proposed[38,40,85,86] that this mechanism serves to exaggerate the impression of body size conveyed by vocalizations. Because vocal tract length is tightly tied to body size in most species,[38,39,87] formant frequencies provide an accurate indication of body size in those species. However, once perceivers begin to rely upon this cue in judging the size of distant or obscured vocalizers, the stage is set for the evolution of a mechanism that fakes large body size by lowering the formants.[88] To the extent that such exaggeration is successful, it will spread through the population to become characteristic of the species as a whole.

Because all primates have a "free" hyoid that is not tightly bound to the skull base by a series of bony links, as in most mammals, their larynx is relatively mobile. Because all species thus far observed cineradiographically lower the larynx during vocalization,[74] it is likely that other primates lower the larynx during vocalization. By the size exaggeration hypothesis, this would be predicted to occur during agonistic vocalizations (such as growls or barks) or in territorial roars. Future observers should be alert to this possibility, as it is sometimes possible to observe laryngeal lowering with the naked eye, and then confirm it via acoustic measurements.[40]

IV. PUTTING IT TO WORK: USING THE PRODUCTION PERSPECTIVE FOR ANALYSIS AND SYNTHESIS

A rich ethological approach to primate communication requires that researchers first study natural calls. This provides a source of hypotheses about call categories, the typical contexts in which calls are produced, and how listeners react to them (e.g., Goodall,[89] Struhsaker,[90] and Hauser[91]). These hypotheses are then tested in perceptual experiments, which can include playing back calls in the field,[92,93] testing calls in a more controlled laboratory setting,[94] or recording neural responses to calls.[95] The ultimate goal of this endeavor is to determine how animals process the complex flow of acoustic information to which they are continually exposed. What acoustic parameters are important? Which ones are ignored? How are these parameters used to categorize calls? The final section of this chapter shows how an understanding of primate vocal production can be used to inform this process. In particular, it focuses on the value of using digital sound synthesis to create natural-sounding primate calls. This allows the experimental manipulation of vocal stimuli to create, for example, call continua that vary along some specific parameter of interest. Though still in its infancy, this approach has great promise.

A. WHY SYNTHESIS?

There are two common approaches to studying auditory function in animals, both of which have corresponding approaches in research on humans. The first, which is favored by psychoacousticians, uses relatively simple signals such as sine waves or noise bursts to explore low-level auditory function. This approach has been used to uncover and explore such key phenomena as equal-loudness contours, critical bands, and the missing fundamental (see Moore[96] for an introduction). Because a linear system can be fully characterized by its response to isolated sine waves (or from its impulse response), this is a reasonable approach to take to initially characterize basic auditory function; however, the auditory system is not linear. It was quickly realized in the 1950s that tone/noise psychophysics would be inadequate to uncover the perceptual mechanisms underlying speech perception, and that richer signals were needed to explore higher level audition.[10] It seems apparent that similar arguments apply to non-human primates as well (e.g., Rauschecker[95]), and that tone/noise psychophysics, while playing an important role, will never suffice for a complete understanding of primate auditory function.

A second common approach, more typical for bioacousticians, is to use unmodified natural calls as stimuli in perceptual experiments. Relative to the previous approach, recorded calls have the advantage of being complex and realistic, thus presumably enlisting the natural perceptual mechanisms that animals use in their communication systems. However, recorded calls have distinct disadvantages as well. First, the vocalizing animal is in control of the call parameters, not the experimenter. This makes the construction of stimulus sets quite difficult, as the researcher must collect a large set of natural calls in order to pick out a few *post hoc* that fall along some continuum (or represent normal exemplars of some category). Second, and more important, it is virtually guaranteed that such a collection of natural calls will also vary in acoustic parameters other than the one of interest, and some of these uncontrolled variables may be significant to the primate listener, but less so (or even inaudible) to the human experimenter. Thus, when using prerecorded calls, we can never be certain of what variables the listener is responding to. It is worthwhile to consider what aspects of human speech perception would have gone undiscovered if only recordings of natural speech had been available. Cues such as formant transitions and voice onset time, and phenomena such as categorical perception, trading relations, prototype magnet effects, duplex perception, and many other central phenomena in speech perception would never had been discovered if researchers had lacked the ability to synthesize speech stimuli.

The solution to these problems is, of course, to synthesize primate calls.[97-99] Modern digital sound synthesis techniques, originally developed for speech but easily modified to work with animal vocalization, allow the creation of synthetic calls that are indistinguishable from natural vocalizations. Such techniques can easily be implemented on standard desktop computers using widely available packages such as Matlab. Although commercial "monkey call synthesis" packages are not available (and seem unlikely to appear), an understanding of call production acoustics, along with the potential pitfalls of the algorithms, puts the creation of natural-sounding synthetic vocalizations within reach of any researcher interested in auditory perception. Following is an overview of one widely available technique, linear prediction, which points out both the pitfalls and power associated with this technique.

B. Linear Prediction in Bioacoustics

Linear prediction (LP) is a powerful spectral modeling technique based on the source–filter theory of speech described above. LP (often called LPC [linear predictive coding] when used for speech compression) became a standard tool for speech engineers during the 1970s.[100,101] Given appropriate vocalizations, linear prediction makes it possible to separate a vocalization into its source and filter components, to modify one or the other, and then put them back together into a highly realistic and natural-seeming synthetic signal. Thus, researchers interested in the perception of vocal tract resonances can modify a specific formant or formants and then recombine the new filter with the original source signal.[99] Researchers interested in pitch perception can modify the LP-derived source using such techniques as phase-vocoding or PSOLA, and then recombine it with the LP-derived filter to obtain natural sounding calls.

Of course, no algorithm can apply to all vocalization types, and linear prediction is no exception. LP is inappropriate for calls where the fundamental frequency is near or exceeds the formant frequencies (high whines or squeaks) or where vocal tract filtering is not evident (e.g., screams). However, linear prediction is appropriate for many calls with relatively low fundamentals or aperiodic sources and clear vocal tract filtering, and the technique has been successfully used both in analysis[38,102,103] and synthesis[97,99] of animal calls. Unfortunately, the value of linear prediction remains relatively unappreciated by researchers in bioacoustics and auditory neuroscience.

The power of linear prediction derives from a very strong limiting assumption about the nature of the signal: that the transfer function contains only resonances (termed *poles*) and not antiresonances (termed *zeros*). A general linear filter can contain either type of spectral feature, but linear prediction assumes an all-pole filter. Fortunately, this assumption is basically correct for speech (with some exceptions, such as nasal sounds such as /m/ and /n/, in which sound resonates in both the nasal and oral cavities). The all-pole model also appears to hold for those mammal vocalizations that have been studied to date (including one primate species). The author[74] studied vocal production in dogs, goats, pigs, and cotton-top tamarin monkeys using cineradiography and found that loud calls in all of these species are made by lowering the larynx into the oral cavity and closing the velum, thus sealing off the nasal cavity and avoiding nasal antiresonances. For calls produced in this manner, the assumption of an all-pole filter appears valid. Other calls (such as pig grunts and dog whines) were produced as purely nasal vocalizations, with the larynx inserted into the nasal passage, so that the oral side-branch that is responsible for the antiresonance in nasalized speech sounds such as /m/ is avoided. Although the trachea, or the complex structure of the nasal passages, might still introduce antiresonances in pure nasals,[104,105] these features appear to be relatively subtle and can be adequately modeled with several poles using linear prediction.[100]

Of course, the production data of Fitch[74] apply only to the species studied, and research must proceed on a case-by-case basis. The primate air sacs described above may introduce antiresonances into the vocal tract filter[106] (although these are not prominent in macaques having a median subhyoid air sac[38]). If so, these calls would require more general models that can include both poles and zeros. In the somewhat arcane terminology of speech engineering, these are often termed ARMA models, because they include both autoregressive (AR, or pole) and moving-average (MA, or zero) components. Unfortunately, the lifting of the all-pole assumption makes it quite difficult to derive accurate ARMA models based solely on the vocal signal. While linear prediction is guaranteed to converge to the optimal model (in a minimum least-squared error sense), no such guarantees are available with unconstrained models such as ARMA. Further, LP is robust to minor deviations from an all-pole system, because zeros can be adequately modeled by multiple poles.[100] Finally, while the correct analysis parameters for linear prediction can be estimated from very basic anatomical data (such as measurements of a skull, see below), the proper number of poles and zeros in an ARMA model is difficult to determine empirically but has a critical effect on the output from the algorithm. Thus, when justified by production data, linear prediction is the preferred technique. The available data, both acoustic and cineradiographic, suggest that the all-pole assumption will be valid for a broad variety of primate vocalizations, underscoring the value of linear prediction in bioacoustics.

C. Analysis and Synthesis Using Linear Prediction

Linear prediction designates a family of spectral estimation techniques; many different algorithms are available to perform LP analysis and different analysis packages implement different algorithms. A complete review of these algorithms or analysis packages is beyond the scope of this chapter; two widely available systems that implement LP and are recommended are Matlab (http://www.mathworks.com/) and Praat (http://www.praat.org). A precise mathematical introduction to LP is given by Markel and Gray,[101] and many practical details relevant to bioacoustics are covered by Owren and Bernacki.[103] This section provides a practical overview of how linear prediction can be used in the synthesis of mammal sounds, highlights the key parameter choices, and stresses some of the potential pitfalls to be avoided. This overview results from the author's years of experience using and teaching linear prediction, first as applied to human speech and later to vertebrate vocalizations, and integrates insights from the literature in speech science, computer music, and vertebrate vocal production.

The essential first step in any LP analysis is deriving an accurate model of the vocal tract transfer function, in which each real formant (in the real vocal tract of the animal) is fitted by a pole (in the mathematical model provided by the LP analysis). Once this is adequately accomplished, we can separate the source from the filter components, and then modify either one before recombining them; or we can cross-synthesize sources and filters from different individuals or even species. This critical and most challenging first step of LP analysis includes the initial modeling of the spectrum of the signal by an all-pole filter. At a minimum, this step requires a high-quality recording (if background noise, such as wind, strong reverberation, or crickets, is present, LP will model the noise along with the animal vocalization of interest; that is, "garbage in gives garbage out"). As mentioned earlier, the vocalization must be also be of a type suited for source–filter analysis in the first place (formants must be present and cleanly outlined by the source). If these minimal criteria are met, the key parameters in an LP analysis are the bandwidth of the initial signal (the sampling rate should be adjusted so that the formants fill the range between 0 Hz and the Nyquist frequency) and the number of poles.

Given a segment of sampled sound and a prespecified number of poles, LP provides the frequencies and amplitudes for the poles that most accurately fit the overall spectrum of the sound. Although the technique is guaranteed to give the optimal solution with a given number of poles, it does *not* guarantee that these poles represent true vocal tract resonances of the vocalizer. This is because LP is critically sensitive to the number of poles chosen for the analysis. For instance, if a human vowel signal possessing four formants is subjected to a two-pole analysis, each pole will be forced to fit two formants and assume an intermediate frequency value and broad bandwidth. These pole frequencies will obviously not correspond to actual formant frequencies. On the other hand, if the same signal is subjected to a 50-pole analysis, each pole might fit one harmonic of the source signal, rather than fitting the formants. Thus, a key to successful LP analysis is the choice of the number of poles. An incorrect choice will produce meaningless results. In my experience, choosing the correct LP parameters requires hours of experimentation and careful listening and a willingness to conclude that a particular signal, or call type, will simply not work.

Choosing the number of poles is best done with some knowledge of the vocal tract length. When the vocal tract length is known, the number of formants (N) can be roughly calculated by:

$$N = \frac{2L}{c} f_c \qquad (6.1)$$

where L is the length of the vocal tract (in m), c is the speed of sound (350 m/sec in the moist 37°C air of a mammalian vocal tract), and f_c is the cutoff frequency (in Hz; e.g., the Nyquist frequency of the sampled signal). Note, however, that often no source energy exists at higher frequencies (e.g., above about 5 kHz in speech) and thus no energy is present in the acoustic signal at higher formant frequencies. Thus, signals should always be appropriately filtered and downsampled so that their spectrum fills the entire frequency range between 0 Hz and the Nyquist frequency (half the sampling rate). For human speech, a Nyquist of around 5 kHz (sampling rate of 10 kHz) often gives good results; for other species, this number may vary widely. Note also that LP often employs several poles as shaping poles (fitting the overall spectrum rather than particular formants), so the number of poles in the LP analysis will typically be greater than the number of formants suspected in the signal based on Eq. (6.1).

As a practical example, let us analyze the pant-threat vocalization of a rhesus macaque, sampled at 22 kHz. The vocal tract length of a rhesus macaque, with lips extended in the grunt or pant-threat position,[107] varies between 5 and 10 cm (for small juvenile and large adult males, respectively), giving between three and seven formants under the 11-kHz Nyquist frequency. Because each pole is specified by two coefficients, and given an extra shaping pole, this gives an appropriate LP order of 8 to 20 coefficients, depending on the size of the animal.[38] Note the lack of standardized terminology for the parameter determining the number of poles; the number of coefficients in the LP filter is twice the number of poles. Researchers must be careful to use the lowest number of poles that achieves a good fit to the spectrum. This is best determined by visual inspection of a frequency–domain plot of the LP impulse response (which shows the combined frequency estimate from all the poles) superimposed over the magnitude spectrum of the original signal.

Once a signal bandwidth and number of poles have been determined that provide an accurate fit to the formants of the vocal tract filter, it is possible to isolate the source signal from the filter. The vocal tract filter function is captured by the LP coefficients (which can now be modified flexibly). To derive the source, we invert this filter and then use it to filter the original source signal. Many speech packages (e.g., Praat) include a special inverse filter function for doing so; in a general-purpose program, such as Matlab, that uses a digital filter framework, this involves simply using the poles, which are the denominator coefficients, as zeros, which are the numerator coefficients. The inverse filtering process, by removing the effect of the vocal tract filter, produces an output that is an estimate of the original source signal. We now have a source signal (representing the output from the larynx, plus a radiation term) and a filter (representing the vocal tract transfer function).

The separated source and filter can now be used for further analysis or modified in various ways. Source pitch or spectral slope can be freely modified, as can individual formant center frequencies or bandwidths or the overall formant spacing. Sources can be combined with filters other than the ones they were originally associated with, allowing cross-synthesis of different call types or even creating hybrid calls by combining the source and filter of different species. Synthesis is completed by passing the source signal (derived by inverse filtering and optionally modified) back through the filter function (derived from LP and optionally modified). If no changes are made to either component, this process will reproduce the original signal almost exactly (to within round-off error), which provides a way of checking that the whole procedure is working correctly (and is strongly recommended). When done correctly, the signals produced by this process are indistinguishable from naturally produced signals (both for humans and those non-human species that have been tested; see, for example, Fitch and Kelley[99]). In summary, linear prediction provides a framework allowing researchers to tear apart a signal into its source and filter components. Source and filter components can then be independently analyzed or modified and finally recombined into highly realistic synthetic calls.

As an example of the types of questions that can be addressed with this technique, rhesus macaques typically assume different facial postures during particular call types. For example, copulation calls or screams are always made with retracted lips, while coos or grunts are made with protruded lips.[35,36] These different articulatory positions modify the vocal tract filter function, both by changing the overall vocal tract length and by subtly changing vocal tract shape. It is currently unknown whether monkeys are sensitive to these acoustic differences or whether the facial postures associated with calls simply represent the visual component of a multimodal signal, the incidental acoustic effects of which are ignored. This question could be answered by cross-synthesizing copulation scream sources with coo or grunt filters (or vice versa) and then using playbacks to determine how such hybrid calls are categorized. Other questions suited to this approach concern judgments of individual identity (do formants carry specific, high-quality cues to individuality, as some researchers[1,44,108] have hypothesized?) or the role of different formants as cues to body size[38] or call category.[97] These questions are interesting in their own right but take on additional importance due to the central importance of formants in human speech. Understanding the perception and categorization of formant-rich vocalizations in non-human primates is thus a crucial component in the quest to understand the evolutionary precursors of speech perception mechanisms.

V. CONCLUSION

This brief review of primate vocal production has provided an overview of recent progress in understanding the physics and physiology of vocal production in non-human primates and of the morphological variability in primate production mechanisms. Although a number of outstanding questions in the field have been highlighted, it should be clear that we now have a firm enough understanding of the basics of primate vocal production to open new realms of investigation to researchers interested in audition. To assist this cross-pollination, this chapter has provided an introduction to linear prediction, a signal processing technique based upon production principles and source–filter theory. Linear prediction and similar techniques provide an important tool that can be used to address previously unapproachable issues in primate vocal perception. In particular, because human speech relies so extensively upon formant perception, more detailed analysis of formant perception in non-human primates should eventually provide important insights into the evolutionary precursors (or lack thereof) of speech and language.[85]

REFERENCES

1. Fitch, W.T. and Hauser, M.D., Vocal production in nonhuman primates: acoustics, physiology, and functional constraints on "honest" advertisement, *Am. J. Primatol.*, 37, 191–219, 1995.
2. Hauser, M.D., *The Evolution of Communication*, MIT Press, Cambridge, MA, 1996.
3. Bradbury, J.W. and Vehrencamp, S.L., *Principles of Animal Communication*, Sinauer Associates, Sunderland, MA, 1998.
4. Zoloth, S.R. et al., Species-specific perceptual processing of vocal sounds by monkeys, *Science*, 204, 870–872, 1979.
5. Zimmermann, E., Castration affects the emission of an ultrasonic vocalization in a nocturnal primate, the grey mouse lemur (*Microcebus murinus*), *Physiol. Behav.*, 60, 693–697, 1996.
6. Zimmermann, E., First record of ultrasound in two prosimian species, *Naturwissenschaften*, 68, 531, 1981.
7. Nowak, R.M., *Walker's Mammals of the World*, Johns Hopkins University Press, Baltimore, MD, 1991.
8. Chiba, T. and Kajiyama, M., *The Vowel: Its Nature and Structure*, Tokyo-Kaiseikan, Tokyo, Japan, 1941.
9. Fant, G., *Acoustic Theory of Speech Production*, Mouton & Co., The Hague, 1960.
10. Liberman, A.M., *Speech: A Special Code*, MIT Press, Cambridge, MA, 1996, p. 458.
11. Titze, I.R., *Principles of Voice Production*, Prentice Hall, Englewood Cliffs, NJ, 1994.
12. Lieberman, P. and Blumstein, S.E., *Speech Physiology, Speech Perception, and Acoustic Phonetics*, Cambridge University Press, Cambridge, U.K., 1988.
13. Fitch, W.T. and Hauser, M.D., Unpacking "honesty": Vertebrate vocal production and the evolution of acoustic signals, in *Acoustic Communication*, A. Simmons, R.R. Fay, and A.N. Popper, Eds., Springer, New York, 2002.
14. Negus, V.E., *The Comparative Anatomy and Physiology of the Larynx*, Hafner Publishing, New York, 1949.
15. Harrison, D.F.N., The anatomy and physiology of the mammalian larynx, Cambridge University Press, New York, 1995.
16. Mueller, J., *The Physiology of the Senses, Voice and Muscular Motion with the Mental Faculties*, Taylor, Walton and Maberly, London, 1848.

17. van den Berg, J., Myoelastic–aerodynamic theory of voice production, *J. Speech Hearing Res.*, 1, 227–244, 1958.
18. Lieberman, P., *Intonation, Perception and Language*, MIT Press, Cambridge, MA, 1967.
19. Titze, I.R., On the mechanics of vocal fold-vibration, *J. Acoust. Soc. Am.*, 60, 1366–1380, 1976.
20. Titze, I.R., Comments on the myoelastic–aerodynamic theory of phonation, *J. Speech Hearing Res.*, 23, 495–510, 1980.
21. Paulsen, K., *Das Prinzip der Stimmbildung in der Wirbeltierreihe und beim Menschen*, Akademische Verlagsgesellschaft, Frankfurt, Germany, 1967, p. 143.
22. Slavitt, D.H., Lipton, R.J., and McCaffrey, T.V., Glottographic analysis of phonation in the excised canine larynx, *Ann. Otol. Rhinol. Laryngol.*, 99, 396–402, 1990.
23. Frazer Sissom, D.E., Rice, D.A., and Peters, G., How cats purr, *J. Zool. (London)*, 223, 67–78, 1991.
24. Remmers, J.E. and Gautier, H., Neural and mechanical mechanisms of feline purring, *Respir. Physiol.*, 16, 351–361, 1972.
25. Hirose, H. et al., An experimental study of the contraction properties of the laryngeal muscles in the cat, *Ann. Otol. Rhinol. Laryngol.*, 78, 297–306, 1969.
26. Peters, G. and M. Hast, Hyoid structure, laryngeal anatomy, and vocalization in felids, *Zeitschrift für Säugetierkunde*, 59, 87–104, 1994.
27. Winter, P., Ploog, D., and Latta, J., Vocal repertoire of the squirrel monkey (*Saimiri sciureus*), its analysis and significance, *Exp. Brain Res.*, 1, 359–384, 1966.
28. Remez, R.E. et al., Speech perception without traditional speech cues, *Science*, 212, 947–950, 1981.
29. Lieberman, P., Primate vocalization and human linguistic ability, *J. Acoust. Soc. Am.*, 44(6), 1574–1584, 1968.
30. Andrew, R.J., Use of formants in the grunts of baboons and other nonhuman primates, *Ann. N.Y. Acad. Sci.*, 280, 673–693, 1976.
31. Nowicki, S., Vocal tract resonances in oscine bird sound production: evidence from birdsongs in a helium atmosphere, *Nature*, 325, 53–55, 1987.
32. Nowicki, S. and Marler, P., How do birds sing?, *Music Percep.*, 5, 391–426, 1988.
33. Amundsen, M., Helium effects on the click frequency spectrum of the Harbor porpoise, *Phocoena phocoena*, *J. Acoust. Soc. Am.*, 90, 53–59, 1991.
34. Rand, A.S. and Dudley, R., Frogs in helium: the anuran vocal sac is not a cavity resonator, *Physiol. Zool.*, 66, 793–806, 1993.
35. Hauser, M.D., Evans, C.S., and Marler, P., The role of articulation in the production of rhesus monkey (*Macaca mulatta*) vocalizations, *Animal Behav.*, 45, 423–433, 1993.
36. Hauser, M.D. and Schön Ybarra, M., The role of lip configuration in monkey vocalizations: experiments using xylocaine as a nerve block, *Brain Lang.*, 46, 232–244, 1994.
37. Owren, M.J. and Bernacki, R., The acoustic features of vervet monkey (*Cercopithecus aethiops*) alarm calls, *J. Acoust. Soc. Am.*, 83, 1927–1935, 1988.
38. Fitch, W.T., Vocal tract length and formant frequency dispersion correlate with body size in rhesus macaques, *J. Acoust. Soc. Am.*, 102, 1213–1222, 1997.
39. Riede, T. and Fitch, W.T., Vocal tract length and acoustics of vocalization in the domestic dog *Canis familiaris*, *J. Exp. Biol.*, 202, 2859–2867, 1999.
40. Fitch, W.T. and Reby, D., The descended larynx is not uniquely human, *Proc. Roy. Soc. Biol. Sci.*, 268, 1669–1675, 2001.
41. Sommers, M.S. et al., Formant frequency discrimination by Japanese macaques (*Macaca fuscata*), *J. Acoust. Soc. Am.*, 91, 3499–3510, 1992.

42. Rendall, D., Rodman, P.S., and Emond, R.E., Vocal recognition of individuals and kin in free-ranging rhesus monkeys, *Animal Behav.*, 51, 1007–1015, 1996.
43. Owren, M.J., Seyfarth, R.M., and Cheney, D.L., The acoustic features of vowel-like grunt calls in chacma baboons (*Papio cyncephalus ursinus*): implications for production processes and functions, *J. Acoust. Soc. Am.*, 101, 2951–2963, 1997.
44. Owren, M.J. and Rendall, D., An affect-conditioning model of nonhuman primate vocal signaling, in *Perspectives in Ethology*, Vol. 12, *Communication*, D.H. Owings, M.D. Beecher, and N.S. Thompson, Eds., Plenum Press, New York, 1997, pp. 299–346.
45. Fitch, W.T., Neubauer, J., and Herzel, H., Calls out of chaos: the adaptive significance of nonlinear phenomena in mammalian vocal production, *Animal Behav.*, 63, 407–418, 2002.
46. Wilden, I. et al., Subharmonics, biphonation, and deterministic chaos in mammal vocalization, *Bioacoustics*, 9, 171–196, 1998.
47. Fee, M.S. et al., The role of nonlinear dynamics of the syrinx in the vocalizations of a songbird, *Nature*, 395, 67–71, 1998.
48. Mende, W., Herzel, H., and Wermke, K., Bifurcations and chaos in newborn infant cries, *Phys. Lett. A*, 145, 418–424, 1990.
49. Herzel, H. et al., Chaos and bifurcations during voiced speech, in *Complexity, Chaos and Biological Evolution*, E. Mosekilde and L. Mosekilde, Eds., Plenum, New York, 1991, pp. 41–50.
50. Titze, I.R., Baken, R., and Herzel, H., Evidence of chaos in vocal fold vibration, in *Vocal Fold Physiology: New Frontiers in Basic Science*, I. Titze, Ed., Singular Publishing Group, San Diego, CA, 1993, pp. 143–188.
51. Lorenz, E.N., Deterministic nonperiodic flow, *J. Atmos. Sci.*, 20, 130–141, 1963.
52. May, R.M., Biological populations with nonoverlapping generations: stable points, stable cycles and chaos, *Science*, 186, 645–647, 1974.
53. May, R.M., Simple mathematical models with very complicated dynamics, *Nature*, 261, 459–467, 1976.
54. Glass, L. and Mackey, M.C., *From Clocks to Chaos: The Rhythms of Life*, Princeton University Press, Princeton, NJ, 1988.
55. Freeman, W.J. and Skarda, C.A., Spatial EEG patterns, nonlinear dynamics and perception: the neo-Sheringtonian view, *Brain Res. Rev.*, 10, 147–175, 1985.
56. Babloyantz, A. and Destexhe, A., Low-dimensional chaos in an instance of epilepsy, *Proc. Natl. Acad. Sci.*, 83, 3513–3517, 1986.
57. Liberman, A.M. and Mattingly, I.G., The motor theory of speech perception revised, *Cognition*, 21, 1–36, 1985.
58. Hauser, M.D., Articulatory and social factors influence the acoustic structure of rhesus monkey vocalizations: a learned mode of production?, *J. Acoust. Soc. Am.*, 91, 2175–2179, 1992.
59. Driver, P.M. and Humphries, D.A., *Protean Behaviour: The Biology of Unpredictability*, Oxford University Press, London, 1988, p. 360.
60. Kelemen, G. and Sade, J., The vocal organ of the howling monkey (*Alouatta palliata*), *J. Morphol.*, 107, 123–140, 1960.
61. Schön Ybarra, M., Morphological adaptations for loud phonation in the vocal organ of howling monkeys, *Primate Rep.*, 22, 19–24, 1988.
62. Hill, W.C.O. and Booth, A.H., Voice and larynx in African and Asiatic Colobidae, *J. Bombay Nat. Hist. Soc.*, 54, 309–321, 1957.
63. Nishimura, T. et al., New methods of morphological studies with minimum invasiveness, *Primate Res.*, 15, 259–266, 1999.

64. Schön Ybarra, M., A comparative approach to the nonhuman primate vocal tract: implications for sound production., in *Current Topics in Primate Vocal Communication*, E. Zimmerman and J.D. Newman, Eds., Plenum, New York, 1995, pp. 185–198.
65. Lampert, H., Zur kenntnis des Platyrhinenkehlkopfes, *Gegenbaurs Morphologisches Jahrbuch*, 1926, 607–654, 1926.
66. Buyse, M.L., *Birth Defects Encyclopedia*, Blackwell Scientific Publications, St. Louis, MO, 1990.
67. Gautier, J.P., Etude morphologique et fonctionnelle des annexes extra-laryngées des cercopithecinae; liaison avec les cris d'espacement, *Biol. Gabonica*, 7(2), 230–267, 1971.
68. Ryan, M.J., *The Túngara Frog: A Study in Sexual Selection and Communication*, University of Chicago Press, 1985.
69. Dudley, R. and Rand, A.S., Sound production and vocal sac inflation in the Túngara frog, *Physalaemus pustulosus* (Leptodactylidae), *Copeia*, 1991(2), 460–470, 1991.
70. Kelemen, G., Anatomy of the larynx and the anatomical basis of vocal performance, in *The Chimpanzee*, Vol. 1, G. Bourne, Ed., S. Karger, Basel, 1969, pp. 165–187.
71. Fay, F.H., Structure and function of the pharyngeal pouches of the walrus (*Odobenus rosmarus* L.), *Mammalia*,. 24, 361–371, 1960.
72. Gregory, W.K., *The Anatomy of the Gorilla: The Studies of Henry Cushier Raven and Contributions by William B. Atkinson [and Others]*, Columbia University Press, New York, 1950.
73. Brandes, R., Über den Kehlkopf des Orang-utan in verschiedenen Altrersstadien mit besonderer Berücksichtigung der Kehlsackfrage, *Morphologisches Jahrbuch*, 69, 1–61, 1931.
74. Fitch, W.T., The phonetic potential of nonhuman vocal tracts: comparative cineradiographic observations of vocalizing animals, *Phonetica*, 57, 205–218, 2000.
75. Kelemen, G., Comparative anatomy and performance of the vocal organ in vertebrates, in *Acoustic Behavior of Animals*, R. Busnel, Ed., Elsevier, Amsterdam, 1963, pp. 489–521.
76. Nemai, J. and Kelemen, G., Beiträge zür Kenntnis des Gibbonkehlkopfes, *Zeitschrift für Anatomie und Entwickelungsgeschichte*, 59, 259–292, 1933.
77. Starck, D. and Schneider, R., Respirationsorgane, in *Primatologia*, Vol. III, Part 2, H. Hofer, A.H. Schultz, and D. Starck, Eds., S. Karger, Basel, 1960.
78. Nemai, J., Das Stimmorgan der Primaten, *Zeitschrift für Anatomie und Entwickelungsgeschichte*, 81, 657–672, 1926.
79. Griffin, D.R., *Listening in the Dark*, Yale University Press, New Haven, CT, 1958.
80. Griffiths, T.A., Modification of *M. cricothyroideus* and the larynx in the Mormoopidae, with reference to amplification of high-frequency pulses, *J. Mammol.*, 59, 724–730, 1978.
81. Suthers, R.A., The production of echolocation signals by bats and birds, in *Animal Sonar: Processes and Performance*, P.E. Nachtigall and P.W.B. Moore, Eds., Plenum, New York, 1988, pp. 23–45.
82. Mergell, P., Fitch, W.T., and Herzel, H., Modeling the role of non-human vocal membranes in phonation, *J. Acoust. Soc. Am.*, 105, 2020–2028, 1999.
83. Napier, J.R. and Napier, P.H., *The Natural History of the Primates*, MIT Press, Cambridge, MA, 1985.
84. Hast, M., The larynx of roaring and non-roaring cats, *J. Anat.*, 163, 117–121, 1989.
85. Fitch, W.T., The evolution of speech: a comparative review, *Trends Cognitive Sci.*, 4, 258–267, 2000.

86. Fitch, W.T., Acoustic exaggeration of size in birds by tracheal elongation: comparative and theoretical analyses, *J. Zool. (London)*, 248, 31–49, 1999.
87. Fitch, W.T., Skull dimensions in relation to body size in nonhuman mammals: the causal bases for acoustic allometry, *Zoology*, 103, 40–58, 2000.
88. Krebs, J.R. and Dawkins, R., Animal signals: mind reading and manipulation, in *Behavioural Ecology*, J.R. Krebs and N.B. Davies, Eds., Sinauer Associates, Sunderland, MA, 1984, pp. 380–402.
89. Goodall, J., *The Chimpanzees of Gombe: Patterns of Behavior*, Harvard University Press, Cambridge, MA, 1986.
90. Struhsaker, T.T., Auditory communication among vervet monkeys (*Cercopithecus aethiops*), in *Social Communication among Primates*, S.A. Altmann, Ed., Chicago University Press, 1967, pp. 281–324.
91. Hauser, M.D., Rhesus monkey (*Macaca mulatta*) copulation calls: honest signals for female choice?, *Proc. Roy. Soc. London*, 254, 93–96, 1993.
92. Seyfarth, R.M., Cheney, D.L., and Marler, P., Monkey responses to three different alarm calls: evidence of predator classification and semantic communication, *Science*, 210, 801–803, 1980.
93. Hauser, M.D., Functional referents and acoustic similarity: field playback experiments with rhesus monkeys, *Animal Behav.*, 55, 1647–1658, 1998.
94. Owren, M.J., Acoustic classification of alarm calls by vervet monkeys (*Cercopithecus aethiops*) and humans. I. Natural calls, *J. Comp. Psychol.*, 104, 20–28, 1990.
95. Rauschecker, J.P., Tian, B., and Hauser, M., Processing of complex sounds in the macaque nonprimary auditory cortex, *Science*, 268, 111–114, 1995.
96. Moore, B.C.J., *An Introduction to the Psychology of Hearing*, Academic Press, New York, 1988.
97. Owren, M.J., Acoustic classification of alarm calls by vervet monkeys (*Cercopithecus aethiops*) and humans. II. Synthetic calls, *J. Comp. Psychol.*, 104, 29–40, 1990.
98. Norcross, J.L., Newman, J.D., and Fitch, W., Responses to natural and synthetic phee calls by common marmosets, *Am. J. Primatol.*, 33, 15–29, 1994.
99. Fitch, W.T. and Kelley, J.P., Perception of vocal tract resonances by whooping cranes, *Grus americana*, *Ethology*, 106(6), 559–574, 2000.
100. Atal, B.A. and Hanauer, S.L., Speech analysis and synthesis by linear prediction of the speech wave, *J. Acoust. Soc. Am.*, 50(2), 637–655, 1971.
101. Markel, J.D. and Gray, A.H., *Linear Prediction of Speech*, Springer–Verlag, New York, 1976.
102. Carterette, E., Shipley, C., and Buchwald, J., Linear prediction theory of vocalization in cat and kitten, in *Frontiers in Speech Communication Research*, B. Lindblom and S. Ohman, Eds., Acacdemic Press, New York, pp. 245–257, 1979.
103. Owren, M.J. and Bernacki, R.H., Applying linear predictive coding (LPC) to frequency-spectrum analysis of animal acoustic signals, in *Animal Acoustic Communication: Sound Analysis and Research Methods*, S.L. Hopp, M.J. Owren, and C.S. Evans, Eds., Springer, New York, pp. 130–162, 1998.
104. Dang, J. and Honda, K., Acoustic characteristics of the human paranasal sinuses derived from transmission characteristic measurement and morphological observation, *J. Acoust. Soc. Am.*, 100, 3374–3383, 1996.
105. Dang, J., Honda, K., and Suzuki, H., Morphological and acoustical analysis of the nasal and the paranasal cavities, *J. Acoust. Soc. Am.*, 96(4), 2088–2100, 1994.
106. Haimoff, E.H., Occurrence of anti-resonance in the song of the siamang (*Hylobates syndactylus*), *Am. J. Primatol.*, 5, 249–256, 1983.

107. Hauser, M.D., Evans, C.S., and Marler, P., The role of articulation in the production of rhesus monkey (*Macaca mulatta*) vocalizations, *Animal Behav.*, 45, 423–433, 1993.
108. Rendall, D.A., Social Communication and Vocal Recognition in Free-Ranging Rhesus Monkeys (*Macaca mulatta*), University of California, Davis, 1996.

7 Developmental Modifications in the Vocal Behavior of Non-Human Primates

Julia Fischer

CONTENTS

I. Introduction 109
II. Impact of Methodological Advances 110
III. Plasticity in the Usage and Comprehension of Calls 113
    A. The Case of Pygmy Marmoset Babbling 113
    B. Dialects and the Question of Vocal Convergence 114
    C. Infants' Responses to Acoustic Gradation 117
IV. Future Research 120
Acknowledgments 121
References 121

I. INTRODUCTION

Understanding the roots of human language is a fascinating topic that has generated as much interest as controversy (see References 1 to 3 for some recent contributions). A key feature of human language is that it is learned, in terms of both its production and comprehension. From a comparative perspective, this raises the question of whether and to what extent other primate species exhibit vocal learning. One of the earliest studies in this field involved an attempt to teach a chimpanzee to speak.[4,5] Vicki, the chimpanzee raised with the Hayeses, for instance, in fact learned to utter a few "words," but the difficulties she had in mastering this task were apparently more striking than her successes.[4]

The finding that apes or monkeys have difficulties acquiring human speech, however, does not refute the possibility that learning plays an important role in the development of their own species-typical vocalizations. Most of the evidence accumulated about vocal development comes from studies on monkeys,[6,7] while hardly anything is known about the vocal development of apes.[8] In contrast, numerous studies have explored apes' ability to employ signs or linguistic symbols,[9] as well as gestures.[10] Thus, there is a serious imbalance in what we know about monkeys vs. apes in these different domains of communication.[8]

Much of the original research on primate vocal development focused on whether the ability to produce certain vocalizations was a learned vs. an innate process.[11–13] More recently, primatologists have realized that it is important to distinguish among the developmental trajectories of the ontogeny of vocal production, call usage, and comprehension of calls.[14,15] In a nutshell, whereas there is little evidence that non-human primates learn to produce their sounds through imitation,[16] learning does seem to play an important role in the usage and comprehension of calls. An extensive review of these earlier studies is given in Seyfarth and Cheney.[14]

The occurrence and possible function of small-scale variability in monkey and ape vocalizations have recently attracted increased attention. The goal of this chapter is to review some recent studies that have targeted such vocal variability during development, as they apply to both production and perception. First, we will examine whether methodological advances in the field of acoustic analyses have altered the assessment of the degree of vocal modification during ontogenetic development. This will be followed by an examination of the occurrence of increased vocal activity ("babbling") in young non-human primates, as well as the extent of and possible mechanisms underlying variation in call production at the group or population level ("dialects"). These studies may shed some light on the functional significance of modifiability in monkey and ape calls. They also provide an opportunity to discuss the distinction between usage and production of calls. The next section turns to recent studies of the development of the comprehension of calls and examines the question of how the ability to attach different meanings to similar sounding calls develops with age.

II. IMPACT OF METHODOLOGICAL ADVANCES

In their 1997 review of vocal development in non-human primates, Seyfarth and Cheney[14] reported that the majority of studies published before 1987 supported the notion that non-human primate calls appear fully formed at birth and show no developmental modification. The majority of studies published after 1987, however, provide evidence of more substantial modification in vocal production. They hypothesized that this shift may be related to the advent of computer-based digital processing of sounds.

Two recent studies in which an earlier investigation was replicated provide a good opportunity to test this hypothesis. In the first such study, Hammerschmidt and colleagues[17] analyzed recordings of young squirrel monkeys, *Saimiri sciureus*. This data set had been gathered in an attempt to replicate the classic paper by Winter et al.[12] on the ontogenetic development of squirrel monkey calls. Winter and colleagues investigated whether species-specific auditory input is necessary for the proper development of the species' vocal repertoire. In this earlier study, squirrel monkey infants raised normally were compared with infants raised in social isolation. A qualitative description of the vocal repertoire yielded no differences between the two groups of infants, suggesting that hearing conspecific vocalizations is not necessary for developing the full range of normal vocalizations. Quantitative analyses of some temporal and spectral characteristic of two call types ("isolation peep" and "cackling") also failed to reveal significant differences in relation to rearing condition. Furthermore, Winter and colleagues[12] found no structural differences in the vocalizations from normally hearing squirrel monkeys and ones deafened shortly after birth. Based on these findings, the authors concluded that all but two call types appear fully formed at birth and do not change during development. Only the isolation peep and twitter exhibited structural changes with age that were probably due to changes in body size.

More recently, Hammerschmidt and colleagues[17] analyzed recordings from four normally raised squirrel monkeys, one Kaspar–Hauser animal reared in social isolation, and one animal that was congenitally deaf. The study investigated changes in 12 different call types over a period of 2 years, using a state-of-the-art acoustic analysis in which a large set of acoustic variables was measured and then submitted to multivariate statistical analysis. With the aid of these more sensitive analytical tools, Hammerschmidt and colleagues found some ontogenetic changes in all of the call types studied, with some calls showing more pronounced changes with age than others. As the call repertoire of the squirrel monkey is extremely diverse, it is not surprising that, for each call type, different sets of acoustic features showed different patterns of developmental modification.[17] However, many of the changes could simply be attributed to growth, such as changes in resonant frequencies or in call duration, a variable that — at least in loud calls — is presumed to be related to lung volume.[18] The majority of changes occurred in the first 4 months of life. Vocal development seemed to be completed at the age of about 10 months. The authors found no substantial differences in the vocalizations of the isolated or the deaf subject, and in this regard confirmed the earlier findings of Winter et al.[12,17] Most importantly, though, results indicated that calls did not become more stereotyped with age but that all calls maintained a high degree of inter- and intra-individual variation. The authors argued that such variation could be the substrate for modifications later in life, a claim that remains to be tested.

In a second similar study, Hammerschmidt and colleagues[19] traced the developmental modification in the "coo" calls of the rhesus macaque (*Macaca mulatta*) and compared their findings to the results of an earlier study by Newman and Symmes.[20] This earlier study reported that young rhesus monkeys reared in isolation produced abnormal-sounding coo calls as adults. In their more recent study, Hammerschmidt et al.[19] recorded 20 subjects from the first week of life until 5 months of age. The animals were raised under two different rearing conditions. Some were housed with their mothers and other animals, whereas others were nursery reared with only age-matched peers and were thus deprived of adult vocalizations to use as potential vocal models. Figure 7.1 shows spectrograms of calls recorded from three subjects at three different ages, illustrating the findings of the acoustic analysis. The analysis revealed a drop in fundamental frequency and more consistent amplitude across the duration of the coo call as the animals grew older. The amplitude consistency reflects whether and to what degree a coo call may reveal gaps during the course of the fundamental frequency. The occurrence of such gaps decreased with age. Whereas a drop in the fundamental frequency correlated significantly with an increase in weight (which in

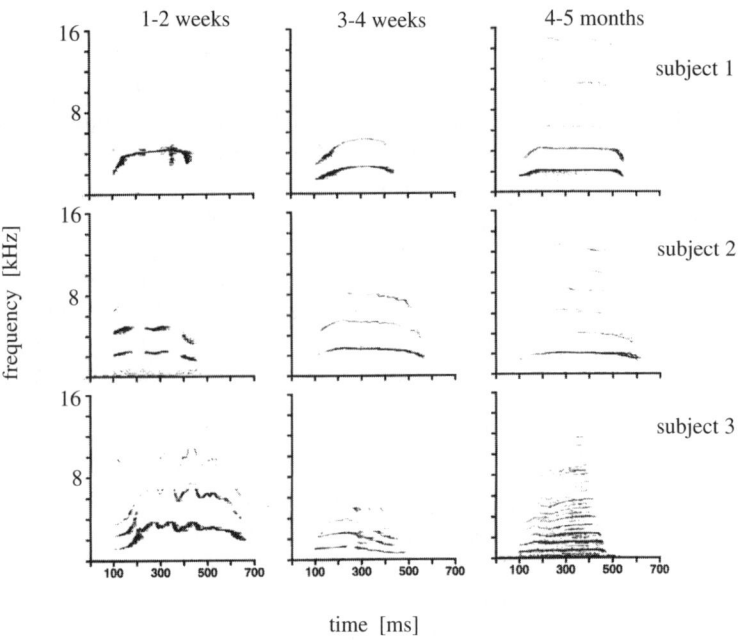

FIGURE 7.1 Spectrograms of coo calls recorded from rhesus macaques at three different ages. Frequency is depicted on the *y*-axis; time on the *x*-axis. (Unpublished material courtesy of Kurt Hammerschmidt.)

turn is related to size), the increase in amplitude consistency indicated that some practice might be required before the mature version of a call can be produced.

Whether or not auditory feedback plays a role in producing coos with a stable fundamental frequency remains an open question. Rearing conditions did not affect the acoustic structure of the coo call, suggesting that exposure to an adult model is not a prerequisite for the formation of species-typical call characteristics. This finding stands in contrast to Newman and Symmes' earlier study.[20] However, these authors lumped coo calls with a tonal structure with others that graded into screaming. The differences in vocal output could therefore rather be related to differences in social experience instead of auditory input, as it seems likely that the recording conditions (social isolation) were perceived differently by isolates and normally reared infants. The results of both the original study and the more recent one suggest that auditory feedback is not a prerequisite. The increase in constancy could either be due to increased experience or to maturation of the neuronal substrate controlling motor output. Interestingly, like the squirrel monkey study described above, the current study also failed to find a significant increase in call stereotypy over time. Instead, throughout the study period, animals exhibited marked intra-individual variation.

To summarize, two recent studies using modern techniques of acoustic and statistical analysis offer conclusions that deviate from earlier claims. In the squirrel monkey study, Hammerschmidt et al.[17] reported a larger degree of acoustic variation than found in previous work, whereas in the rhesus monkey study[19] authors were unable to replicate the finding that rearing conditions affected call structure. Both recent studies support the view that the vast majority of the motor patterns underlying vocal production are largely innate and in place at least from birth, while some features may change in relation to age and practice or maturation. Both studies also suggest that a high degree of variability is maintained throughout development.

It is therefore unclear whether the application of sophisticated analytical tools will fundamentally change current views on vocal development. It is possible that some of these divergent statements can be reconciled if earlier studies are put into their proper historical perspective. For instance, those researchers who described calls as being "adult-like from birth on and showing no signs of modification"[14] would probably have had no trouble estimating the age of a caller just on the basis of the call characteristics, despite the fact that, essentially, the call types are the same across different age classes. At a time when comparatively little evidence had been accumulated yet, the extent as to which vocal production appears to be innate may have come as a surprise, particularly in light of the close relatedness of humans and non-human primates. When it became clear that monkeys neither go through sensitive phases (in terms of their vocal production) nor gradually acquire the structure of their vocalizations, subtle changes became more interesting.

III. PLASTICITY IN THE USAGE AND COMPREHENSION OF CALLS

A. THE CASE OF PYGMY MARMOSET BABBLING

Whereas the previous studies addressed the issue of whether non-human primate calls exhibit plasticity in terms of their acoustic structure, less is known about plasticity in terms of the usage of calls. Several studies suggest that infants vocalize more frequently than adults (reviewed by Snowden and Elowson[21]). Whereas some of this vocal activity may be related to the weaning process,[22,23] other studies suggest that this increased vocal activity may provide vocal practice. One outstanding case is the vocal activity of young pygmy marmosets, *Cebuella pygmaea*. Elowson, Snowdon, and colleagues[24,25] observed that infant pygmy marmosets go through a phase of intense vocal output during which they produce long bouts of calls, most of which resemble calls of the species' repertoire. These call bouts consist of rhythmic and repetitive strings of syllables which occur within the first 2 weeks of life. Vocal activity peaks just prior to weaning at the age of 7 to 8 weeks. The authors likened this behavior to human infant babbling.[24] Human infant babbling typically begins between 6 and 10 months of age and also consists of repetitive strings of syllables.[26,27] Unlike human words, which typically have an external referent, human babbling seems to be content free and is thought to promote control over one's vocal output via auditory feedback. Elowson and colleagues pointed out that both human and pygmy marmoset babbling lack apparent meaning, promote interactions with caregivers, and might provide vocal practice.[21,24]

It may be worth noting that some important differences exist between pygmy marmoset and human infant babbling. For one, human infant babbling can be clearly differentiated from earlier vocal behavior. Furthermore, children assemble the parts that make up their early words from their prelinguistic phonetic repertoire (i.e., from the syllables that form part of the babbling).[27,28] In other words, whereas babbling in children is a crucial part of the child's development of speech, pygmy marmosets, after their babbling phase, decrease the amount of their vocal activity, while the structure of the elements largely remains the same. Finally, not only the pygmy marmoset babbling but also the large majority of adult vocalizations fail to provide information about external objects or events and are, in this sense, content free. Doubtless, numerous parallels do exist between pygmy marmoset and human infant babbling, yet some important differences remain.

B. Dialects and the Question of Vocal Convergence

As noted earlier, vocalizations of at least some primate species retain a high degree of variability throughout adult life, and it was hypothesized that this variability may be the substrate for modifications of the call structure later in life. This issue has been addressed at two levels: first, whether animals tend to converge at a group or population level, and, second, whether dyads tend to produce calls that are similar to each other. If meaningful variation within the range of a species' vocalizations occurs between different populations or groups, a key question is whether they constitute dialects in the sense that the variation can be attributed to experience and not simply to genetic variation. One of the earliest such reports came from Green,[29] who observed that three different populations of Japanese macaques, *Macaca fuscata*, had differences in the acoustic structure of a call produced during feeding times. He suggested that these differences were due to a "behavioral founder effect," where a given subject responded to the food provisioning with a spontaneous vocalization that was subsequently imitated by other group members. In an attempt to document this process in more detail, Sugiura[30] conducted a playback study in which he compared the acoustic structure of calls that were given in response to playbacks of contact calls with those that were given spontaneously. He found that female Japanese macaques in one population (but not another) responded to playback of coo calls by calling back with similar sounding coo calls; that is, these coo calls were acoustically more similar to the playback coo than to coo calls produced spontaneously by the subjects.[30] He concluded that these monkeys must have some control over their output and be able to adjust it accordingly.

But, what exactly is meant by the phrase "control over the vocal output"? On the one hand, animals could truly be imitating one another; that is, they store the auditory input as a sound model and adjust their vocal output so that it matches the model. The acoustic structure of non-human primate calls is determined by oscillation of the vocal folds and sometimes the vocal lip,[31] articulatory gestures that influence the filtering characteristics of the vocal tract,[32–34] and respiration.[35] Current evidence suggests that, in non-human primates, the anterior cingulate cortex serves to control the initiation of vocalizations, facilitating voluntary control over call emission and onset.[36] However, the motor coordination of the vocalization appears to take place in the reticular formation of the lower brainstem,[37] suggesting little voluntary control over the precise structure of the vocal pattern. Following this line of argument, it is difficult to see how such precise control could be achieved. On the other hand, animals may differ in the activation of different motor patterns. Far too little is known about the representation and activation of these motor patterns[38] to resolve this question. In any case, it turns out to be more difficult to distinguish between call usage and call production than it appears at first glance.

Another example for acoustic differences at the group level comes from Barbary macaques, *Macaca sylvanus*. Fischer and colleagues[39] reported that the alarm calls given by members of two populations of Barbary macaques revealed significant variation between sites. Whereas the general structure was the same at both sites, a fine-grained acoustic analysis revealed significant differences between calls from the two populations. Playback experiments in which calls from their own or the other population were broadcast suggested that this observed variation was perceptually salient.[39] Because neither genetic differences nor habitat structure could account for the observed differences between groups, Fischer and colleagues concluded that the general acoustic structure of calls is fixed whereas there is restricted potential for acoustic modification. Again, it remains unclear which mechanisms underlie this observation.

While the evidence for dialects in macaques remains rather weak, and the mechanisms opaque, it could still be the case that apes provide more compelling evidence for modifiability of their vocalizations. Mitani and colleagues[40] examined the differences in the acoustic structure of chimpanzee (*Pan troglodytes*) "pant-hoots" between two different populations. Pant-hoots comprise a series of several elements (dubbed the build-up phase), culminating in a climax element. The authors found differences both in the temporal characteristics of the call and in the frequency characteristics of the climax element. Although genetic or anatomical differences between populations could not be ruled out, the authors assumed that the differences in pant-hoots between groups were mediated by learning. Similarly, Mitani and Gros-Louis[41] investigated the occurrence of call convergence in wild chimpanzees during joint chorusing. In one anecdotal case, they found that calls produced by a pair of closely affiliated male chimpanzees were more similar to each other when they called together than when they called with others. In a recent study, Mitani and colleagues[42] reassessed the evidence for dialects. They recorded pant-hoots from two populations of wild chimpanzees in East Africa. As the structure of pant-hoots varies as a function of the age and sex of the caller,[43] the analysis was restricted to the calls of adult males. This study failed to identify significant variation in the climax elements; however, based on the combination of a number of variables that are related to the duration and the number of elements in the build-up phase, it was possible to assign the majority of calls to their proper populations. In contrast to earlier studies, Mitani and co-workers[42] assumed that this acoustic variation could be related to differences in habitat acoustics and body size of the chimpanzees.

To eliminate genetic relatedness as a variable in chimpanzee call convergence, Marshall and colleagues[44] initiated a study in which they investigated whether two groups of captive chimpanzees differed with regard to the acoustic structure of the pant-hoots. Because the animals came from diverse origins, the influence of a genetic component could be ruled out. The authors found an overall similarity among pant-hoots recorded in captivity and in the wild, suggesting a strong innate component of the call type. Nevertheless, they found significant variation between two captive groups. They also reported that a male in one enclosure introduced a novel element to his pant-hoot that subsequently was also used by five other males from the same colony. This novel element (dubbed a "Bronx cheer") is produced by blowing air through pursed lips. The observation suggests that chimpanzees have some control over their respiratory activity and their lips. However, it is perhaps not surprising that this novel element was not voiced, as studies of the neural control of the vocal apparatus suggest that the control of the tongue and the lips is under voluntary control, while the control over the laryngeal muscles is less developed.[45]

To summarize, several studies suggested that non-human primates have the potential for minor modifications of their call structure. To establish the putative function of these vocal modifications, it is necessary to distinguish between within-group[44,46] and within-dyad convergence.[30,41] Within-group similarity in call structure should be the result of a longer term process, possibly taking place within weeks or months, and could function as a social badge for group recognition.[47] This would facilitate the identification of members in neighboring groups. Identifying members in one's own group could be mediated via knowledge of individual characteristics in call structure, but if groups exceed a certain size then call convergence could also assist in identifying members in one's own group. As far as dyads are concerned, Sugiura[30] proposed that vocal matching functions to indicate that the "matched" coos are given in response to a particular individual's calling. A similar function has been suggested for the matching of signature whistles in dolphins, for which it is believed that vocal matching functions to address a particular individual.[48]

As far as the mechanisms are concerned, several authors suggested that the process of call convergence can be attributed to vocal accommodation.[39,49,50] In speech, humans make minor adjustments in their vocal output so that it sounds more like the speech of the individual with whom they are talking. This subconscious process is called *speech accommodation*.[51] Speech accommodation includes a wide range of subtle adaptations, such as altering the speed at which people talk, the length of pauses and utterances, and the kind of vocabulary and syntax used, as well as intonation and voice pitch. Locke[28] observed similar adjustments in the speech acquisition process of infants and termed it *vocal accommodation*. However, the concept of vocal accommodation has a more descriptive than explanatory value, as the mechanisms underlying both vocal accommodation and matching remain unclear.

In light of the finding that non-human primates have only limited voluntary control over the vocal apparatus,[45,52–54] it remains puzzling how accommodation is accomplished. If vocal accommodation accounts for the observed modification in call structure, then auditory feedback must somehow play a role, despite the fact that it does not seem to be a prerequisite to developing the species-specific repertoire.[55] More precisely, matching requires that the model be rapidly stored, and an individual's own motor output be planned in accordance with the stored template.[56] With regard to long-term effects of call convergence, an alternative mechanism to vocal accommodation has been suggested by Marler,[57] who proposed that some call variants are selectively reinforced by social stimulation (action-based learning). In this case, contextual learning (*sensu* Janik and Slater[58]) would account for the observed changes in call characteristics. Again, however, the precise mechanisms mediating action-based learning are not known.

C. Infants' Responses to Acoustic Gradation

The previous sections on modifications in the production and/or usage of calls showed that non-human primates apparently exhibit only little, albeit significant, plasticity in some of their calls. In contrast, studies that examined the development of the comprehension of, and correct responses to calls, indicated that subjects undergo pronounced changes in development. Most of the earlier work[59,60] has focused on the development of vervet (*Cercopithecus aethiops*) infants' responses to different alarm calls given in response to their main predators, leopards, martial eagles, and pythons.[61,62] Vervet alarm calls are acoustically distinct and elicit different, adaptive responses. For instance, upon hearing an eagle alarm, group members look up into the sky. Playback experiments conducted on a small sample of infants showed that after playback of the different alarm calls, infants 3 to 4 months of age typically ran to their mothers, no matter which call was broadcast. At 4 to 5 months of age, many infants produced an incorrect behavioral strategy; for example, they scanned the ground after hearing an eagle alarm. Detailed inspection of film footage revealed that infants were more likely to respond correctly when they had first looked at an adult. At the age of 6 to 7 months, most infants responded to alarm calls as adults did.[59]

Vervets also attend to the alarm calls of other species, such as the alarm calls of the superb starling (*Spreo superbus*), which are given in response to ground predators such as leopards.[63] Starling alarm calls occur at different rates in different habitats, and playback experiments showed that vervet infants begin to attend to these calls by looking in the direction of the loudspeaker at an age of 6 to 7 weeks.[60] Vervet infants in an area with higher rates of starling alarm calling responded to playbacks of these calls at an earlier age than did infants living in areas with lower rates of starling alarm calling.[60] Additional experiments on a smaller sample of older infants revealed that they responded appropriately to starling alarm calls by running up into a tree at 4 to 5 months of age.[60]

These previous studies have focused on calls that are acoustically distinct; however, the vocal repertoires of many primate species typically consist of a mixture of acoustically graded and discrete calls. Discrete repertoires contain signals with no intermediates between call types, whereas graded signal systems are characterized by continuous acoustic variation between and/or within signal types, with no obvious distinct boundaries that allow a listener to discriminate easily between one signal type and another.[64] For instance, adult Barbary macaques give acoustically different variants of the shrill bark call in response to different predator classes.[65] Typical subtypes of the shrill bark exist, but intermediate variants between these subtypes also occur. Subsequent playback experiments showed that listeners categorized the calls according to the stimulus that elicited the calling. Thus, adult subjects were able to partition the continuous acoustic variation into different categories[66] in a manner that resembled the categorical perception (CP) found in human speech.[67]

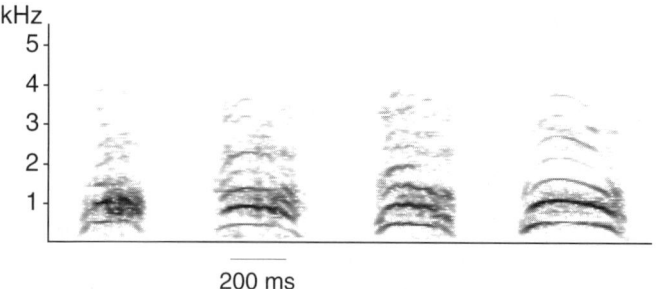

FIGURE 7.2 Spectrograms of female chacma baboon barks recorded from one subject. From left to right: typical alarm bark, intermediate alarm bark, intermediate contact bark, and typical alarm bark. (Redrawn from Reference 69.)

Does the ability to discriminate between variants of the same general call type develop with age? This issue is particularly interesting in light of the suggestion that CP in human speech relies on an innate component.[68] Fischer and colleagues[69] investigated whether a primate's ability to dissect an acoustic continuum into discrete categories is in place from birth on or whether it emerges with experience and age. As the Barbary macaque population currently is under a strict birth-control regime, Fischer et al.[70] extended their studies to baboons (*Papio cynocephalus ursinus*). As an example, they used the bark of adult female baboons. These barks grade from clear, harmonically rich calls into calls with a noisier, harsh structure (Figure 7.2).[71] The clear version is usually given when a caller finds herself in a situation where she is apparently separated from the group or her offspring (contact bark),[72,73] whereas the harsher variant is usually given when a female spots a potential predator (alarm bark). However, there are also intermediate forms between the two subtypes, and these acoustically intermediate calls can occur in both contexts.[71]

Results indicate that infant baboons gradually develop the ability to discriminate between calls that fall along a graded acoustic continuum. The time spent looking toward the (concealed) speaker was measured following playbacks. At 2.5 months of age, infants failed to orient to the speaker after playbacks of either alarm or contact barks. At 4 months of age, infants responded to playback of both alarm and contact barks indiscriminately. At 6 months of age, infants responded strongly to alarm barks but failed to orient to contact barks (Figure 7.3a,b). By this age, therefore, infants reliably discriminate between typical variants of alarm and contact barks. Further experiments showed that infants 6 months and older exhibit a graded series of responses to intermediate call variants (Figure 7.3c). They responded most strongly to typical alarm barks, less strongly to intermediate alarm calls, less strongly still to intermediate contact barks, and hardly at all to typical contact barks. Due to the small sample size, these results remain somewhat equivocal. They may indicate that infants respond in a continuous fashion to continuous variation in perceived urgency in the calls. Alternatively, because infants only responded to intermediate alarm barks, but not to intermediate contact barks, with startle responses, it might also be the case that infants place intermediate contact and alarm barks into two

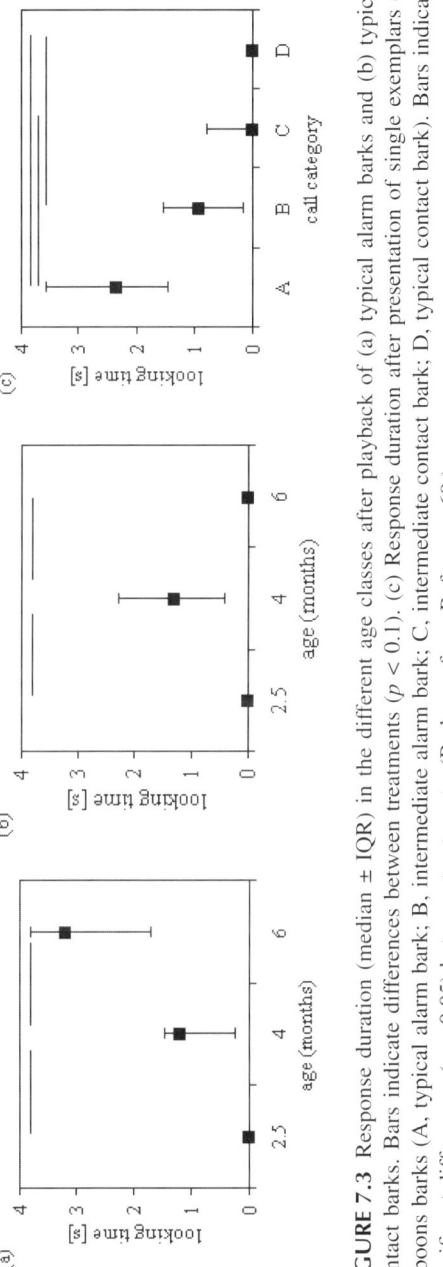

FIGURE 7.3 Response duration (median ± IQR) in the different age classes after playback of (a) typical alarm barks and (b) typical contact barks. Bars indicate differences between treatments ($p < 0.1$). (c) Response duration after presentation of single exemplars of baboons barks (A, typical alarm bark; B, intermediate alarm bark; C, intermediate contact bark; D, typical contact bark). Bars indicate significant differences ($p < 0.05$) between treatments. (Redrawn from Reference 69.)

different categories.[69] Further experiments showed that a lack of responses to clear contact barks was not simply due to habituation to their frequent occurrence; infants responded strongly to the playback of contact barks recorded from their mothers. Some of the infants even interrupted their activity and began to approach the speaker.[69] The finding that infant baboons were able to discriminate between the different call types at about 6 months of age complements the results of the studies mentioned above on vervet infants' responses to different alarm calls.[74]

In sum, the baboon study showed that infants discriminate among graded variants within a call type, and, at the very least, that they can distinguish maternal contact barks from other females' barks. The experiments suggest that infants proceed through a number of different stages. First, they must learn to attend to barks in general; that is, they must distinguish the barks from other sounds. Next, they must associate a particular call type with the external stimulus that evoked it. Presumably, the typical variants of particular call types serve as prototypes because they occur more frequently. If typical variants do indeed function as prototypes, they could then serve as perceptual anchors in the process of dissecting the acoustic continuum into different categories.[75] Whereas the association of a given call with the context of its occurrence clearly forms part of the infant's general cognitive development, the latter should be viewed as part of the infant's perceptual development in the narrower sense.

The view that experience, and not simply maturation, mediates the development of infants' responses is supported by the observations that both vervets and baboons respond to the alarm calls of birds, ungulates, and other primate species.[6,60,63,76] It seems unlikely that these responses have a genetic basis. Finally, in a cross-fostering experiment in which Japanese and rhesus macaque infants were raised by members of the other species, adoptive mothers learned to attend to the calls of their foster offspring even though the infant was a member of another species.[77]

How do infants learn the appropriate responses to a given vocalization? On the one hand, the observation that young vervets who first looked at an adult were more prone to respond appropriately suggests that they may mimic the adult's behavior. On the other hand, juvenile Barbary macaques much more frequently ran away or climbed into trees after playbacks of conspecific alarm calls than adults.[65,78] To date, the evidence for "teaching" and other forms of active information transmission remains anecdotal (for a discussion, see Hauser[7]). McCowan and colleagues[79] suggested that the reason(s) why juveniles respond more readily than adults could be due to infant error or, more plausibly, because it is adaptive to do so. Young monkeys are more vulnerable to dangers because they are smaller and possibly less attentive to their surroundings;[79] thus, not responding to an alarm call could be more costly for young monkeys than for adults.

IV. FUTURE RESEARCH

Compared to the vast amount of information that has accumulated about the development of song in Passerine birds,[80] little is known about early developmental dispositions of non-human primates. In part, this is because social isolation disrupts much of the sociocognitive development in monkeys and is therefore problematic on ethical as well as scientific grounds (this is not to say that social isolation would not disrupt bird social behavior). Nevertheless, even for socially reared infants, there is a conspicuous dearth of information on their early auditory development. For instance, do infant monkeys and apes have a predisposition to attend to their own species' calls over other sounds? When infants learn to attach meaning to particular calls, does this involve general associative mechanisms that mediate the classification of sounds in general, or are special processes at work?

As far as vocal production is concerned, it is the author's belief that we need to develop a better understanding of the relatively strong constraints operating on vocal communication in non-human primates. Why have animals such as dolphins or parrots, but not non-human primates, evolved the ability to imitate vocally? What is the adaptive value of vocal imitation, and what is its function? We know from studies on dolphins that imitation plays a role in the social context,[16] but are there other functions? Finally, we need to develop a better understanding of the co-evolution of caller and listener. For instance, given the relative small degree of modifiability on the caller's side, and the apparent plasticity on the listener's side, could it be the case that smart listeners effectively lift off selective pressure from the caller? This question is of interest with regard to the need to communicate effectively, as well in terms of manipulating one another's behavior.

Many of these issues touch upon the question of the evolution of human language. Clearly, identifying similarities between the vocal behavior of humans and non-human primates is intriguing. In the long term, however, it may be more rewarding to put a thorough understanding of a given species' communicative system first and questions into the similarity to other species' communication second. In other words, animal communication is fascinating in its own right, not only as a road to human linguistic accomplishments.

ACKNOWLEDGMENTS

I gratefully acknowledge funding by the Deutsche Forschungsgemeinschaft (Fi 707/2–1, Fi 707/4–1), and I would like to thank Kurt Hammerschmidt, Mike Tomasello, and Robert Seyfarth for discussion and helpful comments on the manuscript. I am indebted to Asif Ghazanfar for inviting me to contribute to this volume and his careful editing of the chapter.

REFERENCES

1. Jackendoff, R., Possible stages in the evolution of the language capacity, *Trends Cogn. Sci.*, 3, 272, 1999.
2. Fitch, W.T., The evolution of speech: a comparative review, *Trends Cogn. Sci.*, 4, 258, 2000.
3. Tomasello, M., Language is not an instinct, *Cogn. Dev.*, 10, 131, 1995.
4. Hayes, K.J. and Hayes, C., The intellectual development of a home-raised chimpanzee, *Proc. Am. Philos. Soc.*, 95, 105, 1951.

5. Kellogg, W.N., Communication and language in home-raised chimpanzee: gestures, words, and behavioral signals of home-raised apes are critically examined, *Science*, 162, 423, 1968.
6. Cheney, D.L. and Seyfarth, R.M., *How Monkeys See the World*, University of Chicago Press, 1990.
7. Hauser, M.D., *The Evolution of Communication*, MIT Press, Cambridge, MA, 1996.
8. Tomasello, M. and Zuberbuhler, K., Primate vocal and gestural communication, in *The Cognitive Animal*, A. Allen, M. Bekoff, and G.M. Burghardt, Eds., MIT Press, Cambridge, MA, 2002, p. 293.
9. Wallman, J., *Aping Language*, Cambridge University Press, Cambridge, U.K., 1992.
10. Tomasello, M., Call, J., Warren, J., Frost, G.T., Carpenter, M., and Nagell, K., The ontogeny of chimpanzee gestural signals: a comparison across groups and generations, *Evol. Comm.*, 1, 223, 1997.
11. Winter, P., Ploog, D., and Latta, J., Vocal repertoire of the squirrel monkey (*Saimiri sciureus*), its analysis and significance, *Exp. Brain Res.*, 1, 359, 1966.
12. Winter, P.P., Handley, D., Ploog, D., and Schott, D., Ontogeny of squirrel monkey calls under normal conditions and under acoustic isolation, *Behaviour*, 47, 230, 1973.
13. Talmage-Riggs, G., Winter, P., Mayer, W., and Ploog, D., Effect of deafening on vocal behaviour of squirrel monkey (*Saimiri sciureus*), *Folia Primatol.*, 17, 404, 1972.
14. Seyfarth, R.M. and Cheney, D.L., Some features of vocal development in nonhuman primates, in *Social Influences on Vocal Development*, C.T. Snowdon and M. Hausberger, Eds., Cambridge University Press, Cambridge, U.K., 1997, chap. 13.
15. Hauser, M.D., Vocal communication in macaques: causes of variation, in *Evolutionary Ecology and Behavior of Macaques*, J.E. Fa and D. Lindburg, Eds., Cambridge University Press, Cambridge, U.K., 1996.
16. Janik, V.M. and Slater, P.J., Vocal learning in mammals, *Adv. Stud. Behav.*, 26, 59, 1997.
17. Hammerschmidt, K., Freudenstein, T., and Jürgens, U., Vocal development in squirrel monkeys, *Ethology*, 138, 1179.
18. Fitch, W.T. and Hauser, M.D., Vocal production in nonhuman primates: acoustics, physiology, and functional constraints on "honest" advertisement, *Am. J. Primatol.*, 37, 191, 1995.
19. Hammerschmidt, K., Newman, J.D., Champoux, M., and Suomi, S.J., Changes in rhesus macaque coo vocalizations during early development, *Ethology*, 106, 873, 2000.
20. Newman, J.D. and Symmes, D., Vocal pathology in socially deprived monkeys, *Dev. Psychobiol.*, 7, 351, 1974.
21. Snowdon, C.T. and Elowson, A.M., 'Babbling' in pygmy marmosets: development after infancy, *Behaviour*, 138, 1235, 2001.
22. Hammerschmidt, K. and Todt, D., Individual differences in vocalizations of young Barbary macaques (*Macaca sylvanus*): a multi-parametric analysis to identify critical cues in acoustic signalling, *Behaviour*, 132, 381, 1995.
23. Hammerschmidt, K., Ansorge, V., Fischer, J., and Todt, D., Dusk calling in Barbary macaques (*Macaca sylvanus*): demand for social shelter, *Am. J. Primatol.*, 32, 277, 1994.
24. Elowson, A.M., Snowdon, C.T., and Lazaro-Perea, C., Infant 'babbling' in a nonhuman primate: complex vocal sequences with repeated call types, *Behaviour*, 135, 643, 1998.
25. Elowson, A.M., Snowdon, C.T., and Lazaro-Perea, C., 'Babbling' and social context in infant monkeys: parallels to human infants, *Trends Cogn. Sci.*, 2, 31, 1998.

26. Oller, D.K., *The Emergence of the Speech Capacity*, Lawrence Erlbaum, Mahwah, NJ, 2000.
27. Vihman, M.M., *Phonological Development: The Origins of Language in the Child*, Blackwell Publishers, Oxford, 1996.
28. Locke, J.L., *The Child's Path to Spoken Language*, Harvard University Press, Cambridge, MA, 1993.
29. Green, S., Dialects in Japanese monkeys: vocal learning and cultural transmission of locale-specific vocal behavior?, *Z. Tierpsychol.*, 38, 304, 1975.
30. Sugiura, H., Matching of acoustic features during the vocal exchange of coo calls by Japanese macaques, *Animal Behav.*, 55, 673, 1998.
31. Mergell, P., Fitch, W.T., and Herzel, H., Modeling the role of nonhuman vocal membranes in phonation, *J. Acoust. Soc. Am.*, 105, 2020, 1999.
32. Brown, C.H. and Cannito, M.P., Modes of vocal variation in Sykes monkeys (*Cercopithecus albogularis*) squeals, *J. Comp. Psychol.*, 109, 398, 1995.
33. Hauser, M.D., Evans, C.S., and Marler, P., The role of articulation in the production of rhesus monkey, *Macaca mulatta*, vocalizations, *Animal Behav.*, 45, 423, 1993.
34. Hauser, M.D. and Schön Ybarra, M., The role of lip configuration in monkey vocalizations: experiments using xylocaine as a nerve block, *Brain Lang.*, 46, 232, 1994.
35. Häusler, U., Vocalization-correlated respiratory movements in the squirrel monkey, *J. Acoust. Soc. Am.*, 108, 1443, 2000.
36. Jürgens, U., Neuronal control of non-human and human primates, in *Current Topics in Primate Vocal Communication*, E. Zimmermann, J.D. Newman, and U. Jürgens, Eds., Plenum, New York, 1995.
37. Jürgens, U., Neurobiology of primate vocal communication, *Adv. Ethol.*, 36, 17, 2001.
38. Düsterhöft, F., Häusler, U., and Jürgens, U., On the search for the vocal pattern generator: a single-unit recording study, *NeuroReport*, 11, 2031, 2000.
39. Fischer, J., Hammerschmidt, K., and Todt, D., Local variation in Barbary macaque shrill barks, *Animal Behav.*, 56, 623, 1998.
40. Mitani, J.C., Hasegawa, T., Gros-Louis, J., Marler, P., and Byrne, R.W., Dialects in wild chimpanzees?, *Am. J. Primatol.*, 27, 233, 1992.
41. Mitani, J.C. and Gros-Louis, J., Chorusing and call convergence in chimpanzees: tests of three hypotheses, *Behaviour*, 135, 1041, 1998.
42. Mitani, J.C., Hunley, K.L., and Murdoch, M.E., Geographic variation in the calls of wild chimpanzees: a reassessment, *Am. J. Primatol.*, 47, 133, 1999.
43. Marler, P. and Hobbett, L., Individuality in a long-range vocalization of wild chimpanzee, *Z. Tierpsychol.*, 38, 97, 1975.
44. Marshall, A.J., Wrangham, R.T., and Arcadi, A.C., Does learning affect the structure of vocalizations in chimpanzees?, *Animal Behav.*, 58, 825, 1999.
45. Jürgens, U., On the neurobiology of vocal communication, in *Nonverbal Vocal Communication*, H. Papousek, U. Jürgens, and M. Papousek, Eds., Springer–Verlag, Berlin, 1992.
46. Arcadi, A.C., Phrase structure of wild chimpanzee pant hoots: patterns of production and interpopulation variability, *Am. J. Primatol.*, 39, 159, 1996.
47. Guildford, T. and Stamp Dawkins, M., Receiver psychology and the evolution of animal signals, *Animal Behav.*, 42, 1, 1991.
48. Janik, V.M., Whistle matching in wild bottlenose dolphins (*Tursiops truncatus*), *Science*, 289, 1355, 2000.
49. Mitani, J.C. and Brandt, K.L., Social factors influence the acoustic variability in the long-distance calls of male chimpanzees, *Ethology*, 96, 233, 1994.

50. Snowdon, C.T. and Elowson, A.M., Pygmy marmosets modify call structure when paired, *Ethology*, 105, 893, 1999.
51. Giles, H., The dynamics of speech accommodation, *Int. J. Sociol. Lang.*, 46, 1–155, 1984.
52. Deacon, T.W., The neuronal circuitry underlying primate calls and human language, in *Language Origin: A Multidisciplinary Approach*, J. Wind, B. Chiarelli, B. Bichakajian, and A. Nocentini, Eds., Kluwer, Dordecht, Netherlands, 1992.
53. Alipour, M., Chen, Y., and Jürgens, U., Anterograde projections of the cortical tongue area of the tree shrew (*Tupaia belangeri*), *J. Brain Res.*, 38, 405, 1997.
54. Sutton, D., Larson, C., Taylor, E.M., and Lindeman, R.C., Vocalizations in rhesus monkeys: conditionability, *Brain Res.*, 52, 225, 1973.
55. Fichtel, C., Hammerschmidt, K., and Jürgens, U., On the vocal expression of emotion: a multiparametric analysis of different states of aversion in the squirrel monkey, *Behaviour*, 138, 97, 2001.
56. Heyes, C.M., Causes and consequences of imitation, *Trends Cogn. Sci.*, 5, 253, 2001.
57. Marler, P., Song-learning behaviour: the interface with neuroethology, *Trends Neurosci.*, 14, 199, 1991.
58. Janik, V.M. and Slater, P.J., The different roles of social learning in vocal communication, *Animal Behav.*, 60, 1, 2000.
59. Seyfarth, R.M. and Cheney, D.L., The ontogeny of vervet monkey alarm calling behavior: a preliminary report, *Z. Tierpsychol.*, 54, 37, 1980.
60. Hauser, M.D., How infant vervet monkeys learn to recognize starling alarm calls: the role of experience, *Behaviour*, 105, 187, 1988.
61. Seyfarth, R.M., Cheney, D.L., and Marler, P., Vervet monkey alarm calls: semantic communication in a free-ranging primate, *Animal Behav.*, 28, 1070, 1980.
62. Seyfarth, R.M., Cheney, D.L., and Marler, P., Monkey responses to three different alarm calls: evidence of predator classification and semantic communication, *Science*, 210, 801, 1980.
63. Cheney, D.L. and Seyfarth, R.M., Social and nonsocial knowledge in vervet monkeys, *Philos. Trans. Roy. Soc. London Ser. B*, 308, 187, 1985.
64. Marler, P., Social organization, communication and graded signals: the chimpanzee and the gorilla, in *Growing Points in Ethology*, P.P.G. Bateson and R.A. Hinde, Eds., Cambridge University Press, Cambridge U.K., 1976.
65. Fischer, J., Hammerschmidt, K., and Todt, D., Factors affecting acoustic variation in Barbary macaque (*Macaca sylvanus*) disturbance calls, *Ethology*, 101, 51, 1995.
66. Fischer, J., Barbary macaques categorize shrill barks into two call types, *Animal Behav.*, 55, 799, 1998.
67. Harnad, S., *Categorical Perception*, Cambridge University Press, Cambridge, U.K., 1987.
68. Eimas, P.D., Siqueland, E.R., Jusczyk, P., and Vigorito, J., Speech perception in infants, *Science*, 171, 303, 1971.
69. Fischer, J., Cheney, D.L., and Seyfarth, R.M., Development of infant baboons' responses to graded bark variants, *Proc. Roy. Soc. London Ser. B*, 267, 2317, 2000.
70. Hall, K.R. L. and DeVore, I., Baboon social behavior, in *Primate Behavior: Field Studies of Monkeys and Apes*, I. DeVore, Ed., Holt, Rinehart & Winston, New York, 1965, chap. 3.
71. Fischer, J., Hammerschmidt, K., Cheney, D.L., and Seyfarth, R.M., Acoustic features of female chacma baboon barks, *Ethology*, 107, 33, 2001.
72. Cheney, D.L., Seyfarth, R.M., and Palombit, R., The function and mechanisms underlying baboon 'contact' barks, *Animal Behav.*, 52, 507, 1996.

73. Rendall, D., Cheney, D.L., and Seyfarth, R.M., Proximate factors mediating 'contact' calls in adult female baboons and their infants, *J. Comp. Psychol.*, 114, 36, 2000.
74. Seyfarth, R.M. and Cheney, D.L., Vocal development in vervet monkeys, *Animal Behav.*, 34, 1640, 1986.
75. Kuhl, P.K., Human adults and human infants show a 'perceptual magnet effect' for the prototypes of speech categories, monkeys do not, *Percep. Psychophys.*, 50, 93, 1991.
76. Zuberbuhler, K., Interspecies semantic communication in two forest primates, *Proc. Roy. Soc. London Ser. B*, 267, 713, 2000.
77. Owren, M.J., Dieter, J.A., Seyfarth, R.M., and Cheney, D.L., Vocalizations of rhesus (*Macaca mulatta*) and Japanese (*M. fuscata*) macaques cross-fostered between species show evidence of only limited modification, *Dev. Psychobiol.*, 26, 389, 1993.
78. Fischer, J. and Hammerschmidt, K., Functional referents and acoustic similarity revisited: the case of Barbary macaque alarm calls, *Animal Cogn.*, 4, 29, 2001.
79. McCowan, B., Franceschini, N.V., and Vicino, G.A., Age differences and developmental trends in alarm peep responses by squirrel monkeys (*Saimiri sciureus*), *Am. J. Primatol.*, 53, 19, 2001.
80. Catchpole, C.K. and Slater, P.J., *Bird Song*, Cambridge University Press, Cambridge, U.K., 1995.

8 Ecological and Physiological Constraints for Primate Vocal Communication

Charles H. Brown

CONTENTS

I. Introduction 127
II. Ecological Impediments to Vocal Communication 129
   A. Noise 129
   B. Signal Distortion 130
   C. Habitat Acoustics and Neurobiological and Perceptual Studies
      of Call Perception 133
III. Primate Vocal-Fold Biomechanics 136
   A. Misassumptions Regarding the Mechanics of Primate
      Vocal Production 136
   B. Vocal-Fold Biomechanics and Neurobiological and Behavioral
      Studies of Vocal Perception 144
IV. Summary 146
References 146

I. INTRODUCTION

Most neurobiological and behavioral studies of the perception of species-specific vocalizations reflect in their design two assumptions. First, it is usually assumed that primate vocal production mirrors the idealized case for speech. That is, coupled synchronous oscillations of the vocal folds, similar to those underlying the production of vowels by human speakers, serve as the laryngeal source of most calls (or at least the important calls) produced by primates. Second, it is also assumed that the vocalization waveforms incident at the listener's ear are faithful replicas of the emitted signals, perhaps only reduced in signal level due to the distance of propagation and the inverse square law. The goal of this chapter is to review the current research in habitat acoustics and vocal-fold biomechanics that challenges these assumptions. It is hoped that a broader understanding of the mechanisms underlying the distortion of calls in the natural habitat and the mechanics of primate voice production will stimulate innovations in the design of behavioral and neurobiological studies and help advance our understanding of vocal communication and the evolution of speech.

Voicing biomechanics and habitat acoustics are very dissimilar fields; they share no common literature or methodology and, superficially, these disciplines would appear to have no point of intersection and provide no common perspective for understanding primate communication. The converse, however, is true. From the point of view of perceptual systems, the disciplines of voicing biomechanics and habitat acoustics are joined together by posing a single fundamental problem. The problem is: Are the differences between two successive waveforms due to intentional changes in the mechanics of vocal production, or are differences between waveforms due to instabilities in either the mechanisms of vocal production or the propagation of the signal in the habitat?

In the sections that follow, it will be shown that two signals, identical from the standpoint of the vocalizer, may result in two very divergent acoustic waveforms received by the listener. In one case, the differences between successive iterations of the same signal are the product of moment-by-moment variations in the acoustic distortion mechanisms active in the habitat. In the second case, the differences between waveforms are the result of bifurcations in the regime of oscillation of the vocal folds that sometimes may arise spontaneously without any measurable change in subglottal pressure, adduction, or any other parameter of vocal control. Hence, in both cases, perceptual systems must be able to distinguish intended acoustic variation from that generated by uncontrollable phenomena in either sound propagation or vocal production. If the magnitude of acoustic variation produced by intentional changes in the mechanisms of vocal production were large, while those due to other sources were small, then these uncontrolled sources of acoustic variation would be insignificant for studies of communication. However, the data suggest that in some cases the reverse may be closer to the truth; hence, the problem of distinguishing accidental from intended acoustic variation is not a trivial problem for perceptual systems. Accordingly, perceptual systems may be designed to classify as similar two acoustically dissimilar sounds if the differences in waveforms are typically due to instabilities in habitat acoustics or voicing regimes. Conversely, perceptual systems may comparatively amplify the dissimilarity between sound variations that reliably reflect intentional changes in the parameters of vocal control. Finally, we will examine some of the perceptual data supporting the view that prominent acoustic differences between waveforms attributed to source instabilities may be communicatively insignificant, while other classes of subtle acoustic variation may be very salient. These observations suggest that neurons in the primate auditory system may classify as similar two acoustically dissimilar sounds if the differences in the waveforms are typically due to variations in habitat acoustics or reflect spontaneous bifurcations in voicing regimes. Conversely, neurons may be very sensitive to subtle differences between similar waveforms when the differences likely reflect intentional variations in the parameters of vocal production.

II. ECOLOGICAL IMPEDIMENTS TO VOCAL COMMUNICATION

Because the industrialized world emerged so recently in geological time, the impediments to communication created by the modern world are insignificant with respect to the evolution of perceptual systems. Therefore, it is necessary to explore the acoustical properties of selected natural environments to develop a perspective of the impediments to vocal communication that may have influenced the evolution of speech and non-human communication systems. Three habitat types — rain forest, riverine forest, and savanna — comprise the landscape pertinent to acoustic impediments significant to the evolution of both human and non-human primate communication systems.

A. Noise

Listeners are confronted with the task of distinguishing between the utterances of members of their own species (conspecifics) from the calls of sympatric organisms. Many species of primates and birds are exceedingly vocal and use vocalizations to communicate over long distances in relatively noisy habitats.[1] The environmental noise generated by both biotic and nonbiotic sources (dripping condensation, wind-induced vegetation movement, rushing water, falling rocks or limbs, thunder) may readily mask the quietest signals. Furthermore, because the size of the home range and the spacing between competing individuals, troops, or family units may be defined acoustically,[2,3] an evolutionary premium may have been placed on the ability of vocalizers to produce high-amplitude calls. Humans, yelling at the top of their lungs, produce about 1.3 mW of sound power per kg of body weight, while that for Old World monkeys approaches 9 mW/kg;[4] values for birds may exceed 800 mW/kg.[5] Consequently, the amplitude of sounds of biotic origin can be quite striking. Rain forests provide an abundant diversity of niches arrayed both vertically, from the ground to the canopy, and horizontally. Rain forest primates may live in sympatry with perhaps a dozen species of other primates, scores of species of other mammals, hundreds of species of birds, and thousands of species of insects, and some of these vocal sources produce signals audible for distances approaching 2 km.[3] Thus, environmental noise has been a longstanding impediment to oral communication, and noise is not a recent impediment created by the emergence of industrialization.

Though the noise level of the rain forest can be striking, the average ambient noise level of the riverine forest actually exceeds that of the rain forest. Riverine forests are narrow strips of forest nourished by rivers that drain the adjacent savanna. This habitat type has fewer biotic sources of noise, but the noise is heightened due to vegetation movement produced by wind sweeping off the savanna. Overall ambient noise levels are lowest in the savanna habitat, yet some investigators have noted that their greatest subjective impressions of noise were experienced in the savanna, and this noise was produced by the interaction of wind with their pinna and auditory system.[1] Receptive wind noise has been shown to markedly impair hearing in human subjects tested in a wind tunnel.[6] Overall, these observations suggest that, though noise levels differ across habitat types, ambient noise presents a serious and ubiquitous impediment to communication in all habitats relevant to the evolution of speech. Hence, any differences between auditory systems that would render one system more capable of extracting signals at less favorable signal-to-noise ratios would be advantageous and favored by natural selection. Consequently, the emergence of a noise-tolerant auditory perceptual system is likely an evolutionary innovation that is engaged during vocal perception but which evolved prior to the emergence of speech and is not necessarily specialized for speech perception.

B. Signal Distortion

Imagine two theoretical habitats. The first habitat exhibits the acoustic characteristics of an anechoic room without wind and reflections, where sounds of different frequencies are attenuated at essentially the same rate as they are propagated from source to receiver. In this idealized case, the received signal is identical to the emitted signal (except that the amplitude of the received signal is less than that at the source because of spherical spreading), and the only impediment to perception is posed by the ambient noise of the environment. In this habitat, different signals could be readily distinguished from one another through either some sort of time-series template matching or cross-correlation procedure where the whole waveform is sampled, stored, and examined bit by bit. Now consider an alternative theoretical habitat in which the transmission time from source to receiver actually fluctuates from moment to moment. Because sound propagation velocity is no longer stable, neither the frequency nor duration of the received signal need match that at the source. Consider the possibility that different frequency components of complex signals are propagated along different path lengths in this habitat; consequently, signal features such as formant transitions, which occurred synchronously in the source, are now asynchronous in the received waveform. Consider the possibility that, as sound is propagated in this habitat, the rate of attenuation is frequency specific and temporally unstable, and, as a consequence, the relative amplitude of any frequency component of a complex signal at the receiver may be either proportionately greater than or less than that at the source. This second theoretical habitat is clearly much more challenging regarding the viability of auditory communication systems. In the second habitat, it is clear that a template matching or cross-correlation signal discrimination and identification system would be seriously impaired because of the high level of uncertainty of the degree of correspondence between the received and emitted waveforms.

The remainder of this section addresses the proposition that physical measurements of natural habitats relevant to the evolution of primate vocal communication systems exhibit properties resembling the second of these two theoretical habitats, and it provides examples of distortion and some of the processes believed to underlie distortion. The section concludes with a discussion of some implications of these observations for neurobiological and behavioral studies of call perception.

The combined action of acoustic scatter, reverberation, amplitude fluctuation, frequency-specific attenuation, and so forth distort the signal incident at the listener's ear, and the magnitude of distortion produced by these phenomena can be quite striking.[1,7-12] Thus, the listener's perceptual system is confronted with the task of distinguishing signal from noise and potentially estimating the characteristics of the emitted signal from the distorted waveform incident at the listener's ear. The severity and nature of distortion vary with habitat type, and the processes that give rise to distortion can be idealized with synthetic signals. Following are descriptions of the four different phenomena commonly thought to lead to distortion of signals in nature:[12]

> **Case 1: Frequency-Specific Attenuation.** Consider a signal composed of two different frequency components that are attenuated by the environment at different rates. As shown in Figure 8.1A, the high-frequency component of the broadcast signal (represented by a triangular waveform) is initially equal in amplitude to the low-frequency component (represented by a square wave). However, at the receiver, the high-frequency component has been attenuated in amplitude relative to that of the low-frequency component. This example represents the common situation where high-frequency signals, or high-frequency components of signals, are attenuated by the environment at a higher rate than are low-frequency ones.[1,8,11,13,14] Physical measurements show that fluctuations in the amplitude of broadcast tones vary rapidly over time and can easily exceed 10 dB within a 20-sec window. These results suggest that amplitude modulation of a carrier would be a poor strategy for coding information in any of these habitats. With respect to speech and similar signals, this class of distortion has the potential to alter the relative amplitudes of formants such that the receiver is unable to accurately identify the vocal tract transfer function of the emitted waveform.
>
> **Case 2: High-Frequency Scatter.** Consider a second case where the low-frequency component of a complex signal has a sufficiently long wavelength that trees and other obstacles in its path are essentially acoustically transparent (Figure 8.1B). Thus, the low-frequency component is transmitted directly along a straight line from source to receiver, whereas the high-frequency component is scattered by obstacles along the direct path. In this situation, the high-frequency component reaches the receiver only after subsequent reflections from various surfaces in the environment. Even if the high-frequency component of the signal is not strongly attenuated, the temporal relationship between the two frequency components will be distorted. Consequently, some components of complex signals may be transmitted via the direct path, while other components will be scattered and transmitted over longer indirect paths. Acoustical measurements of reverberation in various habitats show that the magnitude of scatter and reflection is frequency specific[1,10] and is significantly more prominent in forested habitats relative to that for the savanna. Reverberation, and frequency-specific differences in the magnitude of scatter, may in theory pose serious problems for coding information via changes in the temporal pattern of the onset and offset of the high- and low-frequency constituents of a vocalization.
>
> **Case 3: Interference between Direct and Reflected Waves.** Consider the case where both the direct and reflected wave of a one-component signal are received by the listener (Figure 8.1C). Reverberation measurements in natural habitats show that this is a common occurrence.[1,10] In natural habitats, interference of this type will cause the received signal to be smeared

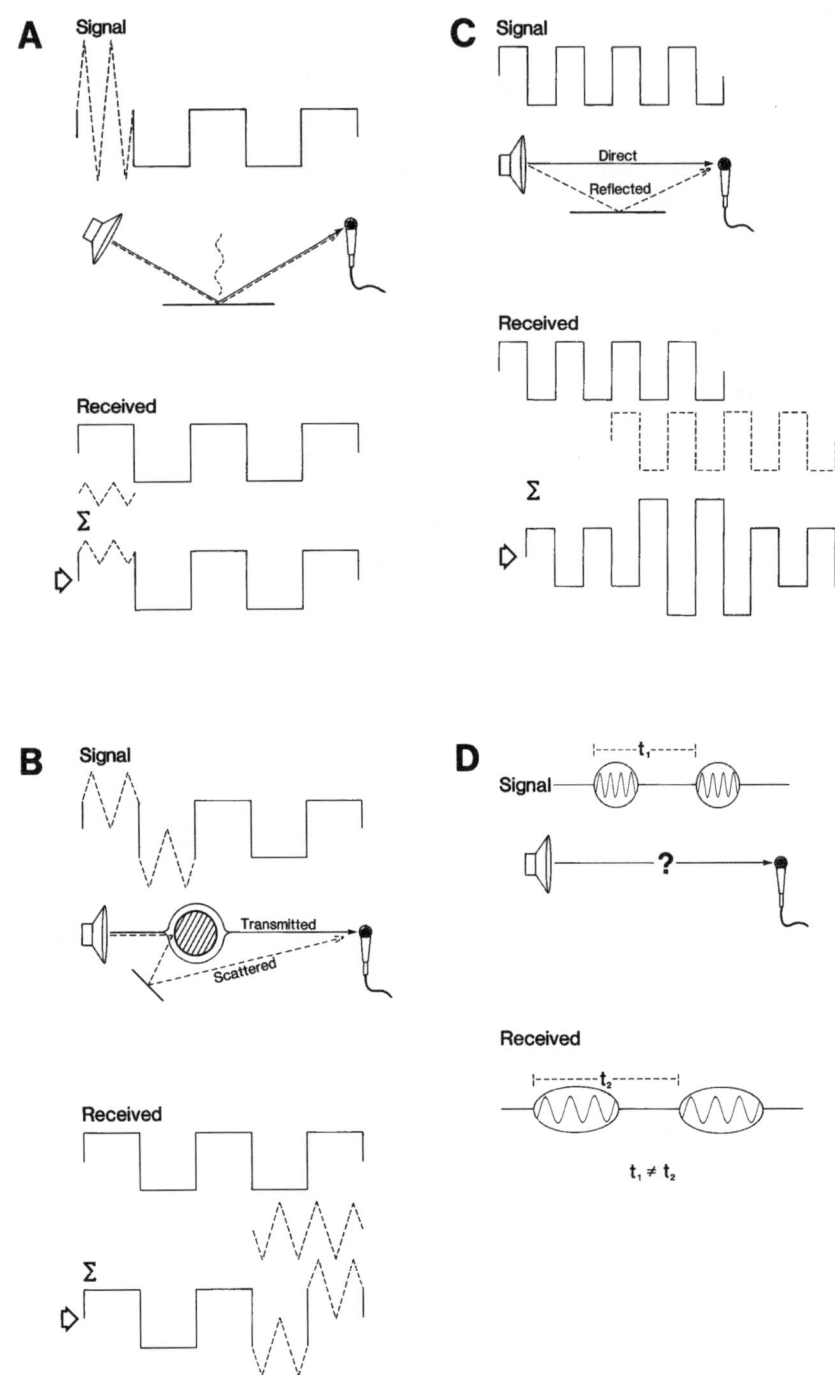

FIGURE 8.1

in time such that it is longer in duration than the broadcast signal, and the envelope of the signal will fluctuate as if the signal was amplitude modulated. In reverberant habitats, coding information through modulating the amplitude of a signal would be prone to distortion by this process.

Case 4: Effective Path-Length Fluctuation. Consider the case where a signal is composed of two successive pulses or elements where the effective path length between the source and receiver changes during the course of sound production (Figure 8.1D). This situation occurs in the common Doppler example[15] where there is relative motion between source and receiver, and it is responsible for the apparent change in the pitch of a train, truck, or car horn as the vehicle approaches and subsequently departs. However, this situation also occurs when either wind velocity or wind direction changes during the course of the emission of the signal. Changes in either wind velocity or wind direction can result in changes in the relative transmission time of successive portions of a signal.

These four cases of distortion may operate simultaneously, and it is not always readily apparent how much each process is contributing to distortion for any given situation. It is clear, however, that habitat-induced distortion is ubiquitous to all environments inhabited by primates. Is habitat-induced distortion just a problem for long-distance communication and relatively insignificant for short-range communication? Over the distances that non-human primates use vocal signals to coordinate social activities, distortion scores are high at both short and long distances, and distortion scores measured at 100 m correlate ($r = 0.97$) with those measured at only one eighth of that distance, 12.5 m.[16] Thus, the evidence suggests that habitat-induced distortion has served as a very serious fundamental impediment to vocal communication at both long and short communication distances and in both open country and forested habitats.

C. HABITAT ACOUSTICS AND NEUROBIOLOGICAL AND PERCEPTUAL STUDIES OF CALL PERCEPTION

The data reviewed here suggest that the natural acoustic world, within which vocal communication systems evolved, has been an extremely challenging acoustic

---

FIGURE 8.1 Four sound distortion phenomena. The signal at the source is shown at the top of each panel. The distortion process is depicted by a drawing, the components of the received signal are then shown, and the addition of the received components are displayed at the bottom of each panel. (A) Frequency-specific attenuation: The high-frequency component (dashed triangular wave) is attenuated by the environment at a higher rate than is the low-frequency component (solid square wave). (B) High-frequency scatter: The high-frequency component is scattered by obstacles along the direct path, while the low-frequency component is not scattered. As a result, the relative time of arrival of the two components as received differs from that initially broadcast. (C) Interference between direct and reflected waves: Path-length differences in the direct and reflected waves of a one-component (square wave) signal result in an apparent increase in signal duration and various possible interference patterns. (D) Effective path-length fluctuation: The air medium accelerates (wind changes velocity or direction) during the course of the emission of the signal. The change in wind direction or velocity is depicted here by a question mark. The received signal is distorted in duration and frequency. (Adapted from Brown and Gomez.[12])

environment. If the received signal is typically seriously degraded relative to the emitted signal, how then is auditory communication possible? Within this framework, it is expected that selection should favor the development of selective attention, or selective processing systems, for those components of waveforms that most reliably differ when the emitted signals differ. That is, perceptual systems should heavily weigh the most consistently reliable sound parameters, those most resistant to distortion by habitat acoustics, and pay less attention to the least stable properties of sounds (i.e., those dimensions along which differences in the received waveform are most apt to be due to fluctuations in habitat acoustics). Similarly, sound production systems should capitalize on any available "stable acoustic islands"[12,13,16–18] by evolving vocalizations that encompass these stable properties. According to this view, the acoustically most prominent components of sounds may be unstable, and perceptual analysis may be diverted from these components to the analysis of much more subtle, but stable components. Finally, critical key signal properties should be stereotypic, so that perceptual systems can compare multiple parameters of received signals to perceptually recover the attributes of the emitted waveform. From this perspective, six predictions follow which have implications for the design of behavioral and neurobiological studies of the perception of primate vocalizations.

Prediction 1: Perceptual systems should not be equally sensitive to all call parameters. That is, both neurobiological and behavioral studies of call perception should find that certain call parameters are more salient than others.

Prediction 2: Perceptual systems should be relatively insensitive to fluctuations in call amplitude. Because amplitude fluctuations are a prominent phenomena in virtually all habitats, perceptual systems should be insensitive to this parameter. Thus, in behavioral multidimensional scaling studies, cluster formation should be independent of changes in signal amplitude.

Prediction 3: Perceptually salient frequency components of calls should be spaced relatively close in frequency, and listeners should be comparatively sensitive to changes in the relative amplitude of these components of spectrally complex calls. This prediction proposes that if the formants (or formant-like resonances) are spaced together closely enough, then frequency-specific attenuation may affect the relative level of all the formants equally (because they are within the same bandwidth of the attenuation phenomenon), and the signal would appear to fluctuate in loudness, but not vary significantly with respect to the perceived vocal tract configuration. In many cases, primates vocalize with high-frequency fundamental frequencies. Hence, a single harmonic of the fundamental frequency may fall within each formant. As a result, as primates retract or protrude their lips or change the location of the jaw (and hence change the shape of the airway), the relative amplitude of the harmonics will correspondingly change. The coo calls of macaque monkeys are a good example of this phenomenon; a single harmonic may fall within a formant, and as the monkey changes its vocal tract configuration the amplitude of the second harmonic, for example, may exceed that of the fundamental.[19] Changes of this sort are quite subtle and easily overlooked; yet, according to prediction 3, changes in the relative amplitude of the harmonics of calls may be important perceptually.

Prediction 4: Perceptual systems should be sensitive to brief punctate call elements or to frequency changes in call elements that occur in brief temporal windows. This prediction is inspired by strategies to minimize degradation induced by reverberation. The arboreal habitat is the natal primate environment, and reverberation degradation can be minimized if vocalizations are segmented into brief pulses or are composed of components that change frequency prior to the arrival of the first main reflection. For example, if a species is adapted to mid-canopy elevations in the rain forest, the principal reflection would be from the forest floor, and the delay time for the reflected wave would be 200–300 msec. In this situation, it would be strategic to package the call into brief pulses that end prior to the delay time expected for the first echo or to change the broadcast fundamental frequency (or transfer function) before the arrival of the echo.

Prediction 5: Perceptual systems should be sensitive to rapid frequency modulations and comparatively insensitive to slower frequency modulations. This prediction is inspired by strategies to distinguish intonation and inflection contours from the path-length-fluctuation distortion process, which acts to change the apparent pitch of the signal. A corollary of this prediction is that perceptual systems should exhibit a preference for stereotypic modulation contours. Thus, changes in the frequency of the received signal that are attributable to fluctuations in the effective path-length should be perceptually insignificant.

Prediction 6: Perceptual systems should exhibit duration/frequency compensation for stereotypic calls. That is, many primates exhibit calls that are composed of a rather long sequence of pulses that are believed to reveal either species or individual identity. The roar of the black and white colobus monkeys likely signals species identity,[20] while the gobble component of the grey-cheeked mangabey's whoop-gobble is believed to communicate individual identity.[21] These and similar calls are subject to distortion by changes in the effective path length, and the duration/frequency compensation prediction states that calls lower in frequency, for example, than those typically issued by a given vocalizer will be perceived as equivalent to the prototype for that subject if they are also correspondingly longer in duration; the converse also holds. This prediction assumes that these calls possess a stereotypic acoustical structure as has been observed in the case of the roars and gobbles described above. In a behavioral multidimensional scaling study, for example, the gobbles recorded from mangabey A should all cluster together, and the gobbles recorded from mangabey B should form a second cluster distinct from that for mangabey A. Exemplars reflecting departures from the prototypical signal equivalent to those produced by the effective path-length distortion phenomenon should not adversely influence the formation of the clusters.

The ideas developed here suggest that selection must operate on both the productive and receptive components of communicative systems to help counter distortion. With respect to vocal production, there is growing evidence that selection may have acted to improve signals so that they more effectively counter the distortion processes typical of the local habitat. In a comparison of the rain forest and savanna habitats, and the vocalizations of four representative species of primates, the results show that, overall, distortion scores were greater in the savanna habitat compared to that in the rain forest.[16] Furthermore, the results show that rain forest monkey calls were distorted less when broadcast in the "appropriate" rain forest habitat compared to that resulting from broadcasts in the "inappropriate" savanna habitat, but savanna monkey calls were distorted about the same in both habitats. This finding suggests that, for some species, selection may have altered the physical form of the vocal repertoire to help counter habitat-induced distortion. Additional evidence of the acoustic adaptation of calls, or call components, to the local habitat has also been reported for birds.[22-28] Selection has acted to alter the physical structure of vocalizations to counter habitat-induced distortion. It is also likely that selection has favored the evolution of receptive, signal-processing mechanisms that would permit listeners to distinguish differences between waveforms due to variations in the signal at its source and those that are due to fluctuations in habitat acoustics.

III. PRIMATE VOCAL-FOLD BIOMECHANICS

Over the past decade, the application of nonlinear dynamics to the analysis of the patterns of vibration of the vocal folds has revolutionized our understanding of the biomechanics of voicing in general,[29,30] and only recently has nonlinear dynamics been introduced to animal bioacoustics.[31] Thus, many researchers are unfamiliar with the terminology of nonlinear dynamics and are equally unprepared for the paradigm shift in thinking that may follow when the implications of this field for animal bioacoustics are understood. Unfortunately, the basic tenets of nonlinear dynamics are complicated and cannot be adequately discussed in the available space of this chapter; consequently, the interested reader is strongly encouraged to read Wilden et al.[31] for an excellent introduction. This section discusses some of the incorrect assumptions made regarding the mechanics of primate vocal production. By drawing attention to mistakes in our own thinking, the author hopes to encourage other investigators to review their understanding of primate vocal behavior and to evaluate the implications of the new perspective of voicing biomechanics for the design of neurobiological and behavioral studies of primate perceptual systems.

A. Misassumptions Regarding the Mechanics of Primate Vocal Production

1. Misassumption: Noisy Calls Are Produced by Turbulence in the Airway

Airflow through a narrow constriction produces turbulence, and, as reflected by the sibilants employed in English (/s/, /ch/, and so forth), the turbulence may result in a prominent noisy hissing sound. It does not necessarily follow, however, that noisy primate calls are produced by turbulence. The vocal folds may oscillate erratically, resulting in irregular periods for each successive oscillation; hence, no fundamental frequency will be apparent in the voice sample. In calls associated with this pattern of vocal-fold oscillation, the broadband noise in the spectrogram of the call is fundamentally the result of the quasi-irregular periods of each glottal cycle, and turbulence may or may not play a role.

In Figure 8.2, the electroglottograph (EGG) waveform of a Sykes's monkey (*Cercopithecus albogularis*) grunt is displayed in association with the sound spectrogram. The EGG waveform is recorded from surface electrodes placed directly over the monkey's larynx. As may be observed from the EGG waveform, both the period and amplitude of each oscillation of the vocal folds are irregular. In the terminology of nonlinear dynamics, this pattern of vocal-fold oscillation is called *deterministic chaos*. Chaos can also be seen in the oscillation patterns of some voice samples obtained from dog and monkey larynges phonated in an excised larynx experiment.[32,33] In this preparation, the larynx is surgically removed from the organism, and the vocal folds are phonated by the flow of compressed air.

Noisy calls, or noisy segments of calls, are prominent in the vocal repertoire of many species of primates. One implication of the chaotic vocal-fold-oscillation phenomenon is that transitions between noisy or clear calls, or between noisy and clear components within a call, are not necessarily due to changes in articulation or shape changes in the airway. Rather, they are due to changes occurring within the larynx itself. This suggests that the regulation of voicing is a very important aspect of primate vocal production.

2. Misassumption: Variations between Successive Calls, or Fluctuations within a Single Vocalization, Are the Acoustic Signature of Changes in the Parameters of Vocal Control

The converse to this assumption is that acoustic variations between successive vocalizations (or fluctuations within a single call) may occur with no measurable change in subglottal pressure, adduction, or any other parameter of vocal control. Analyses of the vibrations of the vocal folds has shown that, under some conditions, two or more patterns (or regimes) of oscillation are equally likely to occur. Also, the vocal folds will spontaneously and unpredictably jump (or bifurcate) from one regime of oscillation to another without any change in the parameters of vocal control. In the nomenclature of nonlinear dynamics, a particular regime of vocal-fold oscillation is due to a specific attractor becoming established in the phase space of the system. For example, coupled synchronous oscillations of the vocal folds are attributed to the limit-cycle attractor. However, under certain conditions, two or more attractors may simultaneously coexist, and the vocal output of this system will unexpectedly jump from one regime of oscillation to another as the output of the system is momentarily dominated by one attractor and then another.[31,34,35]

Figure 8.3A presents an example of the phenomenon of multiple attractors observed in a voice sample recorded from an excised squirrel monkey larynx.[33] In this sample, three different limit-cycle attractors coexisted with fundamental frequencies of 1033 Hz, 1378 Hz, and 1550 Hz, respectively, and the fundamental frequency

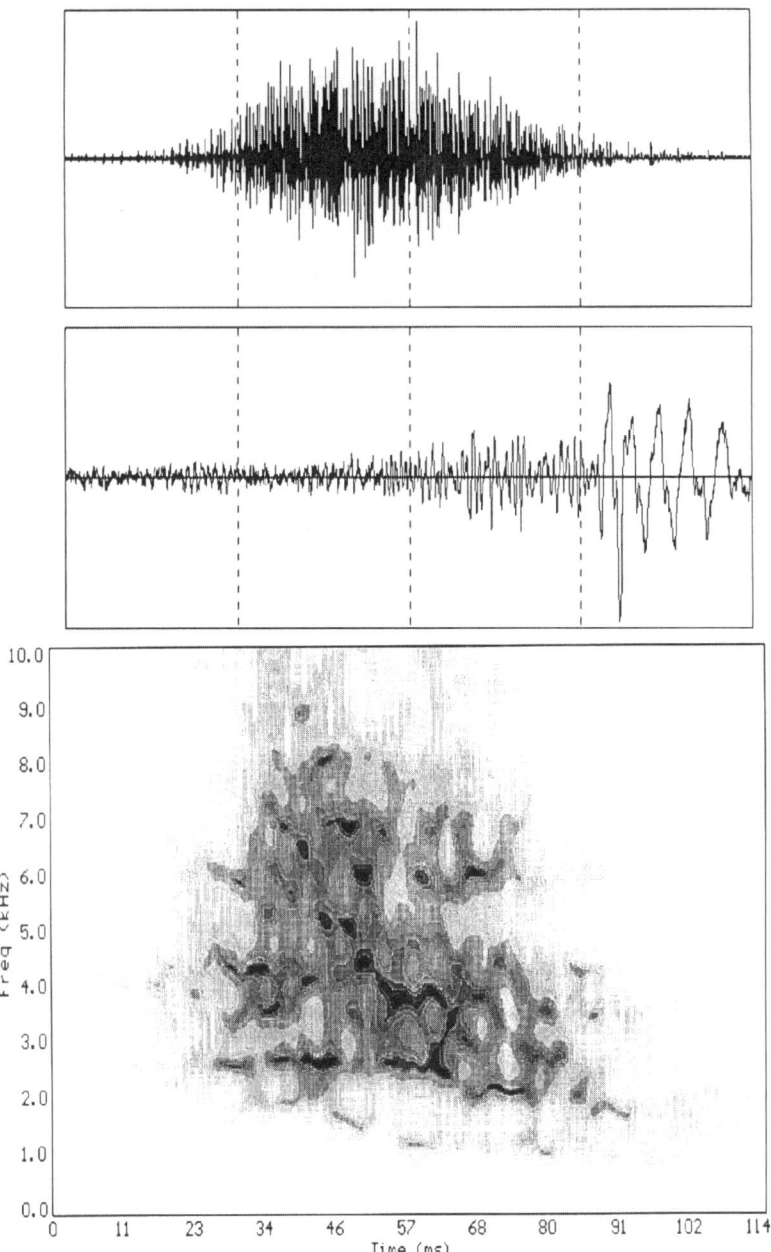

FIGURE 8.2 Grunt vocalization recorded from an adult female Sykes's monkey. Shown are the audio waveform (top), the EGG waveform recorded from surface electrodes positioned directly over the larynx (middle), and the sound spectrogram (bottom).

unpredictable jumped from one of these values to another throughout the sample. The hallmark of multiple attractors is abrupt transitions in the fundamental frequency, not

Ecological and Physiological Constraints for Primate Vocal Communication 139

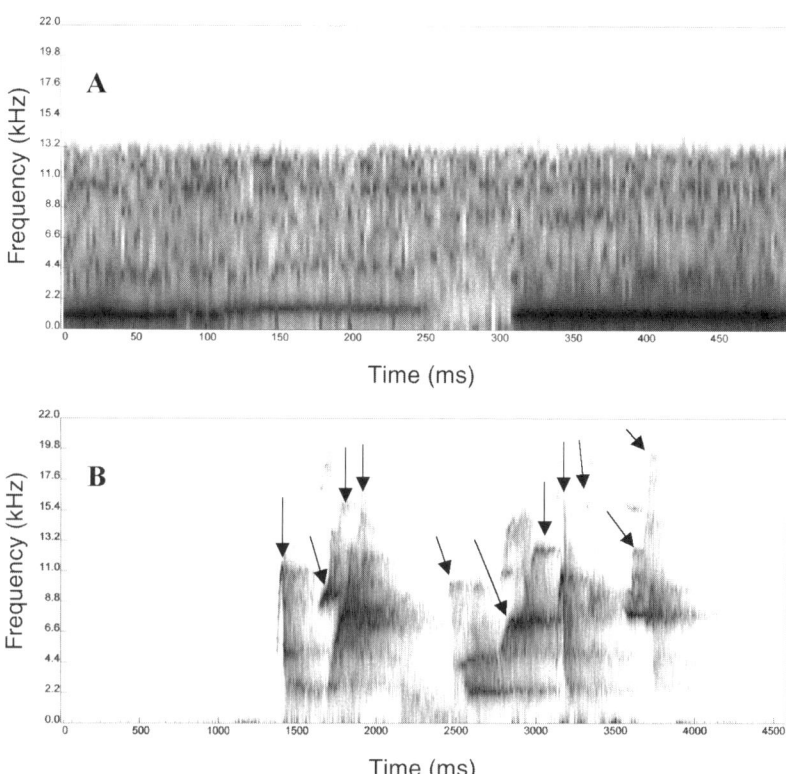

FIGURE 8.3 Sound samples recorded from squirrel monkeys. (A) Spontaneous bifurcations between multiple attractors recorded from an excised larynx: The sound spectrum is displayed in the top panel, the waveform in the middle panel, and the spectrum in the bottom panel. At the onset of the sample the fundamental frequency was 1033 Hz. The frequency jumped to 1550 Hz at 80 msec, to 1033 Hz at 90 msec, and to 1378 Hz at 115 msec. A segment of chaotic oscillations was exhibited between 250 and 310 msec and was followed by a bifurcation to a fundamental frequency of 1033 Hz. (Adapted from Brown et al.[33]). (B) A squirrel monkey chuck call (left-most arrow) is followed by ten successive elements (successive arrows left to right) in a vocal bout. Many of these transitions are believed to be produced by bifurcations in the regime of voicing induced by changes in subglottal pressure, adduction, vocal-fold elongation, and so forth.

gradual and steady changes. Between 250 and 310 msec in this sample, chaotic oscillations of the vocal folds were exhibited, and no fundamental frequency was apparent in this portion of the voice sample. Hence, the phenomenon of the coexistence of multiple attractors may result in abrupt transitions between distinctly different regimes of vocal-fold oscillation with no change in the voice control parameters. We are currently unable to specify where in the primate voice profile range the phenomenon of multiple attractors is likely, or unlikely, to be expressed; more research is necessary to answer this question. Nevertheless, abrupt transitions in the regime of oscillation may occur in natural utterances, and these bifurcations may occur accidentally, or unintentionally. Thus, some acoustic fluctuations between successive utterances (or within a single vocalization) may not be associated with any change in the voice control parameters and consequently were not intentionally produced by the vocalizer. One implication of this phenomenon is that two calls with very different acoustic structures may be the product of nearly identical motor sequences.

3. Misassumption: Acoustically Distinct Calls, or Call Elements Issued within a Single Exhalation, Are the Product of Changes in the Shape of the Airway

Many primates produce rapid sequences of sounds in which a wide variety of acoustic elements are emitted rapidly in succession within a single breath. Because these elements may be so acoustically dissimilar, it is tempting to regard them as being produced by distinctively different vocal gestures involving changes in the configuration of the lips,[36] aspects of the jaw and tongue,[37] and changes in other structures altering the shape of the airway. An example of this phenomenon produced by a male squirrel monkey is shown in Figure 8.3B, in which a squirrel monkey chuck call (left-most arrow), is followed by ten successive elements (successive arrows left to right) in a vocal bout. It is possible that a sequence of abrupt transitions between apparently distinctive acoustic elements is the result of a gradual and steady change in a single parameter of vocal control such as subglottal pressure (or perhaps two linked parameters such as adduction and subglottal pressure). Studies of the excised squirrel monkey larynx show that as subglottal pressure is gradually incremented from the threshold of phonation to higher values, the vocal folds will display a consistent pattern of oscillation for a portion of the voice profile range and then abruptly bifurcate from one regime to another.[33] In the vocal displays of monkeys, these abrupt bifurcations in the regime of oscillation will produce the appearance of a sequence of individually distinctive vocal events involving changes in the shape of the airway. Unlike the case of the coexistence of multiple attractors (misassumption 2), however, the bifurcations between regimes are linked to a change or modulation of a particular parameter of vocal control. One implication of this phenomenon is that vocal elements with very different acoustic structures may be the product of a closely related and coordinated change of one dimension of a motor sequence.

4. Misassumption: With the Exception of Noisy Calls (Discussed in Misassumption 1), Differences in the Frequency Spectrum of Calls in the Vocal Repertoire of Primates Are Due to Changes in the Fundamental Frequency and the Relative Magnitude of the Harmonics of the Fundamental

This assumption is based on the premise that the clear calls produced by primates are generated by coupled synchronous oscillations of the vocal folds, similar to those employed in the production of vowel sounds by humans. While primates may phonate some calls according to this expectation, recent studies have shown that primate calls may also exhibit biphonation, a phenomenon classically attributed to the torus attractor which establishes simultaneous vibrations in two or more uncoupled oscillators.[32,33,38]

Ecological and Physiological Constraints for Primate Vocal Communication 141

In Figure 8.4, biphonation is shown in squeal exemplars recorded from Sykes's monkeys. In these exemplars, as in all instances of biphonation, the frequency peaks are not integer multiples of a single fundamental frequency.

Studies of biphonation in canines and squirrel monkeys have shown that the onset of biphonation is due to an abrupt bifurcation from a regime of oscillation in

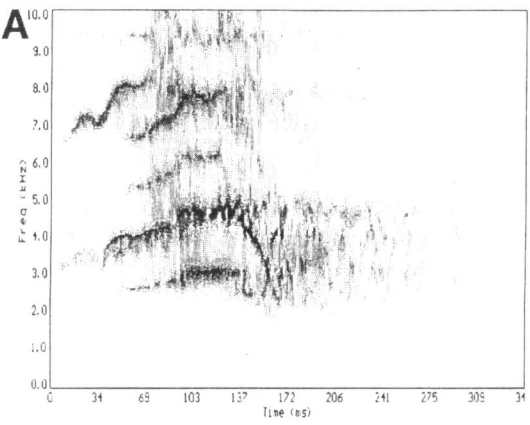

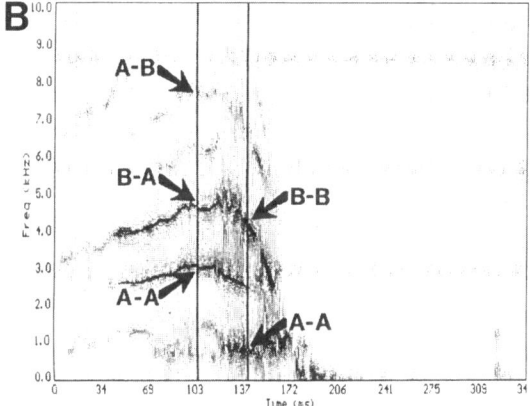

FIGURE 8.4 A Sykes's monkey squeal exemplar produced by biphonation. (A) A sound spectrogram of the acoustic waveform; (B) the EGG waveform recorded from electrodes placed on the larynx during vocalization shown in A. Time-expanded sections of the acoustic and EGG wave shown in A and B are shown for a 5-msec window sampled at 104 msec (C) and 138 msec (D), as indicated by the position of the cursors in B. The EGG wave shows a large-amplitude A wave, and a small amplitude B wave as labeled in C and D. The frequencies of the average A–A, B–B, A–B, B–A cycles are shown in B. Thus, the different frequencies observed in B may be produced by two uncoupled oscillators that differ with respect to the surface area of the tissue in the larynx which make contact when set in vibration. The large-amplitude A wave is attributed to oscillations in the body of the vocal folds (the TA muscle) and the small-amplitude B wave is attributed to oscillations in the vocal membrane, or vocal lip, a medial extension of the mucosal layer of the vocal folds found in some primates. (Adapted from Brown and Cannito.[37]) *(continued)*

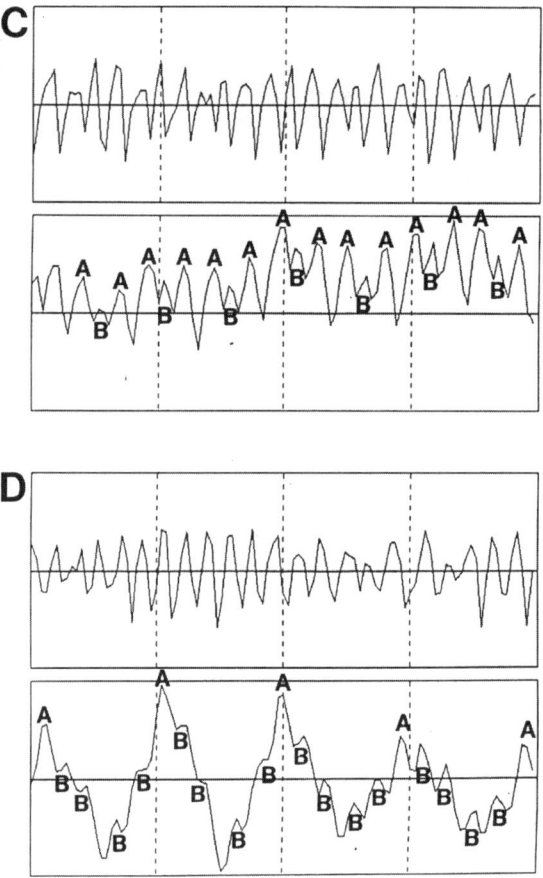

FIGURE 8.4 (Continued from previous page.)

which biphonation was absent to one in which it is present.[30,33] Hence, changes in a natural vocalization, where the change in the frequency composition of the call was linked to the onset of biphonation, is attributed to a bifurcation in the regime of vocal-fold oscillation. It is presently unclear if differences or changes in biphonation patterns (the relative frequencies and amplitudes of the oscillators) are replicable and potentially under volitional control.

5. **Misassumption: The Production of Vocalizations Composed of a Series of Notes, Elements, or Pulses Is Governed by Central Motor Programs which Orchestrate Contractions in the Vocal Musculature that Mimic the Rhythm of the Pattern of the Vocal Sequence**

The number of notes in a trill sequence, for example, and the beginning and ending of each note are associated with a particular sequence of firing of motor neurons that initiate and terminate each individual pulse. In humans, a variant of this concept has been invoked to account for the production of *trillo*, a singing style in which the vocalizer produces a staccato sequence of notes rather than a sustained note. Researchers have hypothesized that rapidly alternating patterns of abduction and adduction of the laryngeal musculature are necessary for the production of *trillo*.[39]

Recent studies of voicing in the excised squirrel monkey larynx have shown that under certain conditions the larynx is capable of producing a waveform that waxes and wanes rhythmically in amplitude, yielding the impression of a series of distinct segments or pulses.[33] This work raises the possibility that at least some pulsed calls are the product of a voicing regime in which the oscillations of the vocal folds rhythmically modulate the amplitude of the signal, and rhythmic contractions of the vocal musculature may not be required for the segmentation of some vocalizations. These observations again suggest that the regulation of voicing is a very important aspect of understanding primate vocal production and that the established voicing regime within the larynx may be responsible for the distinction between sustained and segmented voicing.

6. **Misassumption: Because Noisy Calls Are Produced with an Aperiodic Source, Acoustic Differences between Call Exemplars Are Likely Due to Moment-by-Moment Fluctuations or Instabilities in the Source and Are Communicatively Insignificant**

Though it is true that noise produced by either turbulence or chaotic vibrations of the vocal folds is aperiodic, these calls may reveal variations in the shape of the airway, and these shape variations may be communicatively significant.[40,41] The coughing sound associated with clearing one's throat is a good example. The cough is aperiodic in most people, yet the coughing sounds produced when the jaw is lowered, the lips are retracted, and the teeth are bared differ prominently from those produced when the lips are rounded or the when the mouth is closed. Thus, changes in the shape of the airway change the resonant or formant frequencies of the airway, and these differences are perceptible when the vocal tract is excited by an aperiodic source. The same principle applies to primate vocalizations.[40-45] Figure 8.5 shows the changes in the first formant frequency of a grunt vocalization given by a grey-cheeked mangabey associated with changes in the location of the mandible. As the jaw is lowered, the resonant frequency for the first formant increases, and an equivalent change in formant frequencies associated with jaw lowering is seen in humans.

It is possible that harsh or noisy calls are better at revealing changes in the shape of the airway than are tonal calls. This is because many of the tonal calls emitted by primates are produced with a high fundamental frequency relative to the length of the vocal tract, and the fundamental frequency may actually match the resonant frequency of the first formant. The ability to match the frequency of the first formant with the fundamental frequency is seen in some highly trained opera singers.[46] It is a strategy for producing higher amplitude signals, which would be audible over

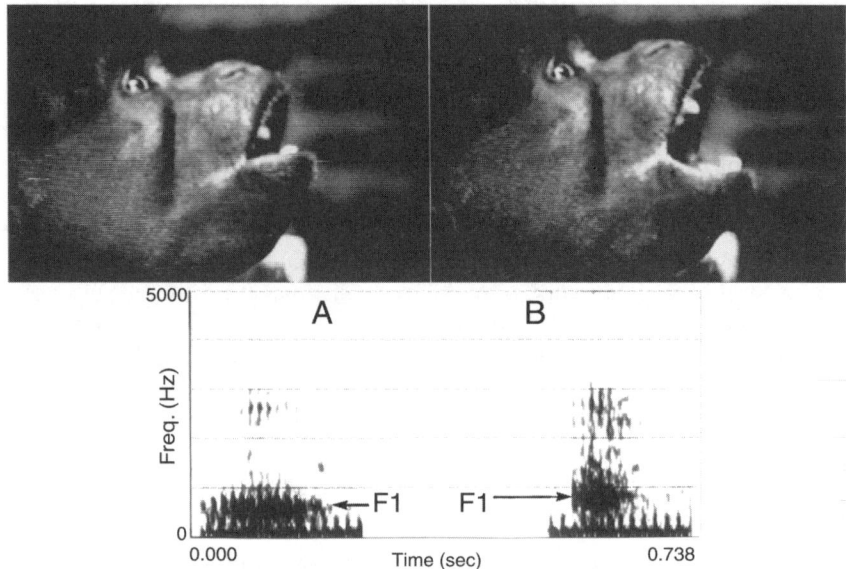

FIGURE 8.5 Two grunts produced in close succession by a grey-cheeked mangabey. The frequency of the first formant (F_1) is denoted by the arrows in the spectrograms; F_0 is about 100 Hz for both grunts. In the second grunt, the mandible is lowered, resulting in an increase in the frequency of F_1. (Adapted from Brown and Cannito.[38])

greater distances (or more audible over the orchestra) but would be poor at revealing changes in the transfer function produced by changing the shape of the airway.

B. Vocal-Fold Biomechanics and Neurobiological and Behavioral Studies of Vocal Perception

In addition to coupled synchronous oscillations, resembling the voicing pattern employed in vowel production, the primate larynx exhibits at least three additional regimes of vocal-fold vibration, including biphonation, staccato phonation, and aperiodic or chaotic phonation. It is quite likely that additional research will reveal other regimes of voicing. In many cases, different call types are associated with bifurcations between different regimes of phonation within the course of a single call or within a single exhalation. Some Sykes's monkey squeals, for example, show bifurcations from the limit-cycle regime to biphonation and then back to the limit-cycle regime within a single utterance.[38] The ability of vocalizers to select a voicing regime and regulate transitions between regimes then becomes an important question for understanding vocal behavior. It is likely, in many instances, that vocalizers intentionally select a regime of phonation and electively execute transitions from one regime to another at will. This perspective is consistent with the observation that many primates tend to produce particular vocalizations in specific social situations.[47-61] It does not follow, however, that all voicing changes are the result of intentional modulations of the voice-control parameters. That is, it is also a possibility that bifurcations from one regime to another occur spontaneously in some portions of the range of voicing without any changes in neural control. The increased variability in the acoustic structure of some calls issued in vocal exchanges at high levels of arousal is not inconsistent with this latter possibility.[47,48,50,51,54–56,61–63] At present, no empirical data exist to provide a good basis for estimating the prevalence of accidental bifurcations for any regime of phonation. However, investigators are beginning to examine the possibility that changes in the spectral pattern of some primate calls occur with limited nervous system control.[64]

The screams of macaques[49,61] and most other primates provide a good example. Many of these calls are neither completely harmonic nor completely noisy, and many calls exhibit segments of harmonic elements accompanied by segments of broadband spectra with energy dispersed over many different frequencies. Different exemplars of these atonal vocalizations[58,65] exhibit distinctive variations between the harmonic and noisy segments, and it is tempting to attribute the striking variations in these calls to changes in the parameters of neural control.

The squeals of Sykes's monkeys illustrate this exact problem. Using acoustical criteria, Brown and Cannito[38] sorted the squeals of Sykes's monkeys into four distinctive categories. In one class of squeals, the variations between exemplars exhibited striking transitions between harmonic and aperiodic segments, while in another class of squeals the aperiodic segment was absent. Brown and Cannito[38] reasoned that the aperiodic segment of these calls was due to turbulence induced by constricting the supralaryngeal airway. They proposed that monkeys could electively produce either periodic harmonic squeals or those in which the harmonic and aperiodic segments were interspersed by regulating the presence or absence of constrictions in the airway. Brown and Cannito's analysis was probably incorrect. More recent observations suggest that their thinking reflects the application of misassumptions 1, 2, and 3. The emerging view of primate voicing suggests that the aperiodic segments of these calls are due to an interval of instability in the periods of the glottal cycle (not produced by turbulence induced by a constriction in the airway). The transitions between periodic and aperiodic segments are due to bifurcations between the limit-cycle regime and chaos (not produced by changes in the shape of the airway). And the pattern of transitions between periodic and aperiodic segments may not have been under tight neural control. That is, the bifurcations between the periodic and aperiodic segments may have been spontaneous or under marginal neural control.

In order to gauge the prevalence of accidental and intentional bifurcations, it is desirable to map the stability of voicing regimes and the transitions between regimes associated with systematic changes in subglottal pressure, vocal-fold length, adduction, and so forth. Unfortunately, bifurcation diagrams of this sort have not been published for any non-human primate. In addition to studies of the biomechanics of primate voicing, behavioral and neurobiological studies of call perception may shed light on this problem. Brown and associates[66] have used multidimensional scaling techniques to measure the perceptual relationships between the different classes of Sykes's monkey squeals described above. The results showed that acoustically dissimilar patterns between the periodic and aperiodic segments of different squeal exemplars were perceived as being nearly identical. That is, even though the calls were different acoustically, they were perceived as being very similar. On the other hand, subtle changes in the contour of the fundamental frequency of periodic squeals were very salient perceptually to conspecific listeners. These results show that the acoustic differences between call exemplars do not predict the perceptual differences between exemplars, and it is possible that prominent differences in the acoustic structure of vocalizations may be perceptually insignificant, if these differences are the result of instabilities in voicing regimes. These observations suggest that the perceptual similarity or dissimilarity between two different vocalizations may be more closely linked to similarities in the mechanics of voice production and voice control parameters than to simple acoustic measures of similarity (such as cross-correlation scores) that are insensitive to the implications of nonlinear dynamics for understanding vocal behavior.

IV. SUMMARY

Distortion measurements for vocalizations broadcast in the natural habitat and biomechanical measurements of primate voicing suggest that perceptual mechanisms may have been exposed to intense selection for distinguishing accidental from intended acoustic variation. Moment-by-moment fluctuations in the acoustic distortion processes active in the habitat result in significant changes between successive broadcasts of the exact same signal. Furthermore, unintended variations between successive iterations of the same vocal gesture will inevitably occur if vocalizations are given under conditions in which the phenomenon of multiple attractors is present. Vocalization classification and recognition processes have likely evolved mechanisms that selectively ignore acoustic variations that are commonly due to these uncontrolled sources of variation and to focus on acoustic parameters that change reliably with changes in voice control parameters. Researchers conducting behavioral and neurobiological studies of vocal perception are expected to encounter instances where the perceptual relationship between two call exemplars is not closely related to acoustic measures of similarity, and the integration of observations from habitat acoustics and the biomechanics of voicing into studies of vocal perception may significantly advance our understanding of primate vocal communication.

REFERENCES

1. Waser, P.M. and Brown, C.H., Habitat acoustics and primate communication, *Am. J. Primatol.*, 10, 135–154, 1986.
2. Brenowitz, E.A., The active space of red-winged black-bird song, *J. Comp. Physiol.*, 147, 511–522, 1982.
3. Brown, C.H., The active space of blue monkey and grey-cheeked mangabey vocalizations, *Animal Behav.*, 37, 1023–1034, 1989.
4. Brown, C.H., The measurement of vocal amplitude and vocal radiation pattern in blue monkeys and grey-cheeked mangabeys, *Bioacoustics*, 1, 253–271, 1989.
5. Brackenbury, J.H., Power capabilities of the avian sound producing system, *J. Exp. Biol.*, 78, 239–251, 1979.
6. Kristiansen, U.R. and Pettersen, O.K.O., Experiments on the noise heard by humans when exposed to atmospheric winds, *J. Sound Vibration*, 58, 285–291, 1978.

7. Michelsen, A., Sound reception in different environments, in *Perspectives in Sensory Ecology*, B.A. Ali, Ed., Plenum, New York, 1978, pp. 345–373.
8. Wiley, R.H. and Richards, D.G., Physical constraints on acoustic communication in the atmosphere: implications for the evolution of animal vocalizations, *Behav. Ecol. Sociobiol.*, 3, 239–263, 1978.
9. Wiley, R.H. and Richards, D.G., Adaptations for acoustic communication in birds: sound transmission and signal detection, in *Acoustic Communication in Birds*, Vol. 1, D.E. Kroodsma, D.E. Miller, and E.H. Miller, Eds., Academic Press, London, 1982, pp. 131–181.
10. Richards, D.G. and Wiley, R.H., Reverberations and amplitude fluctuations in the propagation of sound in the forest: implications for animal communication, *Am. Naturalist*, 115, 381–399, 1980.
11. Brown, C.H. and Waser, P.M., Environmental influences on the structure of primate vocalizations, in *Primate Vocal Communication*, D. Todt, P. Goedeking, and D. Symmes, Eds., Springer-Verlag, Berlin, 1988, pp. 51–66.
12. Brown, C.H. and Gomez, R., Functional design features in primate vocal signals: the acoustic habitat and sound distortion, in *Topics in Primatology*, Vol. 1, *Human Origins*, T. Nishida, W.C. McGrew, P. Marler, M. Pickford, and F.M. B. de Waal, Eds., University of Tokyo Press, Tokyo, 1992, pp. 177–198.
13. Marten, K., Quine, D., and Marler, P., Sound transmission and its significance for animal vocalization. 1. Temperate habitats, *Behav. Ecol. Sociobiol.*, 2, 271–290, 1977.
14. Römer, H. and Lewald, J., High-frequency sound transmission in natural habitats: implications for the evolution of insect acoustic communication, *Am. Naturalist*, 115, 381–399, 1992.
15. Pierce, A.D., *Acoustics: An Introduction to Its Physical Principles and Applications*, The Acoustical Society of America, Woodbury, NY, 1989.
16. Brown, C.H., Gomez, R., and Waser, P.M., Old World monkey vocalizations: adaptation to the local habitat?, *Animal Behav.*, 50, 945–961, 1995.
17. Morton, E.S., Ecological sources of selection on avian sounds, *Am. Naturalist*, 109, 17–34, 1975.
18. Marten, K. and Marler, P., Sound transmission and its significance for animal vocalization. 1. Temperate habitats, *Behav. Ecol. Sociobiol.*, 2, 271–290, 1977.
19. Le Prell, C.G. and Moody, D.B., Perceptual salience of acoustic features in Japanese monkey coo calls, *J. Comp. Psychol.*, 111, 261–274, 1997.
20. Marler, P., Vocalizations of East African monkeys. II. Black and white colobus, *Behaviour*, 42, 175–197, 1972.
21. Waser, P.M., Experimental playbacks show vocal mediation of avoidance on a forest monkey, *Nature*, 255, 56–58, 1977.
22. Anderson, M.E. and Conner, R.N., Northern Cardinal song in three forest habitats in eastern Texas, *Wilson Bull.*, 97, 436–449, 1985.
23. Gish, S.L. and Morton, E.S., Structural adaptations to local habitat acoustics in Carolina wren songs, *Zeitschrift für Tierpsychologie*, 56, 74–84, 1981.
24. Shy, E., The relation of geographical variation in song to habitat characteristics and body size in North American tanagers (*Thraupinae: piranga*). *Behav. Ecol. Sociobiol.*, 12, 71–76, 1983.
25. Waas, J.R., Song pitch-habitat relations in white-throated sparrows: cracks in acoustic windows?, *Can. J. Zool.*, 66, 2578–2581, 1988.
26. Wasserman, F.E., The relationship between habitat and song in the white-throated sparrow, *Condor*, 81, 424–426, 1979.

27. Wiley, R.H., Associations of song properties with habitats for territorial osine birds of eastern North American, *Am. Naturalist*, 138, 973–993, 1991.
28. Ryan, M.J., Cocroft, R.B., and Wilczynski, W., The role of environmental selection in intraspecific divergence of mate recognition signals in the cricket frog, *Acris crepitans*, *Evolution*, 44, 1869–1872, 1990.
29. Titze, I.R., Baken, R., and Herzel, H., Evidence of chaos in vocal fold vibration, in *Vocal Fold Physiology: Frontiers in Basic Science*, I. Titze, Ed., Singular Publishing, San Diego, CA, 1993, pp. 143–188.
30. Berry, D.A., Herzel, H., Titze, I.R., and Story, B.H., Bifurcations in excised larynx experiments, *J. Voice*, 10, 129–138, 1996.
31. Wilden, I., Herzel, H., Peters, G., and Tembrock, G., Subharmonics, biphonation, and deterministic chaos in mammal vocalizations, *Bioacoustics*, 9, 171–196, 1998.
32. Berry, D.A., Herzel, H., Titze, I.R., and Krischer, K., Interpretation of biomechanical simulations of normal and chaotic vocal fold oscillations with empirical eigenfunctions, *J. Acoust. Soc. Am.*, 95, 3595–3604, 1994.
33. Brown, C.H., Alipour, F., Berry, D.A., and Montequin D., Voicing biomechanics and squirrel monkey (*Saimiri boliviensis*) vocal communication, *J. Acoust. Soc. Am.*, 2002 (submitted).
34. Berge, P., Pompeau, Y., and Vidal, C., *Order within Chaos*, Wiley, New York, 1986.
35. Kaplan, D. and Glass, L., *Understanding Nonlinear Dynamics*, Springer–Verlag, Berlin, 1995.
36. Hauser, M.D. and Schön Ybarra, M., The role of lip configuration in monkey vocalizations: experiments using Xylocaine as a nerve block, *Brain Language*, 46, 232–244, 1994.
37. Brown, C.H. and Cannito, M.P., Articulated and inflected primate vocalizations: developing animal models of speech, in *Disorders of Motor Speech: Assessment, Treatment, and Clinical Consideration*, D. Robbin and D. Beukleman, Eds., Brookes Publishing, Baltimore, MD, 1995, pp. 43–63.
38. Brown, C.H. and Cannito, M.P., Modes of vocal variation in Sykes's monkey (*Cercopithecus albigena*) squeals, *J. Comp. Psychol.*, 109, 398–415, 1995.
39. Brown, L.R. and Scherer, R.C., Laryngeal adduction in trillo, *J. Voice*, 6, 27–35, 1992.
40. Fitch, W.T. and Hauser, M.D., Vocal production in nonhuman primates: acoustics, physiology, and functional constraints on "honest" advertisement, *Am. J. Primatol.*, 104, 29–40, 1995.
41. Owren, M.J. and Rendall, D., An affect-conditioning model of nonhuman primate vocal signaling, in *Perspectives in Ethology*, Vol. 12, *Communication*, D.H. Owings, M.D. Beecher, and N.S. Thompson, Eds., Plenum, New York, 1997, pp. 299–346.
42. Lieberman, P., Primate vocalization and human linguistic ability, *J. Acoust. Soc. Am.*, 44, 1574–1584, 1968.
43. Owren, M.J., Acoustic classification of alarm calls by vervet monkeys (*Cercopithecus aethiops*) and humans. II. Synthetic calls, *J. Comp. Psychol.*, 104, 29–40, 1990.
44. Hauser, M.D., Evans, C.S., and Marler, P. The role of articulation in the production of rhesus monkey (*Macaca mulatta*) vocalizations, *Animal Behav.*, 45, 423–433, 1993.
45. Fitch, W.T., Vocal tract length and formant frequency dispersion correlate with body size in rhesus macaques, *J. Acoust. Soc. Am.*, 102, 1213–1222, 1997.
46. Sundberg, J., *The Science of the Singing Voice*, Northern Illinois University Press, Dekalb, 1987.
47. Winter, P., Ploog, P., and Latta, J., Vocal repertoire of the squirrel monkey (*Saimiri sciureus*), its analysis and significance, *Exp. Brain Res.*, 1, 359–384, 1966.

48. Winter, P., Observations of the vocal behavior of free-ranging squirrel monkeys, *Zeitschrift für Tierpsychologie*, 31, 1–7, 1972.
49. Green, S., Variation of vocal pattern with social situation in the Japanese monkey (*Macaca fuscata*): a field study, in *Primate Behavior*, Vol. 4, *Developments in Field and Laboratory Research*, L. Rosenbloom, Ed., Academic Press, New York, 1975, pp. 1–102.
50. Schott, D., Quantitative analysis of the vocal repertoire of squirrel monkeys (*Saimiri sciureus*), *Zeitschrift für Tierpsychologie*, 38, 225–250, 1975.
51. Gautier, J.-P. and Gautier, A., Communication in Old World monkeys, in *How Animals Communicate*, T.E. Sebeok, Ed., Indiana University Press, Bloomington, 1977, pp. 890–964.
52. Seyfarth, R.M., Cheney, D.L., and Marler, P., Monkey responses to three different alarm calls: evidence of predator classification and semantic communication, *Science*, 210, 801–803, 1980.
53. Seyfarth, R.M., Cheney, D.L., and Marler, P., Vervet monkey alarm calls: semantic communication in a free-ranging primate, *Animal Behav.*, 28, 1070–1094, 1980.
54. Smith, H.J., Newman, J.D., and Symmes, D., Vocal concomitants of affiliative behavior in squirrel monkeys (*Saimiri sciureus*), in *Primate Communication*, C.T. Snowdon, C.H. Brown, and M.R. Petersen, Eds., Cambridge University Press, Cambridge, U.K., 1982, pp. 30–49.
55. Boinski, S. and Newman, J.D., Preliminary observations on squirrel monkey (*Saimiri oerstedi*) vocalizations in Costa Rica, *Am. J. Primatol.*, 14, 329–343, 1988.
56. Boinski, S., The coordination of spatial position: a field study of the vocal behavior of adult female squirrel monkeys, *Animal Behav.*, 41, 89–102, 1991.
57. Boinski, S. and Mitchell, C.L., Ecological and social factors affecting the vocal behavior of adult female squirrel monkeys, *Ethology*, 92, 316–330, 1992.
58. Hauser, M.D. and Fowler, C., Declination in fundamental frequency is not unique to human speech: evidence from nonhuman primates, *J. Acoust. Soc. Am.*, 91, 363–369, 1991.
59. Hauser, M.D., A mechanism guiding conversational turn taking in vervet and rhesus macaques, in *Topics in Primatology*, T. Nishida, F.B.M. deWaal, W. McGrew, P. Marler, and M. Pickford, Eds., Tokyo University Press, Tokyo, 1992, pp. 235–248.
60. Boinski, S. and Mitchell, C.L., Wild squirrel monkey (*Saimiri sciureus*) 'caregiver' calls: contexts and acoustic structure, *Am. J. Primatol.*, 35, 129–137, 1995.
61. Gouzoules, H. and Gouzoules, S., Agonistic screams differ among four species of macaques: the significance of motivational-structural rules, *Animal Behav.*, 59, 501–512, 2000.
62. Gouzoules, H. and Gouzoules, S., Design features and developmental modification in pigtail macaque (*Macaca nemestrina*) agonistic screams, *Animal Behav.*, 37, 383–401, 1989.
63. Gouzoules, H. and Gouzoules, S., Body size effects on the acoustic structure of pigtail macaque (*Macaca nemestrina*) screams, *Ethology*, 85, 324–334, 1990.
64. Tokuda, I., Riede, T., Naubauer, J., Owren, M.J., and Herzel, H., Nonlinear prediction of irregular animal vocalizations, *J. Acoust. Soc. Am.*, in press.
65. Tembrock, G., Canid vocalizations, *Behav. Proc.*, 1, 57–75, 1976.
66. Brown, C.H., Sinnott, J.M., and Kressley, R.A., The perception of chirps by Sykes's monkeys (*Cercopithecus albogularis*) and humans, *J. Comp. Psychol.*, 108, 243–251, 1994.

9 Neural Representation of Sound Patterns in the Auditory Cortex of Monkeys

Michael Brosch and Henning Scheich

CONTENTS

I. Introduction 151
II. Auditory Pattern Recognition in Primates 152
III. Effects of Cortical Lesions on Sound Perception 154
IV. Representation of Sound Patterns in Auditory Cortex 155
    A. Representation of Simple Stimuli by Cortical Neurons 155
    B. Response Patterning of Cortical Neurons 159
    C. Representation of the Superposition of Simple Stimuli
       by Cortical Neurons 161
    D. Representation of Sequences of Simple Stimuli
       by Cortical Neurons 161
    E. Neuronal Ensembles in Auditory Cortex 166
V. Concluding Remarks 169
References 169

I. INTRODUCTION

Many behaviorally relevant sounds from the environment — including animal vocalizations, speech, and music — have a very complex spectral composition that typically varies over time. Humans discriminate and identify such auditory objects by selectively attending to spectral and temporal patterns. Vowels, for instance, are distinguished by the spectral profile (formants) of a series of harmonically related frequencies (e.g., see Liberman and Whalen[1]). The consonant–vowel pair /pa/ is readily distinguished from a /ba/ by virtue of the voice onset time (VOT), which is the duration of the silent interval between the initial noise burst and the vowel. Musical tunes are recognized by their melodic and rhythmic structure, which arise from the sequential ordering of tones of different frequency and from the duration of intervals or from the stress of tones, respectively (e.g., see Handel[2]). Many of these patterns are recognized independently of the absolute physical properties of the individual sound segments.

Whereas psychoacoustics has revealed the most important features of sounds that humans use to recognize auditory patterns, our understanding of the neural bases underlying this capacity is still very limited. In this review, we first consider studies that have addressed the perceptual capacities of monkeys and other mammals to discriminate auditory patterns. We then summarize results of lesion studies that indicate potential roles of auditory cortex for auditory object formation. The following section describes how and what features of auditory patterns are encoded by the activity of neurons in the auditory cortex.

II. AUDITORY PATTERN RECOGNITION IN PRIMATES

To understand how monkeys discriminate complex sounds, it is worthwhile to first discuss the ability of monkeys to distinguish simple sounds that vary along only one physical dimension. These tests have been performed on a number of different nonhuman primates, using a variety of behavioral procedures.

Monkeys were found to have intensity-increment difference limens of 1 to 2 dB over a wide range of intensities, which indicates sensitivity similar to that of humans.[3] Monkeys and humans also appear to have similar frequency sensitivity.[4,5] In monkeys, frequency difference limens of pure tones are ~1%[3,6–8] and thus similar to those of untrained human subjects.[9] Even after a year of intensive training, monkeys fail to reach performance levels of well-trained humans.[7,10]

Monkeys, like humans, can discriminate modulated from unmodulated sounds.[11,12] Temporal modulation transfer functions, measured with amplitude-modulated sounds, are similar and reveal that both species are most sensitive at slow temporal modulations of about 10 Hz. Despite this finding, monkeys seem to be much less sensitive to the temporal properties of sounds when compared to humans. Monkeys have several-fold larger frequency-difference limens when they have to discriminate linearly rising frequency sweeps from pure tones;[11] frequency sweeps, centered at 1 kHz, have to vary at least by about 30 Hz in macaques, in contrast to about 2 Hz in humans. Differences are less pronounced at other center frequencies.

The poor temporal sensitivity of primates has further been noted in a study of tone duration discrimination.[13] An increase of the duration of repetitively presented 200-msec standard tones was only detected when the duration was extended by at least 23 to 63%, compared with 8 to 14% in humans. The task became much more difficult for monkeys when tone duration was shortened. In addition, monkeys performed worse than humans for the discrimination of tones differing in rise time.[14] Rise time difference limens were 15 msec for a standard rise time of 5 msec, 30 msec for a standard rise time of 40 msec; thus, they were about one order of magnitude above those of humans.

For more complex sounds, composed of several frequencies, monkeys seem to use cues similar to those of humans. Like humans, monkeys are able to perceptually fuse a complex tone consisting of a series of harmonically related frequencies into a single auditory object with a specific pitch.[15] They even do so when the fundamental frequency is missing. Moreover, monkeys are able to analyze spectral profiles[16,17] and to discriminate vowels.[18–20]

Monkeys can also attend to temporal cues in complex sounds. They can discriminate the synthetic human syllables /ba/ and /pa/, which differ with respect to their VOT.[21] Temporal cues may also be used for the discrimination of conspecific coo calls, which may be distinguished on the basis of the temporal position of the inversion of the sweeping direction of the frequencies[22] or their spectral profile.[23]

Despite the presence of these and other perceptual skills, evidence suggests that the sensitivity of the monkeys' auditory pattern-recognition system is, in various aspects, inferior to that of humans. Sinnott[20] had different Old World monkeys discriminate ten synthetic, steady-state vowels. All monkeys experienced difficulties with spectrally similar pairs of vowels such as /a-ʌ/, /ɛ-œ/, and /U-u/ but performed well on pairs of vowels with spectrally more dissimilar formants. Similar species differences were also observed in a study of VOT[21] and even for the discrimination of recorded and synthetic primate communication calls used by Japanese monkeys.[22,24] Monkeys discriminated these coo calls with a sensitivity that was about half that of humans. Similar species differences were also observed in another study by Sinnott and Adams[21] in which difference limens for VOT for a prevoiced consonant–vowel continuum were measured. For some complex sounds, however, the monkeys' sensitivity is superior to that of humans, such as for the discrimination of bird chirps from monkey chirps.[25]

In the studies reviewed above, monkey subjects have likely based their decisions on absolute values of acoustic signals or have simply attended to changes of a physical quality of individual sound elements. A few studies have addressed the question whether monkeys could also recognize auditory patterns which can only be discriminated by the relationship of individual acoustic elements.

In a series of experiments, d'Amato and co-workers[26] were unable to train monkeys to discriminate tone sequences based on their pattern structure. Brosch et al.,[27] however, have recently succeeded in training two macaques (*Macaca fascicularis*) to discriminate pitch contours. Subjects listened to a sequence of pure tones, which consisted of two or three groups. Tones within the same group had the same frequency, and tones of different groups had different frequencies. Subjects had to respond when the frequency of the tones changed in a downward direction and to refrain from responding when the frequency went up or remained constant; that is, they had to recognize the ordinal relation between the tones and could not use a simple same/different rule. After several thousands of trials, subjects' performance was largely independent of the frequency of the tones in the sequence and of the temporal and ordinal position of the downward frequency shift. The high number of trials suggests that a reason for the negative results reported by d'Amato[26] could be that the frequency generalization was tested only during a short period with transfer tests with no attempt to train the generalization slowly.

Our results demonstrate that macaque monkeys can discriminate the pitch relationships of higher and lower. This implies that monkeys can make use of the concept of ordinal relations between acoustic items. The task, however, appears to be quite difficult for monkeys because they have a strong preference to attend to absolute values of individual acoustic elements, such as absolute pitch. Human subjects, by contrast, can learn to discriminate pitch relationships within a few hundred trials.

In summary, monkeys can be trained to perform auditory discrimination tasks with simple and complex sounds. The behavioral tests suggest that, for many sounds features, monkeys use cues similar to those used by humans, although their sensitivity is often lower. Monkeys exhibit particular problems when they have to attend to the relation between different acoustic items rather than to absolute values of individual items but can be trained on such discriminations. Despite some difficulties, the monkeys seem to provide a suitable animal model to study the neural bases of higher auditory functions, such pattern recognition.

III. EFFECTS OF CORTICAL LESIONS ON SOUND PERCEPTION

During the past decades, a number of lesion studies have addressed the role of auditory cortex in auditory pattern recognition. The findings of these studies are often difficult to interpret because of differences in the location and extent of lesions, differences in behavioral testing procedures, species differences, and differences in the recovery and compensation mechanisms coming into effect.[28] Despite these problems, some general lines of evidence have emerged.

Contrary to earlier notions (e.g., see Neff et al.[29]), complete loss of the auditory cortex has been shown to produce a profound impairment of the capability to detect sounds,[30] rendering some subjects practically deaf during the first week post-lesion. During the following 3 to 7 weeks, a gradual recovery takes place, after which the animals retain a permanent, frequency-dependent threshold elevation of some 10 dB. Comparable transient hearing impairments have been confirmed in the rat by the technique of reversible pharmacological inactivation of the auditory cortex.[31]

Similar patterns of post-lesion loss and recovery have held for the perception of a variety of auditory cues, including the detection of gaps in noise (in the rat, see Kelly et al.[32]), frequency discrimination[33,34] (in the cat, see Strominger et al.[35]), discrimination between noise and pure tone bursts,[18] discrimination between trains of clicks with different rates[36] (in the cat, see Cranford et al.[37]), and discrimination of tones of different duration (1 vs. 4 sec) (in the cat, see Scharlock et al.[38]).

Nevertheless, lesioned animals retain severe impairments in, or complete loss of, the ability to discriminate sounds consisting of more complex features, such as the discrimination of vowels,[18] discrimination of pure tones from frequency sweeps[34] or of the direction of frequency sweeps (in gerbils, see Ohl et al.;[39] in cats, see Kelly and Whitfield[40]), discrimination of different monkey coo calls,[34,41] and discrimination of periodically interrupted noise (at 10 Hz) from steady-state noise.[42] It has also been reported that auditory cortex ablation disrupts the animal's capacity to discriminate different temporal sound patterns[33,43] (for cats, see Diamond et al.[44,45]), although it is possible that subjects merely discriminated these patterns by the absolute pitch or intensity of individual or groups of acoustic items. Lesioned cats no longer discriminated harmonic complex tones by their pitch but responded solely to the individual components of the complex tone.[46]

Lesions restricted to parts of the auditory cortex sometimes affected specific functions. Bilateral lesions of auditory association fields resulted in no effect or transient effects on the ability to discriminate periodically interrupted noise from steady-state noise,[42] discrimination of vowels,[18] discrimination of temporal sound patterns,[33] and discrimination of click trains.[36]

Transient elevation of pure tone thresholds for a few days seems to follow unilateral auditory cortex ablations.[47] If the lesion involved the left hemisphere only, monkeys exhibited an impairment to discriminate coo calls for 5 to 15 days.[41] A right-hemisphere dependence was reported in gerbils for discrimination of the direction of frequency sweeps,[48] although whether the effect is transient or permanent was not studied.

In conclusion, hearing requires the functioning of at least some portions of the auditory cortex. Following auditory cortex ablation, a subject regains some hearing, such as the ability to detect sound and to make some simple sound discriminations. This suggests that auditory perceptions may arise from the activity in the subcortical auditory system, but also that this activity does not seem to be accessible to an intact subject without the action of the auditory cortex. Discrimination and recognition of more complex sounds are not possible without the auditory cortex. This indicates that auditory pattern recognition relies on computations performed in the auditory cortex.

IV. REPRESENTATION OF SOUND PATTERNS IN AUDITORY CORTEX

The three aspects of sound representation — namely, the neuronal code, neuronal ensemble, and receptive field of neurons (e.g., see Eggermont[49]) — are intimately related to one another. The receptive field properties of a neuron, for example, may vary with the features of its firing that are used for its characterization (i.e., with the code the neuron employs). Auditory system researchers have primarily concentrated on two types of coding of single neurons: rate coding and temporal coding. Because it is clear that the representation of sound patterns in the auditory system cannot be understood by considering the activity of single neurons, a few studies have also analyzed how groups, or ensembles, of neurons fire to acoustic signals.

A. Representation of Simple Stimuli by Cortical Neurons

Following the rationale that many sound patterns consist of a single or a sequence of discrete acoustic items consisting of a combination of individual frequencies or noise bands, many studies have used sinusoidal tones, superpositions of such tones, noise bands, frequency sweeps, or periodically modulated signals in studying single auditory neuron responses. The responses to these simple stimuli are then used to define the spectral and temporal receptive field of a neuron. The spectral dimension of the receptive field is often termed the *frequency tuning curve* or *frequency response area*. Its temporal dimension is often referred to as the *modulation transfer function*. The receptive field is then used to distinguish different types of neurons and to predict their responses to complex sound patterns.

Figure 9.1A gives an example of spectral aspects of sounds that are represented by neurons in the auditory cortex. This neuron exhibited a brief period of increased discharge in response to tones within a frequency band of 11.5 to 31.2 kHz, with

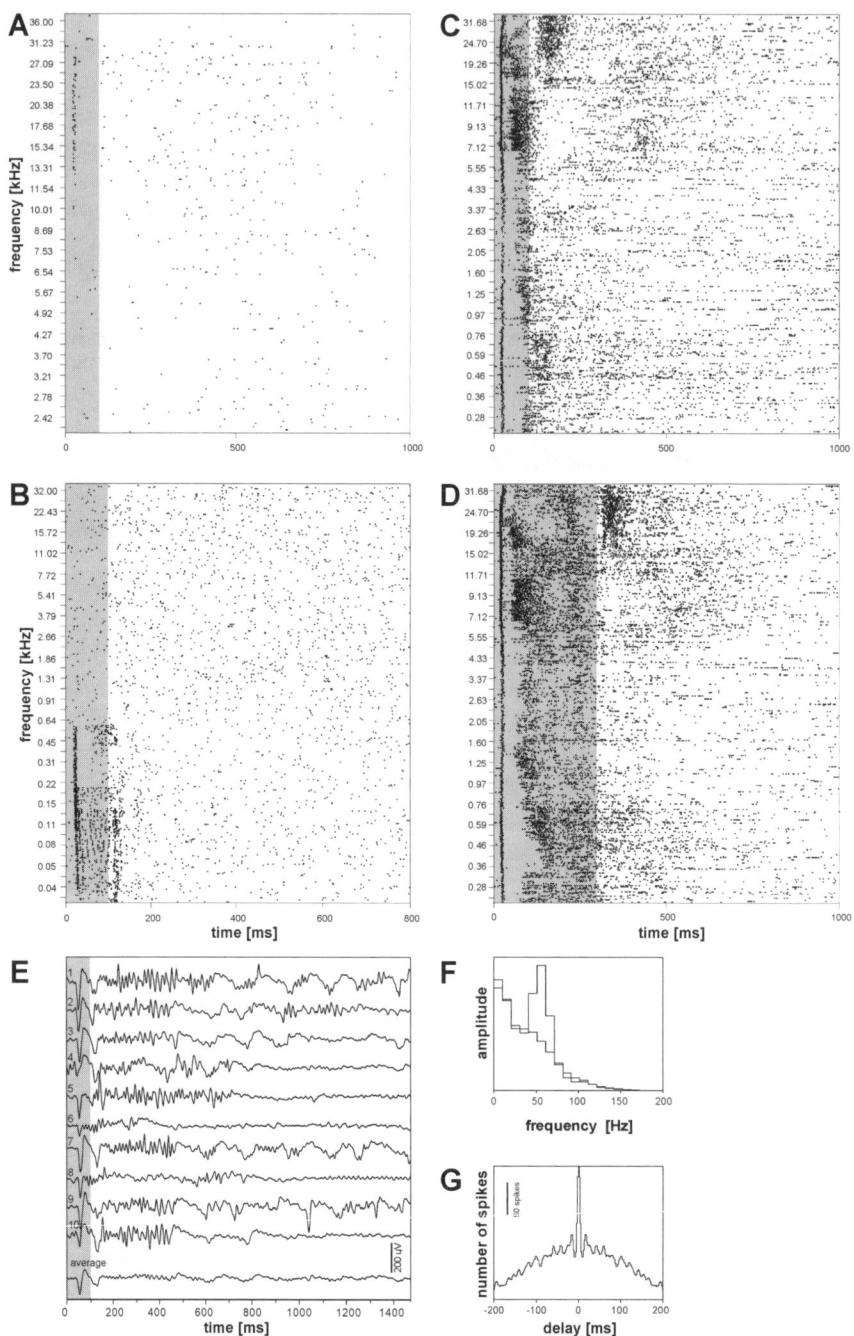

FIGURE 9.1

most discharges occurring at 17.7 kHz. The bandwidth and the best frequency (BF) (the stimulus frequency that elicits the most discharges) are typically used to characterize the spectral part of the receptive field of a neuron. Other spectral aspects of the receptive field have been determined with frequency sweeps[50] (see Clarey et al.[51] for a review of non-primate mammals) and bandpass filtered noise bursts.[52,53]

A change in the number of discharges with stimulus frequency provides an example of rate coding. In the auditory cortex, the integration period seems to be relatively short because many neurons exhibit only a phasic response to acoustic stimuli with a few spikes, after which the discharge rate returns rapidly to prestimulus levels. Because different stimuli can elicit the same firing rate, rate coding of individual neurons is highly ambiguous. One possible solution for the external observer, as well as for the brain, is to integrate within the same topographic region the activity of many neurons that respond to a given stimulus.[54] Topography provides an example of place coding or labeled line coding, in which the position of a neuron in the map provides information about the acoustic stimulus. In the ideal case, the spatiotemporal distribution of active neurons in a map provides a unique representation of a given stimulus.

To date, only a few receptive field properties have been found to be arranged in an orderly fashion in the auditory cortex. Most prominent is the BF (tonotopic) organization, which was first demonstrated by electrophysiological recordings from many sites in the auditory cortex[55] (Figure 9.2A). It can also be seen with brain imaging methods, such as the 14-C-2-deoxyglucose mapping (Figure 9.2B; unpublished results by Scheich and Dieckmann). The tonotopic organization is characterized by the finding that neurons with the same BF are arranged in isofrequency stripes that run parallel to the cortical surface and that the BF increase or decrease is orthogonal to the isofrequency stripes. Several tonotopic maps have been found in the superior temporal plane and gyrus of the monkey.[55-57] Within an isofrequency

---

FIGURE 9.1 Examples of temporal response patterns in the auditory cortex of the macaque. (A) Single unit with a single response component. Here, as in parts B to D, we presented 400 pure tones at 40 frequencies in a randomized order at a rate of 1475 (A, C, D) or 1178 (B) msec. The vertical axis represents the tone frequency, and the horizontal axis is time after tone onset. Presence of the tones is indicated by vertical lines. Occurrence of spikes is marked by dots. (B) Multiunit with phase-locked responses to low-frequency tones. (C, D) Multiunit with several distinct response components. Note that the late response to tones >19 kHz was shifted in latency when the tone duration was changed from 100 msec (C) to 300 msec (D). (E) Intracortically recorded field potentials with stimulus-induced γ-oscillations. Here, only the responses to ten presentations of a 760-Hz tone burst are shown. Note the emergence of spindle-like γ-oscillations after cessation of the tone. The bottom-most trace gives the average of the ten individual traces shown above. (F) Average amplitude spectra of the field potentials shown in part E, calculated from the time window 0 to 300 msec before tone onset (gray line) and 100 to 400 msec after tone onset (black line). Note the stimulus-induced enhancement of field potential frequencies between 41 and 71 Hz in the latter time period. (G) Autocorrelogram of the discharges recorded in parallel with the field potentials shown in part E, calculated from the time window 100 to 400 msec after tone onset. Note the oscillatory structure of the auto correlogram, indicating that the unit fired many spikes at interspike intervals of ~20 msec (corresponding to ~50 Hz). (Adapted from Brosch et al.[82])

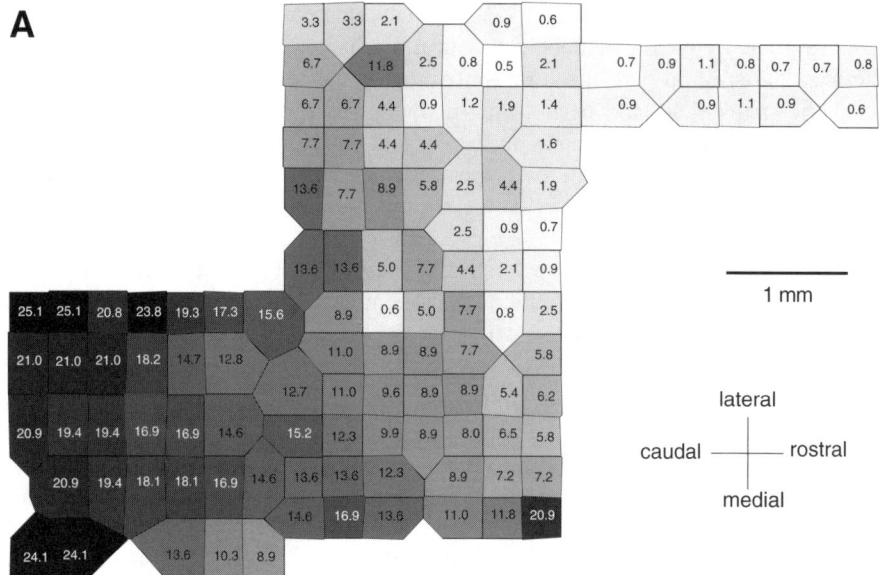

FIGURE 9.2 Tonotopic organization in the auditory cortex. (A) Distribution of BFs (in kHz) in the left auditory cortex of one monkey. The BF at each recording site is represented by a polyeder, the gray value of which codes its BF. Note the orderly low-to-high frequency progression of BF along the rostrocaudal direction, characteristic of the primary auditory cortex (AI). (B) 14-C-2-deoxyglucose labeling in auditory cortex fields AI and R of awake *Macaca fascicularis*. Upper panels show right and left auditory cortex autoradiographic sections cut parallel to the plane of the dorsal surface of the temporal lobe. The animal was stimulated for 45 minutes with 1- and 2-kHz tones at an alteration rate of 4/sec and an intensity of 75 dB SPL. CM: caudomedial field; scale bar = 5 mm. Note that there are two stripes of increased 2-deoxyglucose labeling in each of the fields AI and R, indicating tonotopic organization in agreement with similar findings in rodents.[120] The increased 2-deoxyglucose uptake seen in other areas of the auditory cortex could not be tonotopically interpreted. Lower left: serial reconstructions of several tonotopic slabs through cortex from the case shown above. Lower right: serial reconstructions from the right hemisphere of another animal, which was stimulated with 1- and 4-kHz tones. Note that this stimulation resulted in a labeling in which the outer slabs had a larger separation than for stimulation with 1 and 2 kHz. *(continued)*

stripe, neurons differ with regard to other receptive field properties: some of these have been shown to be arranged orderly, as well, at least in AI (primary auditory complex) of the cat[58] and the gerbil.[59]

Aside from rate and place coding, neurons may also utilize temporal coding, in which information about sound is explicitly encoded by the precise synchronization of neuronal firing to temporal features of the sound. An extreme example of this is shown in Figure 9.1B, in which the neuronal firing was phase locked, with little jitter, to low-frequency pure tones between ~50 and 180 Hz. Very few cortical neurons, however, synchronize their firing to low-frequency tones in this way.[60] In contrast, almost all neurons can synchronize their firing to the envelope of sounds. By examining this ability

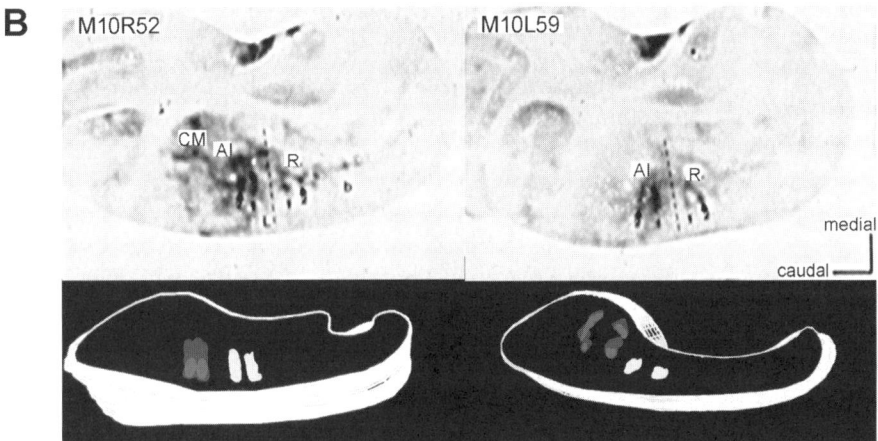

FIGURE 9.2 (continued).

with periodically modulated sounds and click trains, one can measure the temporal modulation transfer function of a neuron — the temporal dimension of receptive fields.

In auditory cortex, the timing of discharges encodes a particular phase of the sound modulation when this sound is modulated at a specific, intermediate period (reviewed in Langner[61]), known as the best modulation frequency (BMF). The vast majority of neurons in anesthetized cats have BMFs < 10 Hz. The cut-off frequency (i.e., the highest modulation to which neurons can phase-lock their firing) is about twice the BMF. It has been argued that temporal modulation transfer functions are highly sensitive to anesthesia such that significantly higher BMFs and cut-off frequencies might be seen in awake animals. In AI and the rostral field (R) of awake squirrel monkeys,[62,63] several researchers have reported BMF values similar to those found in anesthetized cats.[61] By contrast, recordings from AI of the awake rhesus monkey suggest that higher BMFs up to ~200 Hz are also represented by phase-locked responses.[64] The exact value of the BMFs may be of importance because it puts constraints on hypotheses regarding whether neurons in auditory cortex contribute to representation of the rhythmic structure[65] or to the sensation of the roughness of sounds[63] emerging in time ranges about one magnitude of order apart. It should be noted, however, that rapidly modulated sounds could also be encoded by the firing rate of neurons in auditory cortex.[66]

B. RESPONSE PATTERNING OF CORTICAL NEURONS

Units with single phasic responses are commonly found in the auditory cortex of anesthetized monkeys and other mammals, particularly when barbiturates are used as the anesthetic. In lightly anesthetized or awake animals, however, many cortical neurons exhibit more complex temporal responses to acoustic signals (Figure 9.1C,D).

In the authors' lab, more than 80% of the units show more than one discrete excitatory response component, which could occur at latencies of several hundred milliseconds. Complex temporal responses have also been found in a number of other single-unit and multiunit studies in auditory cortex of anesthetized[67] and awake[68–77] monkeys. In our recordings, the majority of the response components are evoked by the onset of stimulation or by steady-state portions of stimuli. A smaller proportion of the response components is elicited by the offset of stimulation, and a small subset of neurons responds exclusively to stimulus offsets. This preference for onsets may underlie the higher perceptual salience of stimulus onsets (e.g., see Zera and Green[78]).

Different latency components of the response of a neuron can often be evoked by tones of different frequency (Miller et al.;[68] Pfingst and O'Connor;[70] unpublished observations by the authors) or by sound-source location.[75] These components can sometimes be related to perception. Indeed, a late response component with a latency of about 100 msec, called the *mismatch negativity* and occurring in both humans[79] and monkeys,[80] could be involved in the automatic detection of stimulus deviancy in a sound sequence. Although the functions of different response components are largely unknown, their existence has wide implications for the representation of sounds in the auditory cortex. First, they can pose problems for attempts to predict the response of neurons to complex sound patterns from responses to simple stimuli. Second, the preferred stimulus of a neuron may be ill defined when it is determined by some arbitrary feature of the neuron's activity, such as the peak firing rate at any time during acoustic stimulation. Third, complex temporal responses challenge the classical view, where a single function is attributed to a single neuron.[81]

During late parts of the response, many neurons exhibit a particular temporal patterning in their activity with no obvious phase-locking to the stimulus.[82] This can best be seen in intracortical recordings of the local slow-wave field potential (Figure 9.1E), in which a period with high-frequency oscillations in the range of 41 to 71 Hz (Figure 9.1F) often emerges after the middle-latency–evoked-potential complex. These so-called γ-oscillations last for several hundred milliseconds and are not seen during ongoing activity. The phase of oscillations is not phase locked to stimulus onset, such that they cannot be seen in the average evoked potential. The γ-oscillations were also present in the neuronal discharges, characterized by an excessive number of interspike intervals between 12 and 22.2 msec (Figure 9.1G).

In the AI and caudomedial (CM) fields of anesthetized monkeys, we observed stimulus-induced γ activity in 465 of 616 multiunits and at 321 of 422 sites at which field potentials were recorded.[82] The γ activity was most pronounced in the frequency range of 60 to 90 Hz and could emerge 100 to 900 msec after stimulus onset, with highest incidence at about 120 msec. The occurrence of γ activity was stimulus dependent. It occurred at highest probability when the neurons were stimulated with a tone at their BF. The probability decreased with increasing difference between the stimulus frequency and the BF. This suggests that, in addition to conventional rate and temporal coding, the acoustic stimuli are also encoded by the rhythmic patterning of neural discharges in the auditory cortex. It is interesting that the frequency of γ-oscillations is above the highest frequency of modulated sounds to which cortical neurons can synchronize their response. This may provide additional coding capacity to cortical cells during the late part of the stimulus-evoked response when discharge rates are generally low.

For the auditory system[82] and previously also for the visual system,[83,84] γ-oscillations have been suggested to contribute to the processing of sensory information by coupling the firing of different neurons. The γ-oscillations are also of wide interest to auditory researchers of human electroencephalograms and magnetoencephalograms who have implicated γ-oscillations with attentional processes,[85] vigilance and consciousness,[86,87] and the temporal binding of successive acoustic events.[88]

C. Representation of the Superposition of Simple Stimuli by Cortical Neurons

Little is known about how neurons in the auditory cortex represent superpositions of simple stimuli. In the cat, one experimental approach has been to analyze neurons with so-called multipeaked tuning curves, which have two or more clearly separated frequency threshold minima in their tuning curve.[89,90] One of these studies[90] found that the average ratio of these minima was close to 1.5, which corresponds to the relation between the second and third harmonics of a harmonic tone complex. Neurons with multipeaked tuning curves have also been observed in the auditory cortex of monkeys (References 50, 73, 76, 91; unpublished observation by the authors); however, neither a preferred frequency relation of 1.5 nor other relations have been reported.

Sutter and Schreiner[90] also tested how multipeaked neurons in the cat responded to the superposition of two pure tones but found little correspondence between tuning curves and the responses to tone combinations. Although neurons often exhibited a strong response when the two tones had frequencies close to the peaks of the tuning curve, these responses were never enhanced over the sum of the two single-tone responses. No such responses have been seen using ripple stimuli.[92,93] Thus, there is not yet any indication for the existence of combination-sensitive neurons in the auditory cortex of nonspecialized mammals similar to those found in the auditory cortex of a bat.[94]

Two studies from the auditory cortex of the monkey have examined how neurons respond to specific combinations of pure tones. The combinations formed harmonic tone complexes, which consisted of the fundamental frequency and a number of higher harmonics or in which the lowest frequencies were selectively deleted. Both tone complexes give rise to the perception of the same pitch, determined by the (missing) fundamental frequency. Using these types of stimuli, a study of AI in anesthetized monkeys[95] described neurons that responded to both complex tones, irrespective of the presence or absence of the fundamental frequency. These neurons did not respond when only one of the higher harmonics was presented at a time. In contrast, in a large sample of neurons in awake monkeys,[73] no neuron responded to the missing fundamental frequency of a tone complex. In almost all neurons, the responses to tone complexes could be explained by how frequency components were located relative to the excitatory and suppressive subfields of the receptive field.

D. Representation of Sequences of Simple Stimuli by Cortical Neurons

Recently, some progress has been made in determining how neurons in auditory cortex respond to sound sequences.[76,96–100] Figure 9.3 illustrates typical results found

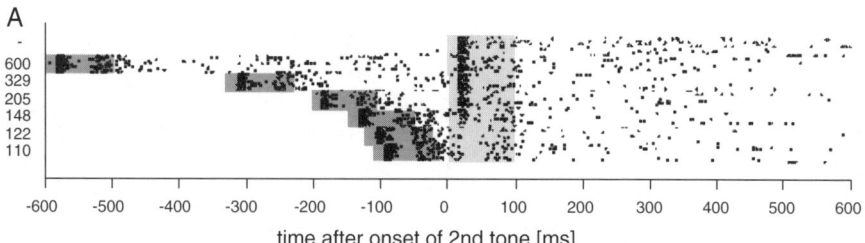
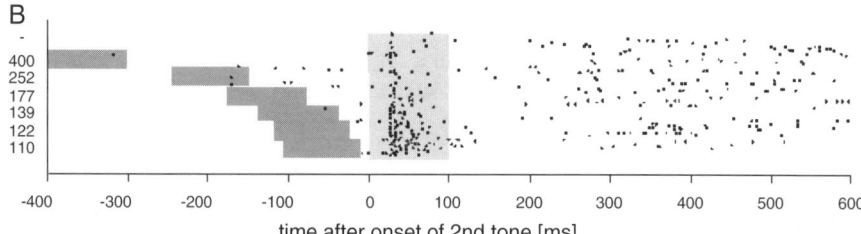
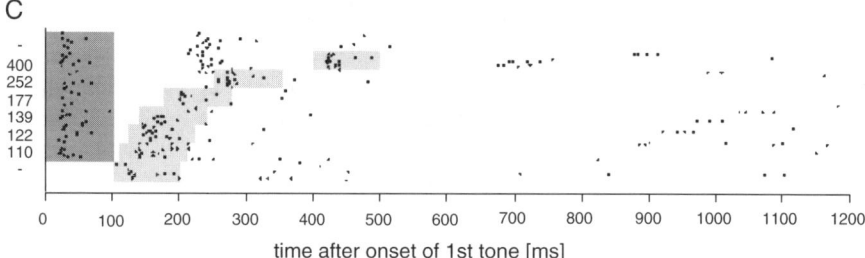
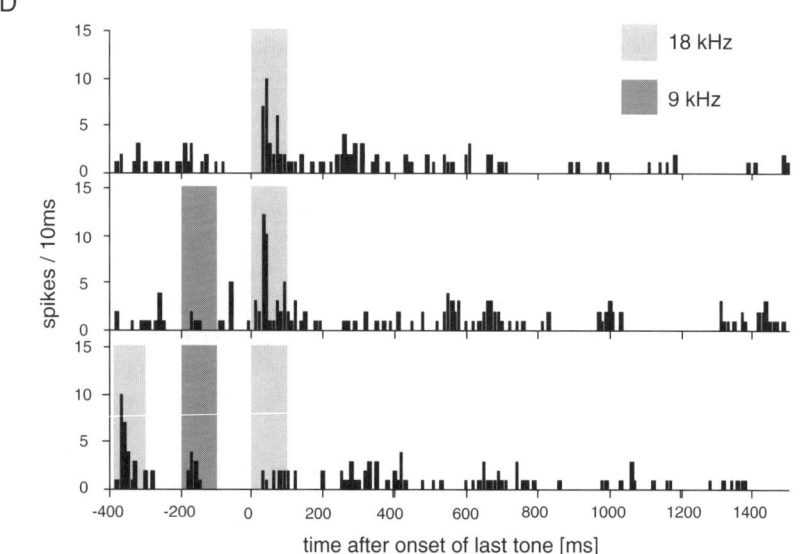

FIGURE 9.3

in such experiments. Stimuli were sequences of two pure tones of varying frequency and stimulus onset asynchrony (SOA). If the first tone causes a response, the second response is attenuated or completely abolished at short SOA. This indicates that the first tone evoked a period of poststimulatory response attenuation during which responses to new tones were suppressed.

Poststimulatory response attenuation was observed in many neurons in the auditory cortex. Response attenuation lasted from 110 to several hundred milliseconds in different neurons in the CM auditory field; median duration of response attenuation was 177 msec (Figure 9.4A). These results are similar to those obtained in AI of the cat,[96,97] where it was found that every neuron exhibited response attenuation and that response attenuation had a minimum duration of about 30 msec in some neurons. Response attenuation in the auditory cortex of monkeys, as well as of cats,[96,97] not only depends on the SOA but also on the spectral separation of the two tones in the sequence. In most neurons, it was strongest and lasted for the longest periods when the first and the second tone had the same frequency (Figure 9.4B).

Another poststimulatory effect, complementary to response attenuation, causes neurons to respond more strongly to a second tone when it is preceded by a first tone (Figure 9.3B). This indicates that the first stimulus evokes a period of poststim-

---

FIGURE 9.3 Responses of neurons in the auditory cortex to tone sequences. (A) Poststimulatory response attenuation: seven stimulus conditions were tested. The top section shows the single-tone condition and the following sections show various two-tone conditions in which the stimulus onset asynchrony (SOA) between the two tones varied between 110 and 600 msec (see ordinate). Frequency of the first tone was 10.7 kHz; that of the second tone, 8 kHz. In the single-tone condition, the tone was at 8 kHz. Presence of tones is indicated by gray shadings. All stimulus conditions were repeated 10 times, resulting in 10 dot rows for each stimulus condition. Tone pairs are graphically aligned with respect to the onset of the second tone. Note that the spike density in response to the second tone varied between the different stimulus conditions. It was high in the single-tone condition and in two-tone conditions with long SOA. It was low in sequences with short SOA. This indicates that the first tone had a poststimulatory influence for a period of approximately 140 msec, during which the response to the second tone was attenuated. (B) Response enhancement: Here, SOA was varied between 110 and 400 msec. The first tone was at 5.2 kHz; the second tone, at 13 kHz. Response enhancement is indicated by finding that the spike density was higher in several two-tone conditions than in the single-tone condition. (C) Retroactive response attenuation: the first tone had a frequency of 8 kHz (dark gray rectangles); the second tone was at 14 kHz (light gray rectangles). The SOA between the two tones varied between 110 and 400 msec (rows 2 to 7). Rows 1 and 8 show the response of this unit when the first or second tone, respectively, was presented alone. Note that the late response to the first tone, visible in the first-tone only and 400-msec SOA condition, disappeared for two-tone sequences with SOA $\leq$ 177 msec. (D) Higher order neighborhood interactions: (upper panel) poststimulus time histogram (PSTH) showing the response to a single 18-kHz tone. (middle panel) PSTH for a two-tone sequence of a 9-kHz and a 18-kHz tone. Note that the first tone did not affect the magnitude of the response to the second tone. (lower panel) PSTH for a three-tone sequence of the 18-, 9-, and 18-kHz tones. Note that the response to the last tone was suppressed by the preceding tones.

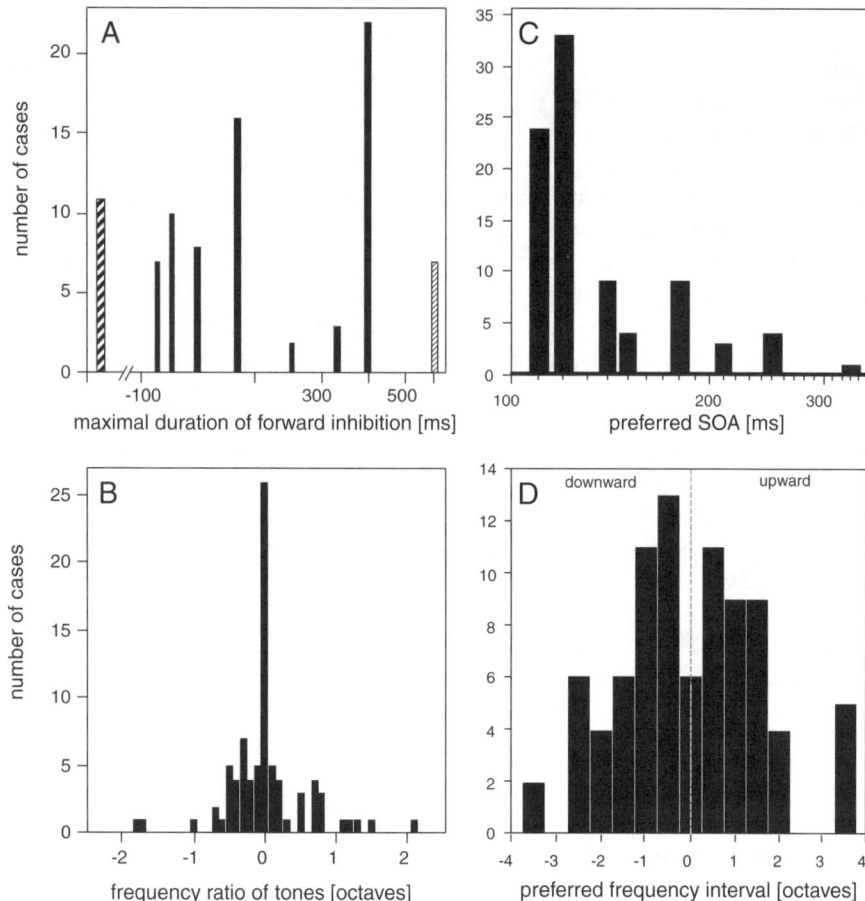

FIGURE 9.4 Characteristics of tone pairs inducing response attenuation and response enhancement. (A) Maximal duration of response attenuation (black bars). Thick hatched bar signifies cases for which no response attenuation was observed. These units likely have a maximal duration of response attenuation shorter than the SOA used in our experiments. Thin hatched bars signify cases for which response attenuation was seen up to the longest SOA that was used in the experiments. It is likely that these units have even longer duration of response attenuation. Data in (A) and (B) are from the field CM. (B) Frequency ratio of the second tone and the first tone that produced maximal response attenuation. Note that response attenuation was primarily strongest when the first and second tone of the sequence had similar frequencies. (C) Distribution of the preferred SOA, defined as the SOA at which response enhancement was maximal. (D) Distribution of the preferred frequency interval. Negative values denote tone pairs in which the first tone was higher than the second tone ("downward"). Positive values indicate tone pairs with an "upward" frequency progression. (Adapted from Brosch et al.[100])

ulatory response enhancement during which the response to the second tone is facilitated. In the AI and CM fields, Brosch et al.[100] observed response enhancement in 67% of the multiunit and in 43% of the single-unit recordings. The number of units exhibiting response enhancement may even be higher when more sequences are tested, as suggested by a recent study in cat auditory cortex.[98] Enhanced responses occurred only when two tones had a minimum SOA of several tens of milliseconds. Enhancement was maximal for SOA of approximately 100 msec (Figure 9.4C) and vanished beyond a SOA of a few hundred milliseconds. Like response attenuation, response enhancement depended on the frequency interval between the two tones. Response enhancement often was maximal when the first tone was about one octave below or above the second tone (Figure 9.4D). Therefore, response enhancement and response attenuation can be observed in the same neuron, although under different frequency and SOA conditions.

Our findings also indicate that neurons in auditory cortex have complex and often nonseparable spectrotemporal receptive fields,[93,101] whose components cannot be determined by separately measuring their spectral and temporal filter properties. Accordingly, the spectral filtering of cortical neurons varies with the temporal context, as much as the temporal filtering varies with the spectral content of consecutive segments. This implies that both rate and synchronization measures of conventionally determined temporal modulation transfer functions (see above) change considerably when they are measured with temporally modulated acoustic signals in which the spectral content of consecutive segments varies as well.

In many neurons in monkey auditory cortex, the enhancement of the response to the second tone commenced after the early response to this tone and then lasted for several tens of milliseconds. This suggests that information about the temporal context in which an acoustic item occurs might be encoded in specific time windows of the neuronal response. We have also obtained first evidence that early and late parts of units with complex temporal responses are differentially affected by a preceding stimulus (unpublished observations by the authors).

In addition to poststimulatory effects, we have also found "prestimulatory" or retroactive effects, in which parts of the response to the first tone are suppressed by a consecutive stimulus (Figure 9.3C). We observed this retroactive response attenuation in all neurons in the auditory cortex of the monkey that exhibited late responses.[99] It interrupts the neuronal processing of the first stimulus by canceling late parts of the response that usually occur in the single-tone condition. The loss or modification of late parts of neuronal response could signal that a stimulus is succeeded by another stimulus and, thus, could provide information of the sequential structure of acoustic signals. Retroactive response attenuation may also have implications for the perception of the first stimulus. For similar SOA ranges, auditory backward recognition masking has been found in psychoacoustic studies and is characterized by an impaired judgment of the pitch, intensity, duration, and location of the first tone.[102,103]

The processing of temporal sound patterns cannot be completely understood only by examining how neurons respond to sequences of two pure tones. It also requires analysis of neuronal responses to more complex sequences. One attempt to address this question is the study of Steinschneider and colleagues,[104] who presented the synthesized syllables /ta/ and /da/, consisting of the sequence of a condensation click, a formant transition, and a harmonic series of tones, characteristic for the steady-state portion of the vowel. It was found that the number of onset responses at a recording site in AI distinguished between voiced and unvoiced syllables; only one response to the /da/ and two responses to the /ta/ occurred. In the latter condition, the responses occurred with a temporal separation equal to the VOT of the /ta/.

Another approach to studying the neural representation of complex sound sequences is to examine how neurons respond to sequences of more than two tones, such as a sequence of three tones (Figure 9.3D). We found that the first two tones of a three-tone sequence modified the responses to the last tone of the sequence in many neurons in AI and CM.[105] In most cases, however, response modifications were not due to genuine higher order neighborhood interactions, as in Figure 9.3D, but rather were a consequence of the linear superposition of nearest neighbor interactions; that is, the response to the last tone was the sum of the poststimulatory effects of the first and second stimuli. In neurons with higher order sequential interactions, the type of interaction depended on the frequency relation of the tones, although it is not clear yet whether any frequency or temporal relations are preferred.

Higher order sequential interactions have been observed by others as well. Steinschneider and colleagues[106] examined neural responses to long sequences of two alternating tones, in which they varied the SOA. For sequences with long SOA, most cortical sites responded well to both tones if they were within the receptive field. For short SOA below about 50 msec, however, only the response to one of the alternating tones persisted while the response to the other tone disappeared, depending on which was the starting tone of the sequence.

E. Neuronal Ensembles in Auditory Cortex

Although many neurons in the auditory cortex respond to simple stimuli or to combinations of them, little evidence currently exists that behaviorally important sound patterns, such as vocalizations, are represented by single neurons in the auditory cortex. Theoretical considerations have also questioned the existence of such highly specific cells (e.g., see Singer and Gray[84]). Therefore, some investigators have considered additional coding schemes based on ensembles of neurons. For the auditory cortex, two concepts have been pursued. The notion of ensemble formation suggests that the information represented by many individual neurons is summed; an example of this provides the population vector, originally proposed for the motor cortex.[107] For the construction of the population vector, it is assumed that each neuron represents a particular property of a movement or of a sensory stimulus and that the information of different neurons is independent from each other. The population vector is then calculated by adding the information represented by different neurons.

A variant of the population vector has been suggested by Wang et al.[108] for the auditory cortex. The authors measured the responses of neurons in AI of marmoset monkeys to conspecific vocalizations. They then constructed a two-dimensional chart in which they plotted the spike train of neurons on the x-axis and arranged different neurons with regard to their BF on the y-axis. The resulting spectrotemporal activity profile resembled the original vocalization in which temporal transients were accentuated. Thus, neurons with specific best frequencies represent the spectral fine structure of the vocalizations, and the timing of the spikes represents the temporal modulation in specific frequency bands.

In other theories of neuronal ensemble coding, the correlated firing between different neurons is crucial. Particularly influential has been the proposition that an ensemble is defined by the synchronization of the firing of neurons. This hypothesis was originally inspired by and formulated for the auditory system,[109,110] although most experimental support has come from the visual system.[83,84] For the auditory system, two variants of this hypothesis have been addressed experimentally. The first variant posits that the information conveyed by different neurons is functionally combined by establishing synchronization between the firing of these neurons.[111] The second variant states that neural synchrony itself represents properties of acoustic signals.[112,113]

Before discussing the experimental findings in detail, it is first important to consider the time scales of neural synchrony and the strength of synchronized firing of neurons in the auditory cortex. Neural synchrony is commonly assessed by computing the cross-correlation of two spike trains. In the auditory cortex of monkeys and other mammals, a number of different shapes of cross-correlograms have been observed.[111,114-119] Most frequent were cross-correlograms with more or less symmetrical peaks, which were centered around the origin of the histogram (Figure 9.5A,C). The widths of these cross-correlograms ranged between a few and several hundred milliseconds,[111,116-118] indicating that neurons fire synchronously for periods of tens of milliseconds. It is interesting that the predominant time scale of neural synchrony in the auditory cortex is similar to the critical temporal window for the detection and processing of individual acoustic events for a complex sound (e.g., see Handel[2]). Therefore, neural synchrony in auditory cortex could provide a mechanism involved in the integration and segregation of different auditory items in an auditory scene.

Another type of neural synchrony could be observed during late parts of the response to acoustic stimuli when many neurons exhibit γ-oscillations (compare Figure 9.3E). In this case, the cross-correlogram had an oscillatory structure with a central and several satellite peaks (Figure 9.5B), indicating that the γ-oscillations of different units were synchronized.[82] For pairs of single units, sometimes correlograms with asymmetric peaks or troughs have been seen,[116,118] which likely indicate a serial flow of information from one neuron to the other.

The number of synchronized neurons as well as the proportion of synchronized neurons is relatively low in the auditory cortex. In the AI and CM of the monkey, peak correlation coefficients ranged between 0.040 and 0.250 (median, 0.077) during ongoing activity and between 0.036 and 0.577 (median, 0.186) during stimulation.[115] Comparable correlation coefficients were observed in cat auditory cortex.[111,116,118,119] Eggermont[116] estimated that in the cat rarely can more than 10% of the spikes in one neuron be accounted for by spikes in another neuron. This indicates that neurons in the auditory cortex are not very tightly synchronized, which may allow them to process different stimulus features widely independently from each other.

If neural synchrony serves as a glue to integrate the information conveyed by different neurons, then specific relationships should exist between the receptive fields of synchronously firing neurons. Such relationships have indeed been found in a number of studies,[80,111,115,116,118,119] including best frequency, overlap of tuning curves, response latencies, and binaural interaction. This suggests that the formation of

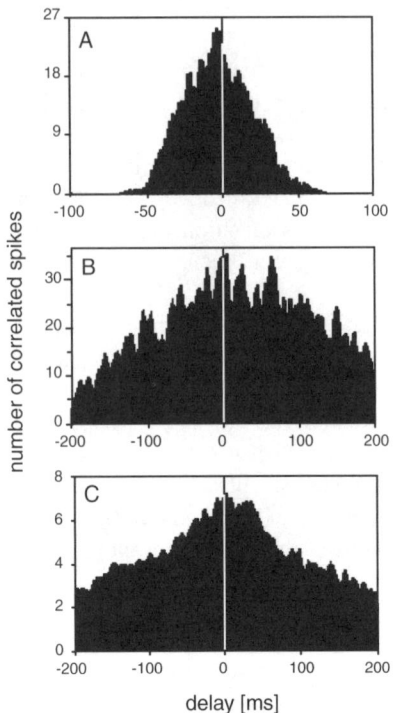

FIGURE 9.5 Examples of the synchronization of neuronal firing of simultaneously recorded spike trains. (A) Cross-correlogram calculated from the initial 100-msec period of the response to a 24.1-kHz tone of 100-msec duration. (B) Cross-correlogram calculated from the late response to a 794-Hz tone of 100-msec duration. Analysis period started 100 msec after tone onset and had a length of 300 msec. (Adapted from Brosch et al.[115]) (C) Cross-correlogram calculated from spontaneous firing.

neuronal ensembles by correlated firing is governed, in part, by the similarity of the receptive fields of neurons. This speculation does not exclude the possibility that neuronal ensembles might also be established according to relations between receptive fields other than similarity.

deCharms and Merzenich[113] have proposed that properties of acoustic signals may be represented by synchronized neural activity. The authors recorded from anesthetized marmoset monkeys to which they presented pure tones with a duration of several seconds. They observed that neurons exhibited a phasic response to the onset of the tones, after which the firing returned to the spontaneous rate. By contrast, neurons were more strongly synchronized during the presence of the stimulus than they were during silence. Therefore, they concluded that the entire presence of the sustained part of the stimulus was represented by synchronously firing neurons in the auditory cortex for periods of 5 to 15 msec, rather than by an overall increase in the firing of the neurons. Unfortunately, they did not report whether the sustained part of the stimulus could also be distinguished from silence by other characteristics of the neuronal activity, such as the temporal patterning of firing. Furthermore, it is not clear whether the lack of increased firing during the sustained portions of the tones can be replicated in awake animals.

V. CONCLUDING REMARKS

In this chapter, we have reviewed the current knowledge of the neural representation of sound patterns in the auditory cortex of monkeys. Behavioral studies have shown that monkeys have many auditory perceptual capabilities in common with humans. Lesion studies have strongly suggested that auditory pattern recognition requires the functioning of the auditory cortex. With almost no exception, electrophysiological recordings in the auditory cortex have focused on relations between acoustic stimuli and neural activity in the auditory cortex but have not included links to the perception of these stimuli. Many of these studies have revealed receptive field properties of individual neurons in the auditory cortex by presenting simple acoustic stimuli such as pure tones, frequency sweeps, and periodically modulated sounds. Some progress has been made in characterizing neuronal responses to superpositions and sequences of simple acoustic stimuli. Also, some evidence has been obtained that neuronal ensembles may contribute to the representation of sound patterns.

Gaining an understanding of how the stimulus–response relation of neurons contributes to auditory perception and behavior requires electrophysiological recordings in behaving animals. Monkeys can be trained on a variety of auditory tasks, recording techniques are well established, and some basic knowledge on the parcellation of auditory cortex and on the receptive field properties are now readily available. Therefore, the monkey may provide a suitable animal model for studying the neural mechanisms underlying higher auditory functions. The advent of new imaging techniques such as magnetic resonance imaging and their use in monkeys may help to find those parts of the auditory cortex that are activated by specific auditory stimulus classes and tasks. In spite of the high interest, especially in speech signals and music, it is clear that the monkey is not yet as powerful a model for auditory research as it is for the study of visual neuroscience and perception.

REFERENCES

1. Liberman, A.M. and Whalen, D.H., On the relation of speech to language, *Trends Cogn. Sci.*, 4, 187, 2000.
2. Handel, S., *Listening: An Introduction to the Perception of Auditory Events*, MIT Press, Cambridge, MA, 1993.
3. Sinnott, J. M., Petersen, M.R., and Hopp, S.L., Frequency and intensity discrimination in humans and monkeys, *J. Acoust. Soc. Am.*, 78, 1977, 1985.
4. Gourevitch, G., Detectibility of tones in quiet and in noise by rats and monkeys, in *Animal Psychophysics*, W.C. Stebbins, Ed., Appleton-Century-Crofts, New York, 1970, p. 67.
5. Serafin, J.V., Moody, D.B., and Stebbins, W.C., Frequency selectivity of the monkey's auditory system: psychophysical tuning curves, *J. Acoust. Soc. Am.*, 71, 1513, 1982.
6. Stebbins, W.C., Hearing of Old World monkeys (Cercopithecinae), *Am. J. Phys Anthropol.*, 38, 357, 1973.

7. Prosen, C.A. et al., Frequency discrimination in the monkey, *J. Acoust. Soc. Am.*, 88, 2152, 1990.
8. Pfingst, B.E., Comparison of spectral and nonspectral frequency difference limens for human and nonhuman primates, *J. Acoust. Soc. Am.*, 93, 2124, 1993.
9. Tyler, R.S., Wood, E.J., and Fernandes, M., Frequency resolution and discrimination of constant and dynamic tones in normal and hearing-impaired listeners, *J. Acoust. Soc. Am.*, 74, 1190, 1983.
10. Olsho, L.W., Koch, E.G., Carter, E.A., Halpin, C.F., and Spetner, N.B., Pure-tone sensitivity of human infants, *J. Acoust. Soc. Am.*, 84, 1316, 1988.
11. Moody, D.B. et al., The role of frequency modulation in the perception of complex stimuli by primates, *Exp. Biol.*, 45, 219, 1986.
12. Moody, D.B., Detection and discrimination of amplitude-modulated signals by macaque monkeys, *J. Acoust. Soc. Am.*, 95, 3499, 1994.
13. Sinnott, J.M., Owren, M.J., and Petersen, M.R., Auditory duration discrimination in Old World monkeys (*Macaca, Cercopithecus*) and humans, *J. Acoust. Soc. Am.*, 82, 465, 1987.
14. Prosen, C.A. and Moody, D.B., Rise-time difference thresholds in the monkey, *J. Acoust. Soc. Am.*, 97, 697, 1995.
15. Tomlinson, R.W. and Schwarz, D.W., Perception of the missing fundamental in nonhuman primates, *J. Acoust. Soc. Am.*, 84, 560, 1988.
16. O'Connor, K.N., Barruel, P., and Sutter, M.L., Global processing of spectrally complex sounds in macaques (*Macaca mullata*) and humans, *J. Comp. Physiol.*, 186, 903, 2000.
17. Sommers, M.S. et al., Formant frequency discrimination by Japanese macaques (*Macaca fuscata*), *J. Acoust. Soc. Am.*, 91, 3499, 1992.
18. Dewson, J.H. 3rd, Pribram, K.H., and Lynch, J.C., Effects of ablations of temporal cortex upon speech sound discrimination in the monkey, *Exp. Neurol.*, 24, 579, 1969.
19. Kuhl, P.K. and Padden, D.M., Enhanced discriminability at the phonetic boundaries for the place feature in macaques, *J. Acoust. Soc. Am.*, 73, 1003, 1983.
20. Sinnott, J.M., Detection and discrimination of synthetic English vowels by Old World monkeys (*Cercopithecus, Macaca*) and humans, *J. Acoust. Soc. Am.*, 86, 557, 1989.
21. Sinnott, J.M. and Adams, F.S., Differences in human and monkey sensitivity to acoustic cues underlying voicing contrasts, *J. Acoust. Soc. Am.*, 82, 1539, 1987.
22. Hopp, S.L. et al., Differential sensitivity of Japanese macaques (*Macaca fuscata*) and humans (*Homo sapiens*) to peak position along a synthetic coo call continuum, *J. Comp. Psychol.*, 106, 128, 1992.
23. Le Prell, C.G. and Moody, D.B., Factors influencing the salience of temporal cues in the discrimination of synthetic Japanese monkey (*Macaca fuscata*) coo calls, *J. Exp. Psychol. Animal Behav. Process.*, 26, 261, 2000.
24. May, B., Moody, D.B., and Stebbins, W.C., Categorical perception of conspecific communication sounds by Japanese macaques, *Macaca fuscata*, *J. Acoust. Soc. Am.*, 85, 837, 1989.
25. Brown, C.H., Sinnott, J.M., and Kressley, R.A., Perception of chirps by Sykes's monkeys (*Cercopithecus albogularis*) and humans (*Homo sapiens*), *J. Comp. Psychol.*, 108, 243, 1994.
26. d'Amato, M.R., A search for tonal pattern perception in cebus monkey: why monkeys can't hum a tune, *Music Perc.*, 5, 452, 1988.
27. Brosch, M. et al., Macaque monkeys discriminate ascending from descending tone sequences, submitted.

28. Finger, S., Levere, T.E., Almli, C.R., and Stein, D.G., *Brain Injury and Recovery*, Plenum, New York, 1988.
29. Neff, W.D., Diamond, I.T., and Casseday, J.H., Behavioral studies of auditory discrimination: central nervous system, in *Handbook of Sensory Physiology*, Vol. V, Part 2, *Auditory System*, W.D. Keidel and W.D. Neff, Eds., Springer–Verlag, New York, 1975, p. 307.
30. Heffner, H.E. and Heffner, R.S., Effect of bilateral auditory cortex lesions on absolute thresholds in Japanese macaques, *J. Neurophysiol.*, 64, 191, 1990.
31. Talwar, S.K., Musial, P.G., and Gerstein, G.L., Role of mammalian auditory cortex in the perception of elementary sound properties, *J. Neurophysiol.*, 85, 2350, 2001.
32. Kelly, J.B., Rooney, B.J., and Phillips D.P., Effects of bilateral auditory cortical lesions on gap-detection thresholds in the ferret (*Mustela putorius*), *Behav. Neurosci.*, 110, 542, 1996.
33. Strominger, N.L., Oesterreich, R.E., and Neff, W.D., Sequential auditory and vsiual discriminations after temporal lobe ablation in monkeys, *Physiol. Behav.*, 24, 1149, 1980.
34. Harrington, I.A., Heffner, R.S., and Heffner, H.E., An investigation of sensory deficits underlying the aphasia-like behavior of macaques with auditory cortex lesions, *NeuroReport*, 12, 1217, 2001.
35. Cranford, J.L., Igarashi, M., and Stramler, J.H., Effect of auditory neocortex ablation on pitch perception in the cat, *J. Neurophysiol.*, 39, 143, 1976.
36. Stepien, L.S., Cordeau, J.P., and Rasmussen T., The effect of temporal lobe and hippocampal lesions on auditory and visual recent memory in monkeys, *Brain*, 83, 470, 1960.
37. Cranford, J.L., Igarashi, M., and Stramler, J.H., Effect of auditory neocortex ablation on identification of click rates in cats, *Brain Res.*, 116, 69, 1976.
38. Scharlock, D.P., Neff, W.D., and Strominger, N.L., Discrimination of tone duration after bilateral ablation of cortical auditory areas, *J. Neurophysiol.*, 28, 673, 1965.
39. Ohl, F.W. et al., Bilateral ablation of auditory cortex in Mongolian gerbil affects discrimination of frequency modulated tones but not of pure tones, *Learn. Mem.*, 6, 347, 1999.
40. Kelly, J.B. and Whitfield, I.C., Effects of auditory cortical lesions on discriminations of rising and falling frequency-modulated tones, *J. Neurophysiol.*, 34, 802, 1971.
41. Heffner, H.E. and Heffner, R.S., Temporal lobe lesions and perception of species-specific vocalizations by macaques, *Science*, 226, 75, 1984.
42. Symmes, D., Discrimination of intermittent noise by macaques following lesions of the temporal lobe, *Exp. Neurol.*, 16, 201, 1966.
43. Jerison, H.J. and Neff, W.D., Effect of cortical ablation in the monkey on discrimination of auditory patterns, *Fed. Proc.*, 12, 237, 1953.
44. Diamond, I.T. and Neff, W.D., Ablation of temporal cortex and discrimination of auditory patterns, *J. Neurophysiol.*, 20, 300, 1957.
45. Diamond, I.T., Goldberg, J.M., and Neff, W.D., Tonal discrimination after ablation of the auditory cortex, *J. Neurophysiol.*, 25, 223, 1962.
46. Whitfield, I.C., Auditory cortex and the pitch of complex tones, *J. Acoust. Soc. Am.*, 67, 644, 1980.
47. Heffner, H.E. and Heffner, R.S., Unilateral auditory cortex ablation in macaques results in a contralateral hearing loss, *J. Neurophysiol.*, 62, 789, 1989.
48. Wetzel, W. et al., Right auditory cortex lesion in Mongolian gerbils impairs discrimination of rising and falling frequency-modulated tones, *Neurosci. Lett.*, 252, 115, 1998.

49. Eggermont, J.J., Between sound and perception: reviewing the search for a neural code, *Hear. Res.*, 157, 1, 2001.
50. Rauschecker, J.P., Cortical processing of complex sounds, *Curr. Opin. Neurobiol.*, 8, 51, 1998.
51. Clarey, J.C., Barone, P., and Imig, T.J., Physiology of thalamus and cortex, in *The Mammalian Auditory Pathway: Neurophysiology*, A.N. Popper and R.R. Fay, Eds., Springer–Verlag, New York, 1992, p. 232.
52. Rauschecker, J.P., Tian, B., and Hauser, M., Processing of complex sounds in the macaque nonprimary auditory cortex, *Science*, 268, 111, 1995.
53. Scott, B.H., Malone, B.J., and Semple, M.N., Physiological delineation of primary auditory cortex in the alert rhesus monkey, *Soc. Neurosci. Abstr.*, 26, 358.20, 2000.
54. Scheich, H., Auditory cortex: comparative aspects of maps and plasticity, *Curr. Opin. Neurobiol.*, 1, 236, 1991.
55. Merzenich, M.M. and Brugge, J.F., Representation of the cochlear partition of the superior temporal plane of the macaque monkey, *Brain Res.*, 28, 275, 1973.
56. Morel, A., Garraghty, P.E., and Kaas, J.H., Tonotopic organization, architectonic fields, and connections of auditory cortex in macaque monkeys, *J. Comp. Neurol.*, 335, 437, 1993.
57. Kaas, J.H., Hackett, T.A., and Tramo, M.J., Auditory processing in primate cerebral cortex, *Curr. Opin. Neurobiol.*, 9, 164, 1999.
58. Schreiner, C.E, Read, H.L., and Sutter, M.L., Modular organization of frequency integration in primary auditory cortex, *Annu. Rev. Neurosci.*, 23, 501, 2000.
59. Ohl, F.W. and Scheich, H., Orderly cortical representation of vowels based on formant interaction, *Proc. Natl. Acad. Sci. USA*, 94, 9440, 1997.
60. Wallace, M.N. et al., Phase-locked responses to pure tones in guinea pig auditory cortex., *NeuroReport*, 18, 3989, 2000.
61. Langner, G., Periodicity coding in the auditory system, *Hear. Res.*, 60, 115, 1992.
62. Muller-Preuss, P., On the mechanisms of call coding through auditory neurons in the squirrel monkey, *Eur. Arch. Psychiatry Neurol. Sci.*, 236, 50, 1986.
63. Bieser A. and Muller-Preuss P., Auditory responsive cortex in the squirrel monkey: neural responses to amplitude-modulated sounds, *Exp. Brain Res.*, 108, 273, 1996.
64. Steinschneider, M. et al., Click train encoding in primary auditory cortex of the awake monkey: evidence for two mechanisms subserving pitch perception, *J. Acoust. Soc. Am.*, 104, 2935, 1998.
65. Eggermont, J.J., Representation of spectral and temporal sound features in three cortical fields of the cat: similarities outweigh differences, *J. Neurophysiol.*, 80, 2743, 1998.
66. Lu, T., Liang, L., and Wang, X., Temporal and rate representations of time-varying signals in the auditory cortex of awake primates, *Nat. Neurosci.*, 4, 1131, 2001.
67. Riquimaroux, H., Neuronal scene analysis? Basic evidence found in the monkey primary auditory cortex, in *Transaction of Technical Committee on Psychological and Physiological Acoustics*, Acoustic Society of Japan, Vol. 29, p. 1, 1994.
68. Miller, J.M., Beaton, R.D., O'Connor, T., and Pfingst, B.E., Response pattern complexity of auditory cells in the cortex of unanesthetized monkeys, *Brain Res.*, 69, 101, 1974.
69. Benson, D.A. and Hienz, R.D., Single-unit activity in the auditory cortex of monkeys selectively attending left vs. right ear stimuli, *Brain Res.*, 29, 307, 1978.
70. Pfingst, B.E. and O'Connor, T.A., Characteristics of neurons in auditory cortex of monkeys performing a simple auditory task, *J. Neurophysiol.*, 45, 16, 1981.

71. Pelleg-Toiba, R. and Wollberg, Z., Tuning properties of auditory cortex cells in the awake squirrel monkey, *Exp. Brain Res.*, 74, 353, 1983.
72. Shamma, S.A. and Symmes, D., Patterns of inhibition in auditory cortical cells in awake squirrel monkeys, *Hear. Res.*, 19, 1, 1985.
73. Schwarz, D.W. and Tomlinson, R.W., Spectral response patterns of auditory cortex neurons to harmonic complex tones in alert monkey (*Macaca mulatta*), *J. Neurophysiol.*, 64, 282, 1990.
74. Hocherman, S. and Yirmiya, R., Neuronal activity in the medial geniculate nucleus and in the auditory cortex of the rhesus monkey reflects signal anticipation, *Brain*, 113, 1707, 1990.
75. Ahissar, M. et al., Encoding of sound-source location and movement: activity of single neurons and interactions between adjacent neurons in the monkey auditory cortex, *J. Neurophysiol.*, 67, 203, 1992.
76. deCharms, R.C., Blake, D.T., and Merzenich, M.M., Optimizing sound features for cortical neurons, *Science*, 280, 1439, 1998.
77. Recanzone, G.H., Response profiles of auditory cortical neurons to tones and noise in behaving macaque monkeys, *Hear. Res.*, 150, 104, 2000.
78. Zera, J. and Green, D.M., Detecting temporal asynchrony with asynchronous standards, *J. Acoust. Soc. Am.*, 93, 1571, 1993.
79. Näätänen, R. et al., "Primitive intelligence" in the auditory cortex, *Trends Neurosci.*, 24, 283, 2001.
80. Javitt, D.C. et al., Detection of stimulus deviance within primate primary auditory cortex: intracortical mechanisms of mismatch negativity (MMN) generation, *Brain Res.*, 26, 192, 1994.
81. Vaadia, E., Gottlieb, Y., and Abeles, M., Single-unit activity related to sensorimotor association in auditory cortex of a monkey, *J. Neurophysiol.*, 48, 1201, 1982.
82. Brosch, M., Budinger, E., and Scheich, H., Stimulus-related gamma-oscillations in primate auditory cortex, *J. Neurophysiol.*, 87, 2715, 2002.
83. Eckhorn, R., Oscillatory and non-oscillatory synchronizations in the visual cortex and their possible roles in associations of visual features, *Prog. Brain Res.*, 102, 405, 1994.
84. Singer W. and Gray, C.M., Visual feature integration and the temporal correlation hypothesis, *Annu. Rev. Neurosci.*, 18, 555, 1995.
85. Tiitinen, H. et al., Selective attention enhances the auditory 40-Hz transient response in humans, *Nature*, 364, 59, 1993.
86. Llinas, R. and Ribary, U., Coherent 40-Hz oscillation characterizes dream state in humans, *Proc. Natl. Acad. Sci. USA*, 90, 2078, 1993.
87. May, P. et al., Long-term stimulation attenuates the transient 40-Hz response, *NeuroReport*, 5, 1918, 1994.
88. Joliot, M., Ribary, U., and Llinas, R., Human oscillatory brain activity near 40 Hz coexists with cognitive temporal binding, *Proc. Natl. Acad. Sci. USA*, 91, 11748, 1994.
89. Abeles, M. and Goldstein, M.H. Jr., Functional architecture in cat primary auditory cortex: columnar organization and organization according to depth, *J. Neurophysiol.*, 33, 172, 1970.
90. Sutter, M.L. and Schreiner, C.E., Physiology and topography of neurons with multi-peaked tuning curves in cat primary auditory cortex, *J. Neurophysiol.*, 65, 1207, 1991.
91. Recanzone, G.H., Guard, D.C., and Phan, M.L., Frequency and intensity response properties of single neurons in the auditory cortex of the behaving macaque monkey, *J. Neurophysiol.*, 83, 2315, 2000.

92. Calhoun, B.M. and Schreiner, C.E., Spectral envelope coding in cat primary auditory cortex: linear and non-linear effects of stimulus characteristics, *Eur. J. Neurosci.*, 10, 926, 1998.
93. Shamma, S., On the role of space and time in auditory processing, *Trends Cogn. Sci.*, 5, 340, 2001.
94. Suga, N., Cortical computational maps for auditory imaging, *Neural Networks*, 3, 3, 1990.
95. Riquimaroux, H. and Hashikawa, T., Units in the auditory cortex of the Japanese monkey can demonstrate a conversion of temporal and place pitch in the central auditory system, *J. de Physiche III*, 4, 419, 1994.
96. Calford, M.B. and Semple, M.N., Monaural inhibition in cat auditory cortex, *J. Neurophysiol.*, 73, 1876, 1995.
97. Brosch, M. and Schreiner, C.E., Time course of forward masking tuning curves in cat primary auditory cortex, *J. Neurophysiol.*, 77, 923, 1997.
98. Brosch, M. and Schreiner, C.E., Sequence selectivity of neurons in cat primary auditory cortex, *Cerebr. Cortex*, 10, 1155, 2000.
99. Brosch, M., Schulz, A., and Scheich, H., Neuronal mechanisms of auditory backward recognition masking in macaque auditory cortex, *NeuroReport*, 9, 2551, 1998.
100. Brosch, M., Schulz, A., and Scheich, H., Processing of sound sequences in macaque auditory cortex: response enhancement, *J. Neurophysiol.*, 82, 1542, 1999.
101. Aertsen, A.M. and Johannesma, P.I., Spectro-temporal receptive fields of auditory neurons in the grassfrog, *Biol. Cybern.*, 38, 223, 1980.
102. Massaro, D.W., Backward recognition masking, *J. Acoust. Soc. Am.*, 58, 1059, 1975.
103. Plack, C.J. and Viemeister, N.F., Intensity discrimination under backward masking, *J. Acoust. Soc. Am.*, 92, 3097, 1992.
104. Steinschneider, M., Arezzo, J. and Vaughan, H.G. Jr., Speech evoked activity in the auditory radiations and cortex of the awake monkey, *Brain Res.*, 252, 353, 1982.
105. Schulz, A., Brosch, M., and Scheich, H., Neuronal responses to three tone sequences in the auditory cortex of macaca fascicularis, *Soc. Neurosci. Abstr.*, 25, p157.4, 1999.
106. Fishman, Y.I., Reser, D.H., Arezzo, J. C., and Steinschneider, M., Neural correlates of auditory stream segregation in primary auditory cortex of the awake monkey, *Hear. Res.*, 151, 167, 2001.
107. Georgopoulos, A.P., Schwartz, A.B., and Kettner, R.E., Neuronal population coding of movement direction, *Science*, 233, 1416, 1986.
108. Wang, X. et al., Representation of a species-specific vocalization in the primary auditory cortex of the common marmoset: temporal and spectral characteristics, *J. Neurophysiol.*, 74, 2685, 1995.
109. Reitböck, H.J., A multi-electrode matrix for studies of temporal signal correlations within neural assemblies, in *Synergetics of the Brain*, E. Basar, H. Flohr, H. Haken, and A.J. Mandell, Eds., Springer–Verlag, New York, 1983, p. 174.
110. von der Malsburg, C. and Schneider, W., A neural cocktail-party processor, *Biol. Cybern.*, 54, 29, 1986.
111. Brosch, M. and Schreiner, C.E., Neural discharge correlations are related to receptive field properties in cat primary auditory cortex, *Eur. J. Neurosci.*, 11, 1, 1999.
112. Abeles, M., *Corticonics: Neural Circuits of Cerebral Cortex*, Cambridge University Press, Cambridge, U.K., 1991.
113. deCharms, R.C. and Merzenich, M.M., Primary cortical representation of sounds by the coordination of action-potential timing, *Nature*, 381, 610, 1996.
114. Bieser, A., Processing of twitter-call fundamental frequencies in insula and auditory cortex of squirrel monkeys, *Exp. Brain Res.*, 122, 139, 1998.

115. Brosch, M. and Scheich, H., Representation of stimulus relations by correlated firing in the auditory cortex of the monkey, *Abstr. Assoc. Res. Otolaryngol.*, 23, 2002.
116. Eggermont, J.J., Neural interaction in cat auditory cortex: dependence on recording depth, electrode separation and age, *J. Neurophysiol.*, 68, 1216, 1992.
117. Eggermont, J.J., Functional aspects of synchrony and correlation in the auditory nervous system, *Conc. Neurosci.*, 4, 105, 1993.
118. Eggermont, J.J., Neural interaction in cat auditory cortex: effects of sound stimulation, *J. Neurophysiol.*, 71, 246, 1994.
119. Eggermont, J.J., Sound-induced synchronization of neural activity between and within three auditory cortical areas, *J. Neurophysiol.*, 83, 2708, 2000.
120. Scheich, H., Heil, P., and Langner, G., Functional organization of auditory cortex in the mongolian gerbil (*Meriones unguiculatus*). II. Tonotopic 2-deoxyglucose, *Eur. J. Neurosci.*, 5, 898, 1993.

10 Representation of Sound Location in the Primate Brain

*Kristin A. Kelly, Ryan Metzger,
O'Dhaniel A. Mullette-Gillman,
Uri Werner-Reiss, and Jennifer M. Groh*

CONTENTS

I. Introduction 177
II. Technical Considerations 178
III. Brain Stem 180
    A. Superior Olivary Complex 180
    B. Inferior Colliculus 180
        1. Anatomy 180
        2. Lesion Studies 181
        3. Representation of Sound Location 182
IV. Thalamus: Medial Geniculate Body 185
V. Auditory Cortex 186
    A. Anatomy 186
    B. Lesion Studies 186
    C. Representation of Sound Location 187
VI. Sensorimotor and Motor Areas 187
    A. Parietal Cortex 187
    B. Frontal Eye Fields and Superior Colliculus 188
VII. Summary and Conclusions 189
Acknowledgments 190
References 190

I. INTRODUCTION

Knowing where things are in space is essential to our existence. The visual, auditory, and cutaneous senses all contribute to the perception of stimulus location, but they acquire spatial information in radically different ways. Positional information is literally built into the neural wiring for vision and touch: stimuli at different positions in the environment activate receptors at different positions on the retina or on the body surface. In contrast, the topographic organization of the cochlea produces a map for the frequency content of a sound, not its direction. Sound source position must be inferred from the direction-dependent filtering of the sound by the external ear (spectral cues) and from differences in sound arrival time and pressure level (or intensity) across the two ears (binaural difference cues).

Animals differ in the extent to which they rely on sound localization to survive. Bats and owls specialize in sound localization for hunting prey and navigating through the nighttime world. The neural basis of sound localization has been explored extensively in these species (for review, see Knudsen[1] and Simmons[2]). Humans and many other primates occupy a different ecological niche; for diurnal creatures, hearing can work in tandem with vision to guide orientation to stimuli and movements through the world. To understand how this happens, we need to know not only how the brain's auditory pathway represents sound location, but also how auditory signals may interact with visual signals. In this chapter, we will review what is known about neural representations of sound location along the auditory pathway, emphasizing the features of these representations that have implications for the coordination with visual information.

One key issue for interactions between hearing and vision concerns the fact that primates can move their eyes with respect to their heads. For vision, the locus of retinal activity evoked by a stimulus at a given location will depend on the angle of gaze with respect to the visual scene. In contrast, binaural and spectral cues depend only on the location of the sound source with respect to the head and ears; the orientation of the eyes is obviously irrelevant to the acoustical input to the ears. Thus, the auditory system is thought to compute the locations of sounds in a head (or ear)-centered reference frame, whereas the visual system employs a reference frame that is retinotopic or eye-centered. For the brain to coordinate visual and auditory information, it must reconcile these discrepant reference frames. As we will see, the actual frame of reference employed at various stages in the brain to represent sound location is not necessarily head centered after all, but incorporates information about the position of the eyes as well.[3–8]

II. TECHNICAL CONSIDERATIONS

As we review the literature on the neural basis of sound localization, certain differences in experimental parameters will be of particular consequence. One major issue is the species of animal involved. Although we are motivated by a desire to understand how the human brain localizes sounds, the methodology of single-unit recording is not normally suited for use in humans. Neuroimaging techniques such as functional magnetic resonance imaging (fMRI) or positron emission tomography (PET) could in principle be used to help identify brain areas that might be involved in sound localization. However, they currently lack the necessary anatomical and temporal precision to reveal how the neurons in the auditory pathway encode sound location. Accordingly, we will concentrate on single-unit recording studies in non-human primates to provide a portrait of neural coding for sound location in a reasonably close relative of humans. However, this body of work is sufficiently small that we will include studies involving other species from time to time. That such substitutions are necessary highlights the rather severe gaps in our knowledge of the primate auditory pathway.

A second issue, and one that is confounded with the species, concerns the behavioral state of the animal. Primate experiments are often conducted with alert animals, but, due to practical considerations, studies in cats and other species usually involve anesthesia. Anesthesia affects the responses of auditory neurons,[9–12] and different types of anesthesia act differently.[13–15] Furthermore, the behavioral state of an alert animal can affect auditory processing; performance of a behavioral task can affect neural responses to auditory stimuli.[11,16–23]

The nature of the acoustic stimuli is also an important issue. Sounds may be presented either through earphones or in the "free field." Stimuli presented with earphones can simulate sounds originating at different spatial locations in the environment. The perception that the sound has an azimuthal location derives from the incorporation of interaural timing and/or level differences of the sound stimuli delivered to each ear. Because spectral cues are omitted, listeners typically experience such stimuli as originating from locations inside the head (for discussion, see Keller et al.[24]). This type of sound is known as a *dichotic stimulus*.

Dichotic stimulation has advantages and disadvantages. It is the method of choice for exploring the specific contributions of timing vs. level differences to neural responses; however, in the real world, interaural timing and level differences co-vary. Varying one without varying the other is an anomalous situation and may evoke neural responses that are anomalous as well. In addition, the weakness of the perceptual experience of sound location associated with this type of acoustic stimulation is worrisome and speaks to the importance of spectral cues in the computation of sound location (for a review, see Keller et al.[24]).

More recent advances in dichotic methods have achieved a higher level of realism. Virtual space stimuli, like conventional dichotic stimuli, are delivered through earphones, but the sounds are prefiltered with a previously measured head-related transfer function. This process approximates the frequency- and direction-dependent modulation of the sound waves that would normally be accomplished by the subject's head and external ears. Interaural timing and level differences are also incorporated into the head-related transfer function. These virtual space stimuli sound more realistic than traditional dichotic stimuli, undoubtedly because of the added contribution of the spectral cues. This technique also allows the presentation of stimuli that mimic moving sounds.[25] This novel technique is increasing in popularity due to its obvious utility.

Free-field stimuli are sounds delivered from speakers located in the environment. Because they are real-world stimuli, they provide all the cues that are normally used under natural conditions. Such stimuli, like virtual-space stimuli, provide the best chance of uncovering spatial sensitivity in neurons because the binaural and spectral cues are all present and providing consistent directional information. The main source of concern with this technique arises from the difficulty in controlling echoes in a free-field setting. Although the presence of echoes complicates efforts to correlate neural responses with specific components of the auditory stimulus, such echoes are a normal part of the environment and contribute to the ecological validity of the setting.

Finally, the issue of quantifying sensitivity to sound location deserves mention. Unlike neurons in the visual and somatosensory pathways, auditory neurons tend not to have clearly circumscribed receptive fields with an obvious excitatory (or inhibitory) island of activity in a sea of unresponsiveness. Instead, neurons vary widely in the type of spatial sensitivity that they show. Broad tuning, to the point at which neurons may respond readily to sounds at every location tested, is often the rule rather than the exception. Many researchers have gone to considerable effort to employ objective and quantitative criteria for classifying spatial tuning profiles. A consensus for the best method of quantifying spatial tuning has yet to emerge. Indeed, there is disagreement concerning the most suitable metric for assessing neural activity; variations in both spike count and latency as a function of sound location have been identified (for discussion, see Furukawa[26]). Nevertheless, these quantitative descriptions are critical for understanding the spatial information present in the population profile at each stage of the auditory pathway. Whenever possible, we will detail the methods that have been used so that valid comparisons between experiments can be made.

III. BRAIN STEM

A. SUPERIOR OLIVARY COMPLEX

We begin with the superior olivary complex (SOC), because it is the first point in the auditory pathway (Figure 10.1) where *ascending* signals from the two ears converge. As such, this structure has received considerable attention for its potential role in sound localization. Although both the cochlear nuclei and the cochlea itself may be subject to binaural influences via descending pathways,[27] spatial sensitivity has not been explored in these structures. Indeed, even at the level of the SOC, very little work has been done in primates (e.g., see Ryan et al.[28]). Studies in cats and other mammals have concentrated on two main subdivisions, the medial (MSO) and lateral (LSO) superior olives (for reviews see References 29 to 33). Neurons in the MSO are sensitive to the interaural timing differences of dichotic auditory stimuli, and neurons in the LSO are sensitive to interaural level differences (cat MSO,[33] cat LSO,[34] dog LSO[35]). The representation of sound location is lateralized: MSO neurons tend to prefer timing differences that correspond to contralateral sound locations, whereas the LSO neurons respond better to ipsilateral stimuli.

The MSO and LSO neurons are also exquisitely sensitive to the frequency of tonal stimuli (as they are throughout the auditory pathway). The two areas appear to specialize in frequency ranges for which the respective types of spatial cues are most informative: MSO neurons tend to prefer the low frequencies for which timing differences are highly relevant, while LSO neurons tend to have best frequencies in a higher range, where intensity differences are larger and more informative.

Nothing is known about the frame of reference employed by neurons in these areas. Because SOC is located so close to the auditory periphery, a head-centered frame of reference is certainly to be expected, but this question has not yet been addressed experimentally.

B. INFERIOR COLLICULUS

1. Anatomy

Located in the midbrain tectum, the inferior colliculus (IC) is situated at an intermediate point in the primate auditory pathway. The IC receives ascending

Representation of Sound Location in the Primate Brain

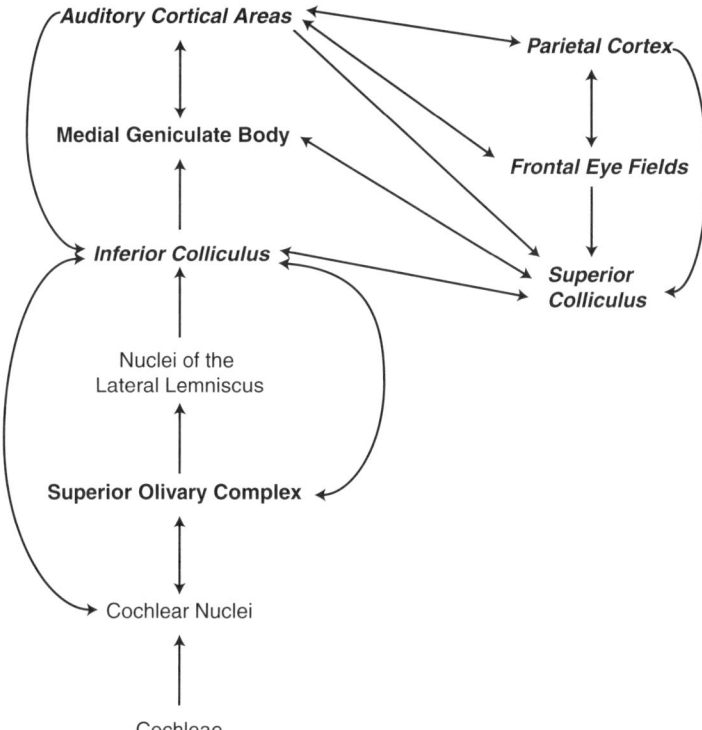

FIGURE 10.1 Overview of the connections within the auditory pathway and between auditory and selected sensorimotor areas. Areas discussed in this chapter are shown in bold. Areas showing at least some dependence on eye position are shown in italics. For review, see References 42, 80, and 122–125, as well as References 38, 82, and 126–133.

inputs (from the SOC, dorsal cochlear nucleus, and nuclei of the lateral lemniscus[36,37]), descending inputs (from the auditory cortex[38–40]), and local collaterals within the IC,[41] as well as projections from the contralateral IC.[37] Efferent projections mainly exit the IC via the brachium of the IC to terminate in the medial geniculate body (MGB), but descending projections also innervate lower auditory structures, as well as projections entering the commissure of the IC to terminate primarily in the contralateral IC.[37,42] The IC is divided into three main subdivisions, commonly called the central (ICC), external (ICX), and pericentral (ICP) nuclei. The largest and best studied of these in the cat is the ICC. In the primate and the cat, the ICC appears tonotopically organized in a manner congruent with the laminar organization of this nucleus.[43–48]

2. Lesion Studies

Ablation of the IC impairs sound localization abilities. In one of the first studies focusing on the effects of IC lesions, Masterton et al.[49] found that cats were unable to lateralize dichotic stimuli after bilateral IC ablations. Transection of the major input and output pathways to and from the IC also impairs sound localization (cats[50,51]). Unilateral lesions of the IC produce deficits in the cat's ability to localize sound sources in the contralateral hemifield.[50,52]

3. Representation of Sound Location

Early studies of coding for sound location in IC neurons concentrated primarily on sensitivity to interaural timing and/or level differences in anesthetized cats. Rose et al.[53] were the first to suggest that some IC neurons have a "characteristic delay," whereby a neuron would respond preferentially to a dichotic stimulus with a specific interaural delay. The results and interpretations of this classic study have been largely corroborated by subsequent studies using broadband noise[54,55] and tones.[56,57] The overwhelming majority of units sensitive to interaural timing differences responded better when the sound delivered to the contralateral ear led the sound delivered to the ipsilateral ear.[57,58] Neurons sensitive to interaural level differences usually respond best when the sound delivered to the contralateral ear is louder, corresponding to a location in the contralateral hemifield.[54,58–60]

Studies investigating responses to free-field or virtual space sounds in the IC of anesthetized cats have identified two broad classes of cells. *Omnidirectional* (also called *nondirectional*) cells have been defined either qualitatively as responding vigorously to sounds over a large range of sound locations[61] or quantitatively as modulating their responses by less than 50% of the maximum discharge rate as a function of sound location.[62–64] The proportion of omnidirectional cells has varied considerably between studies, with findings ranging from as many as 47%[61] to as few as 6%.[64]

The other major category of neurons in cat IC has been termed *directionally sensitive*.[61] This category is usually defined based on the degree of modulation by sound location. Neurons showing a greater than 50% difference in responsiveness between the best and the worst sound locations are said to be directionally selective.[62] Qualitatively, this class of neurons is described as having a clear peak in activity for sounds at a particular (range of) azimuthal location(s). Responses fall off considerably outside this region. The function may be a monotonic plateau or step-shaped function, with one clear border, or it may show tuning in which the responses fall off on both sides of a distinct best azimuth.[61–64] Occasionally more than one best azimuth occurs.[63] As predicted from studies using dichotic stimuli, the majority of directionally sensitive neurons prefer sounds located in the contralateral hemifield. Sensitivity to elevation is also present among IC neurons.[65]

Varying stimulus intensity can influence the range of azimuths to which a neuron is most responsive. Expansion of receptive fields often results from increasing the stimulus intensity;[66,67] however, some neurons remain insensitive to intensity modulation[61,63] (see also Irvine and Gago[60]). Neurons in which the peak or border of the spatial tuning is only minimally affected by sound level are termed *azimuth selective*.[61–64] Note, however, that this definition allows the inclusion of neurons whose overall discharge rate may vary considerably as a function of sound pressure level, provided the location eliciting the strongest responses does not change dramatically.

Another factor that greatly influences the spatial sensitivity of cat IC neurons is the position of the pinnae (particularly the contralateral pinna[61,68]). Each pinna can modulate sounds (particularly high-frequency stimuli), depending on where they originate with respect to the acoustical axis of the ear (for the rhesus monkey, see Spezio et al.[69]), and this must be taken into account when examining the spatial tuning of IC neurons.[61,66,68]

Some evidence for topographic organization of spatial tuning has been reported in the mammalian IC.[70,71] General patterns in the organization of neurons sensitive to interaural timing and level differences, as well as azimuth-sensitive neurons, have also been demonstrated. Low-characteristic (or low-best) frequency neurons are found more commonly in dorsolateral portions of the cat ICC, where afferent projections from the MSO terminate. Neurons in this region, like those in the MSO, are sensitive to interaural timing differences.[57] In contrast, high-characteristic frequency cells are found more ventromedially, receiving abundant input from the LSO. Like LSO neurons, neurons in this area display sensitivity to interaural level differences.[60] It should be noted, however, that IC neurons have complex binaural and monaural response properties. As such, their responses to sounds are not solely dependent upon SOC inputs, but they can arise from complex interactions among other excitatory, inhibitory, and intrinsic inputs, often resulting in *de novo* response patterns.[72] Aitkin and Schuck[73] have suggested a topographical organization for stimulus azimuth in low-characteristic-frequency, azimuth-sensitive neurons, where neurons sensitive to peripherally located tones were located more rostromedially, and those sensitive to tones close to midline were located more caudolaterally. For high-frequency ICC neurons, azimuth-selective cells appear to be located more rostrally, although this gradient was not apparent when noise bursts were used to elicit responses.[62]

Neurons in primate IC are also sensitive to the locations of free-field stimuli and respond best to contralateral locations.[74] However, the frame of reference of this representation is *not* head centered. Groh et al.[3] examined the effects of eye position on responses to sounds in this structure. Monkeys fixated different positions with their eyes while their heads were held stationary. Sounds were delivered from speakers at a range of locations under free-field conditions. The responses of about a third of IC neurons to acoustic stimuli were significantly modulated as a function of eye position.[3] This eye position influence was comparable in magnitude to the effects of sound location *per se*; on average, neural firing rates changed by about the same amount when sound location was held constant and the eyes moved between two positions vs. when the eyes remained in one position and sounds were presented from two different spots. The influence of eye position renders the frame of reference of IC neurons a hybrid between head- and eye-centered coordinates (Figure 10.2). The computational purpose served by a hybrid reference frame is unclear. It may represent an intermediate stage between head- and eye-centered reference frames,[75,76] although coordinate transformations can in principle be accomplished without proceeding through an intermediate or hybrid stage.[77]

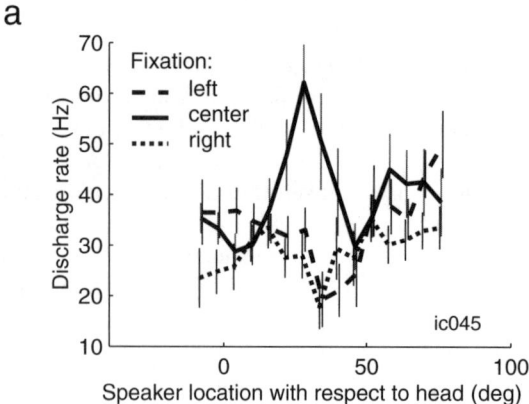
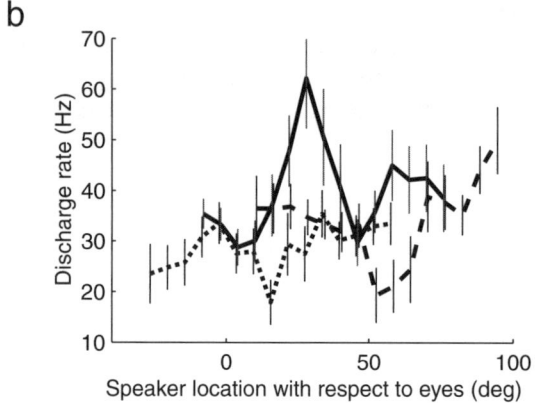
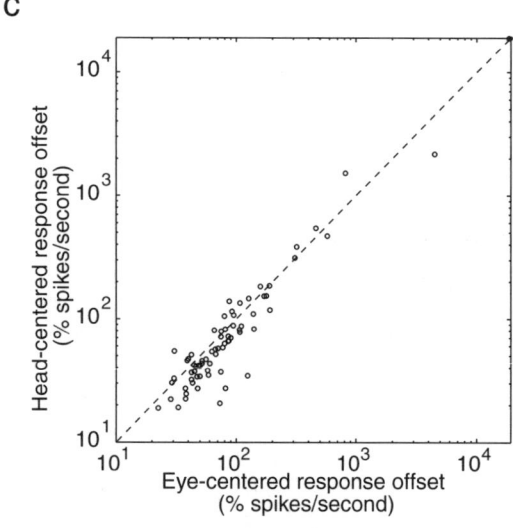

FIGURE 10.2

IV. THALAMUS: MEDIAL GENICULATE BODY

The medial geniculate body (MGB) receives ascending input from the IC and sends projections to auditory cortical regions, which provide a reciprocal descending influence as well. Neurons sensitive to interaural timing and intensity differences have been reported with dichotic stimuli in anesthetized squirrel monkeys.[78] A bias toward sounds located in the contralateral hemifield was evident, but the precise details are as yet unknown.

Several more detailed studies have explored coding for sound location in cat MGB. Barone et al.[79] report that 57% of MGB neurons modulate their responses by at least 75% of the maximum response rate as a function of free-field sound location (i.e., the minimum discharge rate is less than 25% of the maximum discharge rate). Among this population of neurons, several different types of directional sensitivity have been described: bounded response functions, unbounded response functions, and multipeaked response functions. Neurons with bounded response functions have a discharge rate that exceeds 75% of the maximum response for a discrete region of azimuthal locations. To both the left and right of this region, the response rate remains below that level. Neurons with unbounded response functions respond at near peak levels in a region that extends out to 90 degrees of eccentricity on one side. Multipeaked neurons have more than one region of space where sounds elicit responses that exceed 75% of the maximum firing rate. Bounded neurons were about 63% more prevalent than unbounded neurons (the proportion of multipeaked neurons was not reported). The best azimuthal range was centered in the contralateral hemifield about 57% of the time, whereas ipsilateral best ranges occurred in only 14.5% of neurons. The remaining best azimuthal ranges were near the midline or involved multiple peaks.[79]

Nothing is known about the reference frame in this area. Because this structure receives ascending input from the IC, one might predict that these neurons, like those in primate IC, will show sensitivity to the position of the eyes in the orbits, but the answer to this question awaits experimental evidence.

FIGURE 10.2 Effects of eye position on neural activity in primate inferior colliculus (IC). (a) Responses of an IC neuron as a function of sound location (with respect to the head) while the monkey fixated a visual stimulus located 18 degrees left, 18 degrees right, or straight ahead. (b) The same data realigned taking into account the position of the sound with respect to the eyes. (c) A measure of the relative alignment of the neural responses in head- vs. eye-centered frames of reference suggests that the population of neurons employs an intermediate reference frame that is neither head nor eye centered. The "response offset" was the average difference between the evoked responses at the different eye positions when the data were aligned in head- vs. eye-centered coordinates — in other words, the average vertical separation, unsigned, between curves such as those shown in panels a and b, expressed as a percentage of the mean overall response. Only the parts of the curves that overlapped in both frames of reference were included in this analysis. If the IC employed a head-centered frame of reference, then the response curves should align better when plotted in head-centered coordinates, and the points should lie primarily below the dashed line. The opposite pattern should apply in the case of an eye-centered frame of reference. See Groh et al.[3] for further details. (From Reference 3. With permission.)

V. AUDITORY CORTEX

A. ANATOMY

The auditory cortex is perhaps the best-studied area of the primate auditory pathway. The primate auditory cortex consists of several broad regions: the core, the belt, the parabelt, and additional association areas all located in the superior temporal region. The core and belt are located within the lateral sulcus and the parabelt is located on the superior temporal gyrus. The core is further divided into three subregions: AI, R, and RT. AI (primary auditory cortex) is located most caudally, R is rostral with respect to AI, and RT is located at the rostrotemporal end of the core. The belt regions (caudomedial, CM; caudolateral, CL; middle lateral, ML; anterolateral, AL; rostrotemporal medial, RTM; and rostromedial, RM) surround the core and, together with the parabelt, are considered to be auditory association areas. The subdivisions of the parabelt are less clearly defined but appear to consist of distinct caudal and rostral regions lateral to the belt. Projections from ventral MGB synapse primarily on the auditory core, which subsequently projects to the belt regions. The parabelt areas in turn receive the majority of their inputs from the belt regions adjacent to them. The parabelt areas provide input to several other neighboring structures, including the temporoparietal association cortex (Tpt) and caudal superior temporal polysensory region (cSTP).[80–83]

B. LESION STUDIES

That auditory cortex plays a functional role in sound localization has been demonstrated by several lesion studies. Unilateral lesions impair sound localization in the contralateral hemifield for both macaques[84] and squirrel monkeys.[85] Bilateral lesions in macaque auditory cortex also disrupt performance of sound localization tasks.[86,87] Humans with bilateral lesions of auditory cortex still detect sound, but the source of the sound cannot be localized.[88]

A variety of methods have been used to assess the deficits in sound localization performance. Thompson and Cortez[85] trained squirrel monkeys to walk up to speakers in the testing environment to receive a liquid reward from a water-spout affixed below each speaker. Seven speaker locations separated by 30 degrees were used, and training on this task was conducted preoperatively. The effects of unilateral auditory cortical lesions on the accuracy of the behavior were assessed starting 7 days postoperatively. Correct performance dropped from about 90 to 30% on the side contralateral to the lesion. Ipsilateral to the lesion, correct performance was maintained at approximately 80%.

Heffner and colleagues[86,87] assessed the effects of bilateral auditory cortex lesions on the ability of Japanese macaques to learn and/or perform several different types of sound localization tasks. They reported the most profound deficits for tasks involving physical approach to sound sources. The impairments were most severe when the stimulus to be localized was brief; click trains or trains of noise bursts were more easily localized. Performance of tasks that involved conditioned responses to sounds, such as pressing the lever nearest a sound or releasing contact with a water-spout when the location of a sound changed, were also impaired but the deficits were less severe than in the approach task.

C. Representation of Sound Location

Most neurons in the core and belt regions respond to sounds at a broad range of azimuthal locations, usually including both hemifields.[23,89–91] Recanzone et al.,[91] using linear regression of response rate vs. sound location as a measure of spatial sensitivity, report that approximately 80% of neurons in both AI and CM are sensitive to either the azimuth or the elevation of sound sources. Neurons in both areas are more likely to be sensitive to azimuth than to elevation. The proportion of neurons sensitive to azimuth is slightly higher in CM than in AI. In both core and belt areas, the majority of recorded units respond preferentially to the most contralateral speaker, although neurons that prefer ipsilateral locations do exist.[23,89–91]

Spatial sensitivity has also been reported in the lateral part of the belt (CL, ML, and AL) in anesthetized monkeys. Tian et al.[92] defined spatial sensitivity as responses that vary by at least a factor of two as a function of the azimuthal location of the sound.[92] They report that approximately 60% of spatially sensitive neurons respond best to sounds in the contralateral field in all three areas.

Spatial sensitivity within the parabelt region has not been explored, but many neurons in two areas that receive input from the parabelt region, Tpt and cSTP, are broadly tuned for the locations of sounds. In both areas, more than half of the auditory cells modulated by sound location respond preferentially to speakers located in the contralateral hemifield. Neurons that respond to nonauditory stimuli (e.g., visual and somatosensory stimuli) have also been reported in these areas.[93,94]

Behavioral context has been shown to affect the responsiveness of some core/belt neurons. Performance of an auditory task can either increase or decrease the responsiveness of neurons.[11,19–23] Only minor differences are seen when performance of a sound localization task is compared with performance of a sound detection task. Benson et al.[23] reported that the majority of cells (92%) responded identically to either condition. Of the 8% of cells that did show a response preference, the majority had firing rates that were greater for the localization task.

Recently, eye position has been shown to modulate the auditory responses of more than half of the neurons in the auditory cortex.[95] The influence of eye position is very similar to the effect found in the IC.[3] Across the population, cell responses vary nonmonotonically as a function of horizontal eye position.[96] The eye position influence suggests that the frame of reference in the auditory cortex, like that in the IC, is a complex hybrid of head- and eye-centered coordinates.

VI. SENSORIMOTOR AND MOTOR AREAS

A. Parietal Cortex

Unlike the structures discussed thus far, the parietal cortex is not usually thought of as an auditory area. Instead, it is believed to play a role in higher order processing of sensory stimuli of multiple modalities, with an emphasis on the spatial aspects of these stimuli. Control over spatial attention and orientation to sensory stimuli have both been ascribed to this region (e.g., see Anderson et al.[97] and Colby and

Goldberg[98]). Typically, neurons respond to stimuli within specific areas of extrapersonal space when the monkey plans motor responses such as eye movements or limb movements to the locations of those stimuli.

Although the parietal cortex has been studied most extensively with visual stimuli, some neurons respond to auditory stimuli as well. Neurons appear to be sensitive to the location of the sound source,[99] though features such as the shape and size of the receptive fields have not been as thoroughly explored as they have been for visual stimuli (e.g., see Platt and Glimcher[100]). In the lateral intraparietal cortex (LIP), the responses to sound sources often depend on both the position of the sound with respect to the head as well as eye position, with some neurons potentially encoding sound location in an eye-centered reference frame.[6] Neurons in nearby ventral intraparietal cortex (VIP) that are responsive to both visual and auditory stimuli tend to show registry between their receptive fields for each modality.[101]

The context in which the sound is presented can have a major impact on how neurons in the parietal cortex respond to sounds. The most vigorous responses to sounds in LIP neurons occur when the monkey is planning to make an eye movement to that sound source.[17,18] Neurons responsive to sounds have also been reported in the neighboring medial intraparietal cortex (MIP) and the dorsal part of the parieto-occipital area (PO). Interestingly, these neurons are thought to encode the locations of sounds in an eye-centered, rather than head-centered, reference frame, even when the behavioral task involves limb movements but no eye movements. This is a particularly interesting situation because this eye-centered frame of reference is native neither to the sensory stimulus nor to the motor response.[7]

B. Frontal Eye Fields and Superior Colliculus

The functional role of the frontal eye field (FEF) and its subcortical relative, the superior colliculus (SC), is relatively clear: these two brain areas are involved in the generation of saccadic eye movements. Microstimulation in either structure triggers naturalistic saccades with very short latency,[102–105] and combined lesions of both structures eliminate saccades completely and irreversibly.[106] For these reasons, both of these areas are thought of primarily as oculomotor structures. However, sensory responses to visual, auditory, and even tactile stimuli also occur in these areas (e.g., see References 4, 5, and 107–110). These responses are sensory rather than motor in that they occur if and only if the stimulus is presented, they are time locked to the onset of the stimulus, and they are not inevitably followed by a saccade (although the response may be enhanced if the animal is planning a saccade).[111]

Coding for sound location and its reference frame has been studied more extensively in the SC than in the FEF. The majority of primate SC auditory neurons have receptive fields; they respond to sounds only if they are located within a circumscribed region of space (although this region of space can be quite large). These receptive fields move when the eyes move.[4,5] Intriguingly, the receptive fields do not always move as far as they should to maintain a consistent and unambiguous eye-centered reference frame; frequently, the receptive field displaces by only about half of the amount of the change in eye position. Less is known about auditory sensory responses in FEF, but motor activity associated with saccades to auditory stimuli is prevalent and encodes the locations of sounds with respect to the eyes.[8]

VII. SUMMARY AND CONCLUSIONS

The auditory-responsive areas of the brain collectively bear the responsibility for encoding the locations of sounds. Certain key features of this representation emerge consistently:

- Neurons all along the auditory pathway are sensitive to sound location. The proportion of space-sensitive cells varies depending on the methods employed. Even within brain areas, different researchers have reported different proportions of neurons showing spatial sensitivity.
- Sensitivity to sound location is often very broad; neurons frequently respond above baseline levels across a wide range of locations.
- Both monotonic (e.g., linear, sigmoidal, step-shaped, plateau-shaped, unbounded) and nonmonotonic (e.g., bounded, peaked) response-location functions have been reported. Quantifying the relative proportions of these categories in the different brain areas is difficult.
- On the whole, the representation for sound location is lateralized, beginning as early as the SOC. LSO neurons respond preferentially to ipsilateral sounds, but the remainder of the auditory pathway response preferentially to sounds located in the contralateral hemifield. Note, however, that because neurons are typically broadly tuned, responses to ipsilateral sounds may be quite vigorous as well.

Certain features are noticeably absent from this list. A well-organized auditory "map" of space, like that found time and again in homologous areas of the visual pathway, does not appear to exist (although some topographical organization has been found). Further, little evidence indicates that certain areas of the primate auditory pathway specialize in coding sound location; no consistent trend toward increasing (or, for that matter, decreasing) spatial sensitivity emerges as signals ascend to "higher" brain areas, as has been suggested in other species.[112] The problem of how to quantify the spatial response functions of broadly tuned neurons with different types of spatial sensitivity is a thorny one and has impeded attempts to compare the spatial signals present in different areas of the auditory pathway.

How can accurate perception of sound location be mediated by neurons with such broad tuning for sound location? The absence of sharply tuned neurons leaves few alternatives; at least some of these broadly tuned neurons must indeed participate in serving this function. Further, Baldi and Heiligenberg[113] demonstrated that broad tuning of sensory neurons is not necessarily disadvantageous: a population of broadly tuned receptors can accurately encode a stimulus parameter (such as location) with ease. Indeed, relying on a population of neurons may be optimal when the responses of individual neurons are highly variable.[22,26,114–118] However, if the variability of individual neurons is correlated, as has been reported,[89] then pooling will yield diminishing returns[119] (see also Furukawa et al.[26]). The consistency with which neurons respond as a function of sound location is thus worthy of further investigation.

Any nonspatial features that affect the responsiveness of neurons in the auditory pathway must also be taken into consideration.[120] Sound level is such a feature, and the effects of sound level on spatial tuning have received some attention already.[61-64,66,67,121] Increases in sound level can either increase or decrease the responses of individual neurons and can therefore affect the range of locations that evoke activity. Similarly, sound frequency will affect the identity, size, and discharge vigor of the active neural population. Some form of compensation or normalization for the effects of these nonspatial parameters will be necessary before sound location can be read out from the population of acoustically responsive neurons in the auditory pathway (e.g., see Furukawa et al.[26]).

ACKNOWLEDGMENTS

We wish to thank Abigail M. Underhill for her valuable assistance, and Yale Cohen for helpful comments. This work was supported by grants to JMG from the Sloan Foundation, the McKnight Foundation, the Whitehall Foundation, the John Merck Scholars Program, the ONR Young Investigator Program, the Fondation EJLB, and NIH NS 17778-19. Correspondence and requests for materials should be addressed to Jennifer M. Groh (jennifer.m.groh@dartmouth.edu; www.cs.dartmouth.edu/~groh).

REFERENCES

1. Knudsen, E.I., Neural derivation of sound source location in the barn owl: an example of a computational map, *Ann. N.Y. Acad. Sci.*, 510, 33–38, 1987.
2. Simmons, J.A., A view of the world through the bat's ear: the formation of acoustic images in echolocation, *Cognition*, 33, 155–199, 1989.
3. Groh, J.M., Trause, A.S., Underhill, A.M., Clark, K.R., and Inati, S., Eye position influences auditory responses in primate inferior colliculus, *Neuron*, 29, 509–518, 2001.
4. Jay, M.F. and Sparks, D.L., Sensorimotor integration in the primate superior colliculus. II. Coordinates of auditory signals, *J. Neurophysiol.*, 57, 35–55, 1987.
5. Jay, M.F. and Sparks, D.L., Auditory receptive fields in primate superior colliculus shift with changes in eye position, *Nature*, 309, 345–347, 1984.
6. Stricanne, B., Andersen, R.A., and Mazzoni, P., Eye-centered, head-centered, and intermediate coding of remembered sound locations in area LIP, *J. Neurophysiol.*, 76, 2071–2076, 1996.
7. Cohen, Y. and Andersen, R., Reaches to sounds encoded in an eye-centered reference frame, *Neuron*, 27, 647–652, 2000.
8. Russo, G.S. and Bruce, C.J., Frontal eye field activity preceding aurally guided saccades, *J. Neurophysiol.*, 71, 1250–1253, 1994.

9. Kuwada, S., Batra, R., and Stanford, T.R., Monaural and binaural response properties of neurons in the inferior colliculus of the rabbit: effects of sodium pentobarbital, *J. Neurophysiol.*, 61, 269–282, 1989.
10. Gaese, B.H. and Ostwald, J., Anesthesia changes frequency tuning of neurons in the rat primary auditory cortex, *J. Neurophysiol.*, 86, 1062–1066, 2001.
11. Pfingst, B.E., O'Connor, T., and Miller, J.M., Response plasticity of neurons in auditory cortex of the rhesus monkey, *Exp. Brain Res.*, 29, 393–404, 1977.
12. Fitzpatrick, D.C., Kuwada, S., and Batra, R., Neural sensitivity to interaural time differences: beyond the Jeffress model, *J. Neurosci.*, 20, 1605–1615, 2000.
13. Astl, J., Popelar, J., Kvasnak, E., and Syka, J., Comparison of response properties of neurons in the inferior colliculus of guinea pigs under different anesthetics, *Audiology*, 35, 335–345, 1996.
14. Cheung, S.W., Nagarajan, S.S., Bedenbaugh, P.H., Schreiner, C.E., Wang, X.Q., and Wong, A., Auditory cortical neuron response differences under isoflurane versus pentobarbital anesthesia, *Hearing Res.*, 156, 115–127, 2001.
15. Dodd, F. and Capranica, R.R., A comparison of anesthetic agents and their effects on the response properties of the peripheral auditory-system, *Hearing Res.*, 62, 173–180, 1992.
16. Ryan, A. and Miller, J., Effects of behavioral performance on single-unit firing patterns in inferior colliculus of the rhesus monkey, *J. Neurophysiol.*, 40, 943–956, 1977.
17. Grunewald, A., Linden, J.F., and Andersen, R.A., Responses to auditory stimuli in macaque lateral intraparietal area. I. Effects of training, *J. Neurophysiol.*, 82, 330–342, 1999.
18. Linden, J.F., Grunewald, A., and Andersen, R.A., Responses to auditory stimuli in macaque lateral intraparietal area. II. Behavioral modulation, *J. Neurophysiol.*, 82, 343–358, 1999.
19. Beaton, R. and Miller, J.M., Single cell activity in the auditory cortex of the unanesthetized, behaving monkey: correlation with stimulus controlled behavior, *Brain Res.*, 100, 543–562, 1975.
20. Hocherman, S., Benson, D.A., Goldstein, M.H., Jr., Heffner, H.E., and Hienz, R.D., Evoked unit activity in auditory cortex of monkeys performing a selective attention task, *Brain Res.*, 117, 51–68, 1976.
21. Benson, D.A. and Hienz, R.D., Single-unit activity in the auditory cortex of monkeys selectively attending left vs. right ear stimuli, *Brain Res.*, 159, 307–320, 1978.
22. Miller, J.M., Sutton, D., Pfingst, B., Ryan, A., Beaton, R., and Gourevitch, G., Single cell activity in the auditory cortex of rhesus monkeys: behavioral dependency, *Science*, 177, 449–451, 1972.
23. Benson, D.A., Hienz, R.D., and Goldstein, M.H., Jr., Single-unit activity in the auditory cortex of monkeys actively localizing sound sources: spatial tuning and behavioral dependency, *Brain Res.*, 219, 249–267, 1981.
24. Keller, C.H., Hartung, K., and Takahashi, T.T., Head-related transfer functions of the barn owl: measurement and neural responses, *Hearing Res.*, 118, 13–34, 1998.
25. Jacobson, G., Poganiatz, I., and Nelken, I., Synthesizing spatially complex sound in virtual space: an accurate offline algorithm, *J. Neurosci. Meth.*, 106, 29–38, 2001.
26. Furukawa, S., Xu, L., and Middlebrooks, J.C., Coding of sound-source location by ensembles of cortical neurons, *J. Neurosci.*, 20, 1216–1228, 2000.
27. Illing, R.B., Kraus, K.S., and Michler, S.A., Plasticity of the superior olivary complex, *Microsc. Res. Technol.*, 51, 364–381, 2000.

28. Ryan, A.F., Miller, J.M., Pfingst, B.E., and Martin, G.K., Effects of reaction time performance on single-unit activity in the central auditory pathway of the rhesus macaque, *J. Neurosci.*, 4, 298–308, 1984.
29. Boudreau, J.C. and Tsuchitani, C., Cat superior olive S-segment discharge to tonal stimuli, in *Contributions to Sensory Physiology*, W. Neff, Ed., Academic Press, New York, 1970, pp. 143–213.
30. Goldberg, J., Physiological studies of the auditory nuclei of the pons, in *Handbook of Sensory Physiology: Auditory System*, W. Keidel and W. Neff, Eds., Springer–Verlag, Berlin, 1975, pp. 109–144.
31. Irvine, D., The auditory brainstem: a review of the structure and function of auditory brainstem processing mechanisms, in *Progress in Sensory Physiology*, D. Ottoson, Ed., Springer–Verlag, Berlin, 1986, pp. 1–279.
32. Yin, T. and Kuwada, S., Neuronal mechanisms of binaural interaction, in *Dynamic Aspects of Neocortical Function*, Edelman, G.M., Gall, W.E., and Cowan, W.M., Eds., Wiley, New York, 1984, pp. 263–313.
33. Yin, T.C. and Chan, J.C., Interaural time sensitivity in medial superior olive of cat, *J. Neurophysiol.*, 64, 465–488, 1990.
34. Tsuchitani, C. and Boudreau, J., Stimulus level of dichotically presented tones and cat superior olive S-segment cell discharge, *J. Acoust. Soc. Am.*, 46, 979–988, 1969.
35. Goldberg, J.M. and Brown, P.B., Response of binaural neurons of dog superior olivary complex to dichotic tonal stimuli: some physiological mechanisms of sound localization, *J. Neurophysiol.*, 32, 613–636, 1969.
36. Moore, R.Y. and Goldberg, J.M., Ascending projections of the inferior colliculus in the cat, *J. Comp. Neurol.*, 121, 109–135, 1963.
37. Moore, R.Y. and Goldberg, J.M., Projections of the inferior colliculus in the monkey, *Exp. Neurol.*, 14, 429–438, 1966.
38. FitzPatrick, K.A. and Imig, T.J., Projections of auditory cortex upon the thalamus and midbrain in the owl monkey, *J. Comp. Neurol.*, 177, 573–555, 1978.
39. Diamond, I.T., Jones, E.G., and Powell, T.P., The projection of the auditory cortex upon the diencephalon and brain stem in the cat, *Brain Res.*, 15, 305–340, 1969.
40. Andersen, R.A., Snyder, R.L., and Merzenich, M.M., The topographic organization of corticocollicular projections from physiologically identified loci in the AI, AII, and anterior auditory cortical fields of the cat, *J. Comp. Neurol.*, 191, 479–494, 1980.
41. Oliver, D.L., Kuwada, S., Yin, T.C., Haberly, L.B., and Henkel, C.K., Dendritic and axonal morphology of HRP-injected neurons in the inferior colliculus of the cat, *J. Comp. Neurol.*, 303, 75–100, 1991.
42. Huffman, R.F. and Henson, O.W., The descending auditory pathway and acousticomotor systems: connections with the inferior colliculus, *Brain Res. Rev.*, 15, 295–323, 1990.
43. FitzPatrick, K.A., Cellular architecture and topographic organization of the inferior colliculus of the squirrel monkey, *J. Comp. Neurol.*, 164, 185–207, 1975.
44. Webster, W.R., Serviere, J., Crewther, D., and Crewther, S., Iso-frequency 2-DG contours in the inferior colliculus of the awake monkey, *Exp. Brain Res.*, 56, 425–437, 1984.
45. Morest, D.K. and Oliver, D.L., The neuronal architecture of the inferior colliculus in the cat: defining the functional anatomy of the auditory midbrain, *J. Comp. Neurol.*, 222, 209–236, 1984.
46. Merzenich, M.M. and Reid, M.D., Representation of the cochlea within the inferior colliculus of the cat, *Brain Res.*, 77, 397–415, 1974.

47. Rose, J.E., Greenwood, D., Goldberg, J., and Hind, J., Some discharge characteristics of single neurons in the inferior colliculus of the cat. I. Tonotopical organization, relation of spike counts to tone intensity, and firing patterns of single elements, *J. Neurophysiol.*, 26, 294–320, 1963.
48. Semple, M.N. and Aitkin, L.M., Representation of sound frequency and laterality by units in central nucleus of cat inferior colliculus, *J. Neurophysiol.*, 42, 1626–1639, 1979.
49. Masterton, R.B., Jane, J.A., and Diamond, I.T., Role of brain-stem auditory structures in sound localization. II. Inferior colliculus and its brachium, *J. Neurophysiol.*, 31, 96–108, 1968.
50. Strominger, N.L. and Oesterreich, R.E., Localization of sound after section of the brachium of the inferior colliculus, *J. Comp. Neurol.*, 138, 1–18, 1970.
51. Casseday, J.H. and Neff, W.D., Auditory localization: role of auditory pathways in brain stem of the cat, *J. Neurophysiol.*, 38, 842–858, 1975.
52. Jenkins, W.M. and Masterton, R.B., Sound localization: effects of unilateral lesions in central auditory system, *J. Neurophysiol.*, 47, 987–1016, 1982.
53. Rose, J.E., Gross, N.B., Geisler, C.D., and Hind, J.E., Some neural mechanisms in the inferior colliculus of the cat which may be relevant to localization of a sound source, *J. Neurophysiol.*, 29, 288–314, 1966.
54. Geisler, C.D., Rhode, W.S., and Hazelton, D.W., Responses of inferior colliculus neurons in the cat to binaural acoustic stimuli having wide-band spectra, *J. Neurophysiol.*, 32, 960–974, 1969.
55. Yin, T.C., Chan, J.C., and Irvine, D.R., Effects of interaural time delays of noise stimuli on low-frequency cells in the cat's inferior colliculus. I. Responses to wide-band noise, *J. Neurophysiol.*, 55, 280–300, 1986.
56. Yin, T.C. and Kuwada, S., Binaural interaction in low-frequency neurons in inferior colliculus of the cat. III. Effects of changing frequency, *J. Neurophysiol.*, 50, 1020–1042, 1983.
57. Kuwada, S. and Yin, T.C., Binaural interaction in low-frequency neurons in inferior colliculus of the cat. I. Effects of long interaural delays, intensity, and repetition rate on interaural delay function, *J. Neurophysiol.*, 50, 981–999, 1983.
58. Caird, D. and Klinke, R., Processing of interaural time and intensity differences in the cat inferior colliculus, *Exp. Brain Res.*, 68, 379–392, 1987.
59. Blatchley, B.J. and Brugge, J.F., Sensitivity to binaural intensity and phase difference cues in kitten inferior colliculus, *J. Neurophysiol.*, 64, 582–597, 1990.
60. Irvine, D.R. and Gago, G., Binaural interaction in high-frequency neurons in inferior colliculus of the cat: effects of variations in sound pressure level on sensitivity to interaural intensity differences, *J. Neurophysiol.*, 63, 570–591, 1990.
61. Aitkin, L.M., Gates, G.R., and Phillips, S.C., Responses of neurons in inferior colliculus to variations in sound-source azimuth, *J. Neurophysiol.*, 52, 1–17, 1984.
62. Aitkin, L.M. and Martin, R.L., The representation of stimulus azimuth by high best-frequency azimuth-selective neurons in the central nucleus of the inferior colliculus of the cat, *J. Neurophysiol.*, 57, 1185–1200, 1987.
63. Aitkin, L.M., Pettigrew, J.D., Calford, M.B., Phillips, S.C., and Wise, L.Z., Representation of stimulus azimuth by low-frequency neurons in inferior colliculus of the cat, *J. Neurophysiol.*, 53, 43–59, 1985.
64. Delgutte, B., Joris, P.X., Litovsky, R.Y., and Yin, T.C., Receptive fields and binaural interactions for virtual-space stimuli in the cat inferior colliculus, *J. Neurophysiol.*, 81, 2833–2851, 1999.

65. Aitkin, L. and Martin, R., Neurons in the inferior colliculus of cats sensitive to sound-source elevation, *Hearing Res.*, 50, 97–105, 1990.
66. Semple, M.N., Aitkin, L.M., Calford, M.B., Pettigrew, J.D., and Phillips, D.P., Spatial receptive fields in the cat inferior colliculus, *Hearing Res.*, 10, 203–215, 1983.
67. Moore, D.R., Hutchings, M.E., Addison, P.D., Semple, M.N., and Aitkin, L.M., Properties of spatial receptive fields in the central nucleus of the cat inferior colliculus. II. Stimulus intensity effects, *Hearing Res.*, 13, 175–188, 1984.
68. Moore, D.R., Semple, M.N., Addison, P.D., and Aitkin, L.M., Properties of spatial receptive fields in the central nucleus of the cat inferior colliculus. I. Responses to tones of low intensity, *Hearing Res.*, 13, 159–174, 1984.
69. Spezio, M.L., Keller, C.H., Marrocco, R.T., and Takahashi, T.T., Head-related transfer functions of the rhesus monkey, *Hearing Res.*, 144, 73–88, 2000.
70. Binns, K.E., Grant, S., Withington, D.J., and Keating, M.J., A topographic representation of auditory space in the external nucleus of the inferior colliculus of the guinea-pig, *Brain Res.*, 589, 231–242, 1992.
71. Schnupp, J.W. and King, A.J., Coding for auditory space in the nucleus of the brachium of the inferior colliculus in the ferret, *J. Neurophysiol.*, 78, 2717–2731, 1997.
72. Kuwada, S., Batra, R., Yin, T.C., Oliver, D.L., Haberly, L.B., and Stanford, T.R., Intracellular recordings in response to monaural and binaural stimulation of neurons in the inferior colliculus of the cat, *J. Neurosci.*, 17, 7565–7581, 1997.
73. Aitkin, L. and Schuck, D., Low frequency neurons in the lateral central nucleus of the cat inferior colliculus receive their input predominantly from the medial superior olive, *Hearing Res.*, 17, 87–93, 1985.
74. Groh, J.M. and Underhill, A.M., Coding of sound location in primate inferior colliculus, *Soc. Neurosci. Abstr.*, 27, 60.1, 2001.
75. Zipser, D. and Andersen, R.A., A back-propagation programmed network that simulates response properties of a subset of posterior parietal neurons, *Nature*, 331, 679–684, 1988.
76. Pouget, A. and Sejnowski, T.J., Spatial transformations in the parietal cortex using basis functions, *J. Cogn. Neurosci.*, 9, 222–237, 1997.
77. Groh, J.M. and Sparks, D.L., Two models for transforming auditory signals from head-centered to eye-centered coordinates, *Biol. Cybern.*, 67, 291–302, 1992.
78. Starr, A. and Don, M., Responses of squirrel monkey (*Saimiri sciureus*) medial geniculate units to binaural click stimuli, *J. Neurophysiol.*, 35, 501–517, 1972.
79. Barone, P., Clarey, J.C., Irons, W.A., and Imig, T.J., Cortical synthesis of azimuth-sensitive single-unit responses with nonmonotonic level tuning: a thalamocortical comparison in the cat, *J. Neurophysiol.*, 75, 1206–1220, 1996.
80. Kaas, J.H. and Hackett, T.A., Subdivisions of auditory cortex and processing streams in primates, *Proc. Natl. Acad. Sci. USA*, 97, 11793–11799, 2000.
81. Pandya, D.N., Anatomy of the auditory cortex, *Rev. Neurol. (Paris)*, 151, 486–494, 1995.
82. Hackett, T.A., Stepniewska, I., and Kaas, J.H., Prefrontal connections of the parabelt auditory cortex in macaque monkeys, *Brain Res.*, 817, 45–58, 1999.
83. Kaas, J.H. and Hackett, T.A., Subdivisions of auditory cortex and levels of processing in primates, *Audiol. Neurootol.*, 3, 73–85, 1998.
84. Heffner, H., The role of macaque auditory cortex in sound localization, *Acta Otolaryngol. Suppl.*, 532, 22–27, 1997.
85. Thompson, G. and Cortez, A., The inability of squirrel monkeys to localize sound after unilateral ablation of auditory cortex, *Behav. Brain Res.*, 8, 211–216, 1983.

86. Heffner, H. and Masterton, B., Contribution of auditory cortex to sound localization in the monkey (*Macaca mulatta*), *J. Neurophysiol.*, 38, 1340–1358, 1975.
87. Heffner, H.E. and Heffner, R.S., Effect of bilateral auditory cortex lesions on sound localization in Japanese macaques, *J. Neurophysiol.*, 64, 915–931, 1990.
88. Engelien, A., Huber, W., Silbersweig, D., Stern, E., Frith, C.D., Doring, W., Thron, A., and Frackowiak, R.S., The neural correlates of 'deaf-hearing' in man: conscious sensory awareness enabled by attentional modulation, *Brain*, 123, 532–545, 2000.
89. Ahissar, M., Ahissar, E., Bergman, H., and Vaadia, E., Encoding of sound-source location and movement: activity of single neurons and interactions between adjacent neurons in the monkey auditory cortex, *J. Neurophysiol.*, 67, 203–215, 1992.
90. Recanzone, G.H., Spatial processing in the auditory cortex of the macaque monkey, *Proc. Natl. Acad. Sci. USA*, 97, 11829–11835, 2000.
91. Recanzone, G.H., Guard, D.C., Phan, M.L., and Su, T.K., Correlation between the activity of single auditory cortical neurons and sound-localization behavior in the macaque monkey, *J. Neurophysiol.*, 83, 2723–2739, 2000.
92. Tian, B., Reser, D., Durham, A., Kustov, A., and Rauschecker, J.P., Functional specialization in rhesus monkey auditory cortex, *Science*, 292, 290–293, 2001.
93. Leinonen, L., Hyvarinen, J., and Sovijarvi, A.R., Functional properties of neurons in the temporo-parietal association cortex of awake monkey, *Exp. Brain Res.*, 39, 203–215, 1980.
94. Hikosaka, K., Iwai, E., Saito, H., and Tanaka, K., Polysensory properties of neurons in the anterior bank of the caudal superior temporal sulcus of the macaque monkey, *J. Neurophysiol.*, 60, 1615–1637, 1988.
95. Trause, A.S., Werner-Reiss, U., Underhill, A.M., and Groh, J.M., Effects of eye position on auditory signals in primate auditory cortex, *Soc. Neurosi. Abstr.*, 26, 1977, 2000.
96. Werner-Reiss, U., Kelly, K.A., Underhill, A.M., and Groh, J.M., Eye Position Tuning in Primate Auditory Cortex, paper presented at the Society for Neuroscience 31st Annual Meeting, Vol. 27, p. 60.2, 2001.
97. Andersen, R.A., Snyder, L.H., Bradley, D.C., and Xing, J., Multimodal representation of space in the posterior parietal cortex and its use in planning movements, *Annu. Rev. Neurosci.*, 20, 303–330, 1997.
98. Colby, C.L. and Goldberg M.E., Space and attention in parietal cortex, *Annu. Rev. Neurosci.*, 22, 319–349, 1999.
99. Mazzoni, P., Bracewell, R.M., Barash, S., and Andersen, R.A., Spatially tuned auditory responses in area LIP of macaques performing delayed memory saccades to acoustic targets, *J. Neurophysiol.*, 75, 1233–1241, 1996.
100. Platt, M.L. and Glimcher, P.W., Response fields of intraparietal neurons quantified with multiple saccadic targets, *Exp Brain Res.*, 121, 65–75, 1998.
101. Schlack, A., Sterbing, S., Hartung, K., Hoffmann, K.-P., and Bremmer, F.F., Spatially congruent auditory and visual responses in macaque area VIP, *Soc. Neurosci. Abstr.*, 26, 2000.
102. Schiller, P.H. and Stryker, M., Single-unit recording and stimulation in superior colliculus of the alert rhesus monkey, *J. Neurophysiol.*, 35, 915–924, 1972.
103. Schiller, P.H., True, S.D., and Conway, J.L., Paired stimulation of the frontal eye fields and the superior colliculus of the rhesus monkey, *Brain Res.*, 179, 162–164, 1979.
104. Robinson, D.A., Eye movements evoked by collicular stimulation in the alert monkey, *Vision Res.*, 12, 1795–1807, 1972.
105. Stanford, T.R., Freedman, E.G., and Sparks, D.L., Site and parameters of microstimulation: evidence for independent effects on the properties of saccades evoked from the primate superior colliculus, *J. Neurophysiol.*, 76, 3360–3381, 1996.

106. Schiller, P.H. and Conway, J.L., Deficits in eye movements following frontal eye-field and superior colliculus ablations, *J. Neurophysiol.*, 44, 1175–1189, 1980.
107. Groh, J.M. and Sparks, D.L., Saccades to somatosensory targets. III. Eye-position-dependent somatosensory activity in primate superior colliculus, *J. Neurophysiol.*, 75, 439–453, 1996.
108. Jassik-Gerschenfeld, D., Somesthetic and visual responses of superior colliculus neurones, *Nature*, 208, 898–900, 1965.
109. Goldberg, M.E. and Wurtz, R.H., Activity of superior colliculus in behaving monkey. I. Visual receptive fields of single neurons, *J. Neurophysiol.*, 35, 542–558, 1972.
110. Stein, B.E. and Arigbede, M.O., Unimodal and multimodal response properties of neurons in the cat's superior colliculus, *Exp. Neurol.*, 36, 179–196, 1972.
111. Wurtz, R.H. and Mohler, C.W., Enhancement of visual responses in monkey striate cortex and frontal eye fields, *J. Neurophysiol.*, 39, 766–772, 1976.
112. Fitzpatrick, D.C., Batra, R., Stanford, T.R., and Kuwada, S., A neuronal population code for sound localization, *Nature*, 388, 871–874, 1997.
113. Baldi, P. and Heiligenberg, W., How sensory maps could enhance resolution through ordered arrangements of broadly tuned receivers, *Biol. Cybern.*, 59, 313–318, 1988.
114. Brugge, J.F. and Merzenich, M.M., Responses of neurons in auditory cortex of the macaque monkey to monaural and binaural stimulation, *J. Neurophysiol.*, 36, 1138–1158, 1973.
115. Glass, I. and Wollberg, Z., Lability in the responses of cells in the auditory cortex of squirrel monkeys to species-specific vocalizations, *Exp. Brain Res.*, 34, 489–498, 1979.
116. Manley, J.A. and Mueller-Preuss, P., Response variability in the mammalian auditory cortex: an objection to feature detection?, *Fed. Proc.*, 37, 2355–2359, 1978.
117. Manley, J.A. and Muller-Preuss, P., Response variability of auditory cortex cells in the squirrel monkey to constant acoustic stimuli, *Exp. Brain Res.*, 32, 171–180, 1978.
118. Funkenstein, H.H. and Winter, P., Responses to acoustic stimuli of units in the auditory cortex of awake squirrel monkeys, *Exp. Brain Res.*, 18, 464–488, 1973.
119. Zohary, E., Shadlen, M., and Newsome, W., Correlated neuronal discharge rate and its implications for psychophysical performance, *Nature*, 370, 140–143, 1994.
120. Groh, J.M., Converting neural signals from place codes to rate codes, *Biol. Cybern.*, 85, 159–165, 2001.
121. Middlebrooks, J.C., Xu, L., Eddins, A.C., and Green, D.M., Codes for sound-source location in nontonotopic auditory cortex, *J. Neurophysiol.*, 80, 863–881, 1998.
122. Ehret, G., The auditory midbrain, a 'shunting yard' of acoustical information processing, in *The Central Auditory System*, G. Ehret and R. Romand, Eds., Oxford University Press, New York, 1997.
123. Sparks, D.L. and Hartwich-Young, R., The deep layers of the superior colliculus, in *The Neurobiology of Saccadic Eye Movements*, R.H. Wurtz and M.E. Goldberg, Eds., Elsevier, New York, 1989, pp. 213–255.
124. Rauschecker, J.P. and Tian, B., Mechanisms and streams for processing of 'what' and 'where' in auditory cortex, *Proc. Natl. Acad. Sci. USA*, 97(22), 11800–11806, 2000.
125. Masterton, R.B. and Imig, T.J., Neural mechanisms for sound localization, *Annu. Rev. Physiol.*, 46, 275–287, 1984.
126. Thompson, G.C. and Thompson, A.M., Olivocochlear neurons in the squirrel monkey brainstem, *J. Comp. Neurol.*, 254(2), 246–258, 1986.
127. Kaas, J.H., Hackett, T.A., and Tramo, M.J., Auditory processing in primate cerebral cortex, *Curr. Opin. Neurobiol.*, 9(2), 164–170, 1999.

128. Rauschecker, J.P., Tian, B., Pons, T., and Mishkin, M., Serial and parallel processing in rhesus monkey auditory cortex, *J. Comp. Neurol.*, 382(1), 89–103, 1997.
129. Luethke, L.E., Krubitzer, L.A., and Kaas, J.H., Connections of primary auditory cortex in the New World monkey, *Saguinus*, *J. Comp. Neurol.*, 285(4), 487–513, 1989.
130. Fitzpatrick, K.A. and Imig, T.J., Auditory cortico-cortical connections in the owl monkey, *J. Comp. Neurol.*, 192(3), 589–610, 1980.
131. Lewis, J.W. and Van Essen, D.C., Corticocortical connections of visual, sensorimotor, and multimodal processing areas in the parietal lobe of the macaque monkey, *J. Comp. Neurol.*, 428(1), 112–137, 2000.
132. Cavada, C. and Goldman-Rakic, P.S., Posterior parietal cortex in rhesus monkey. II. Evidence for segregated corticocortical networks linking sensory and limbic areas with the frontal lobe, *J. Comp. Neurol.*, 287(4), 422–445, 1989.
133. Asanuma, C., Andersen, R.A., and Cowan, W.M., The thalamic relations of the caudal inferior parietal lobule and the lateral prefrontal cortex in monkeys: divergent cortical projections from cell clusters in the medial pulvinar nucleus, *J. Comp. Neurol.*, 241(3), 357–381, 1985.

11 The Comparative Anatomy of the Primate Auditory Cortex

Troy A. Hackett

CONTENTS

  I. Introduction 199
 II. Organization of Primate Auditory Cortex 202
III. Core 204
     A. Architecture 204
     B. Connections 206
 IV. Belt 208
     A. Architecture 209
     B. Connections 210
  V. Parabelt 210
     A. Architecture 211
     B. Connections 212
 VI. Summary and Conclusions 212
Acknowledgments 214
References 214
Appendix A 221

I. INTRODUCTION

In the sensory systems of the mammalian brain, one of the clearest organizing principles is the existence of topographic maps of the sensory receptors. In the neocortex, orderly representations of the receptor surface are characteristic of the primary sensory areas and some secondary areas, as well. The location and spatial extent of a given map are correlated with distinctive patterns of interneuronal connections and architectonic features of the tissue. Together, these structural and functional properties contribute to the identification of individual areas, and ultimately to the assembly of areas that comprise the cortical system dedicated to a particular sensory modality. For many regions of the brain, however, the organizational picture is uncertain because substantive data are scarce. Thus, the identification of cortical fields devoted to sensory processing remains the subject of ongoing investigation, as it has for over a century.

With respect to audition, the number of cortical areas identified or proposed varies substantially in number, ranging from 1 in some marsupials to over 12 in the macaque monkey.[1] Of these areas, only the primary auditory field, AI, is considered homologous across taxonomic groups. Additional homologies between other areas are likely, but corroborating data are lacking at present. This is partly because research efforts have tended to emphasize the structural and functional characterization of AI over other areas. Thus, the number of auditory areas identified may underestimate the actual number for some species. Comparisons across taxa are further complicated by the variable expression of specialized cortical areas.[2] So, the extent to which findings in one species can be generalized to another is uncertain.

For primates, including humans, the identification of auditory-related fields in cortex has progressed in stages. The first major contribution was made by the British physician, David Ferrier, who studied the functional organization of cerebral cortex in monkeys using electrical stimulation and brain lesions. In one report, Ferrier[3] claimed that bilateral ablation of the superior temporal gyrus in macaque monkeys resulted in deafness, and that electrical stimulation of the same region produced auditory sensations. Ferrier's findings were not broadly accepted at the time,[4] despite independent reports of regional specialization of function by Paul Broca[5] and Karl Wernicke[6] in the brains of aphasics. These early studies were followed by a series of now-classic neuroanatomical studies that detailed architectonic variations in the superior temporal region of monkeys, great apes, and humans.[19–23] In spite of substantial differences between these works, a common organizational theme emerged that forms the cornerstone of current models of primate auditory cortical organization: simply stated, the primate auditory cortex is composed of a central primary, or *core*, region flanked by an array of nonprimary, or *belt*, fields on the superior temporal plane and gyrus (Figure 11.1). It is important to recognize that, in most other mammals, only one primary area is usually recognized (i.e., AI). In primates, however, two or three distinct areas with primary-like features have been identified, all of which are included in a region referred to herein as the *core*. The location of the core is found to be coextensive with an acoustically responsive region on the superior temporal plane of monkeys and chimpanzees in the first electrophysiological maps of primate auditory cortex.[24–27] In the acoustically responsive region, higher stimulus frequencies were represented caudomedially, while lower frequencies were represented rostrolaterally. Sites judged to be well outside of the core, or immediately adjacent fields, generally were not responsive to acoustic stimulation (see Figure 11.2).

These landmark studies were followed by periodic reports describing connections of the superior temporal region[28–31] and recordings from auditory fields in monkeys and humans.[32–36] Interest in the functional organization of primate auditory cortex increased dramatically in the 1970s, marked by important anatomical and physiological studies in macaque monkeys. Pandya and Sanides[37] and Mesulam and Pandya[38] related patterns of corticocortical and thalamocortical connections to architectonic subdivisions[39] of the entire superior temporal region. Merzenich and Brugge[40] identified two tonotopically organized fields (AI, RL) and several adjacent fields on the superior temporal plane. These physiologically identified fields were then related to cytoarchitectonic subdivisions. Despite minor differences in the respective parcellations (Figure 11.1), the collective findings provided support for

The Comparative Anatomy of the Primate Auditory Cortex

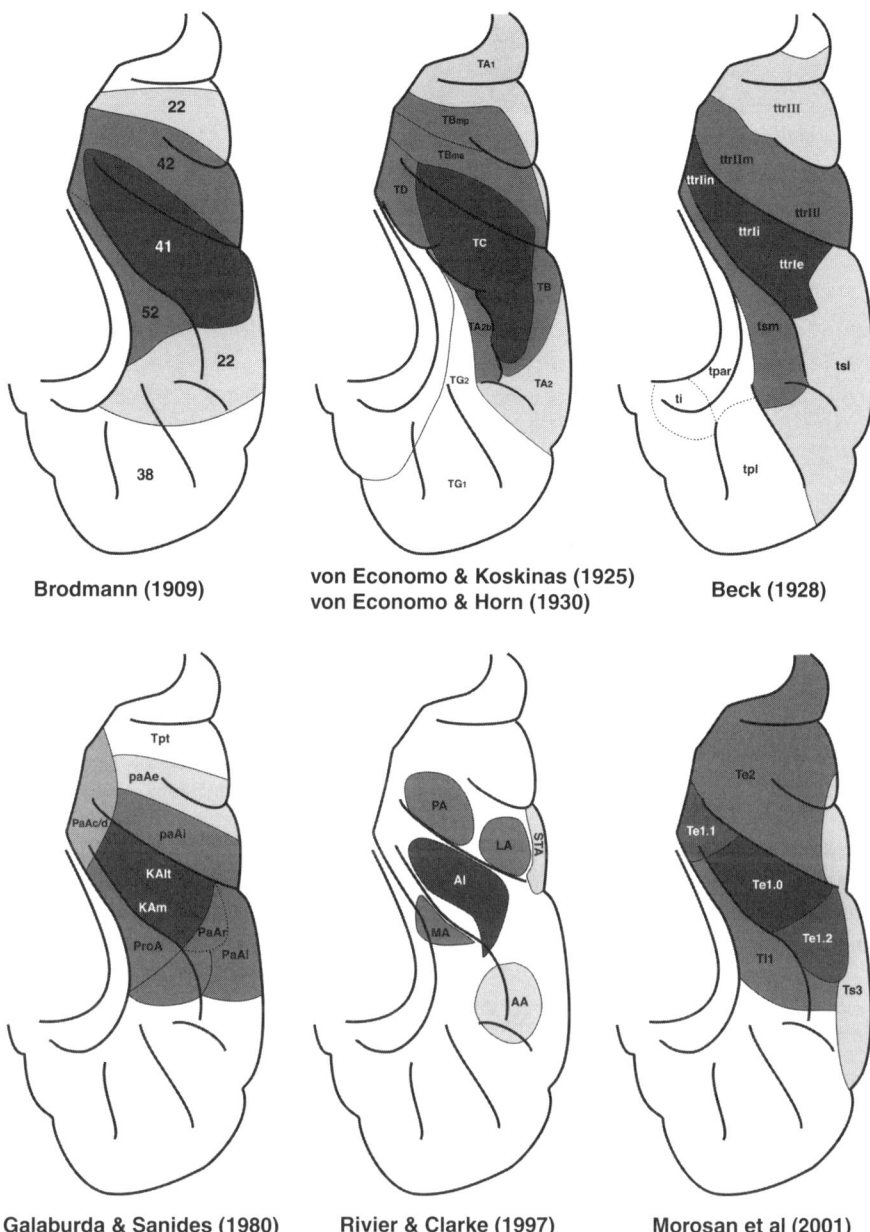

FIGURE 11.1 Parcellations of the human superior temporal cortex. In each panel, the locations of auditory cortex subdivisions are approximated on a standardized drawing of the superior temporal plane. Only the dorsal edge of the superior temporal gyrus is visible. Areas comprising the core region are darkly shaded. Belt areas are moderately shaded, and putative parabelt areas are lightly shaded.

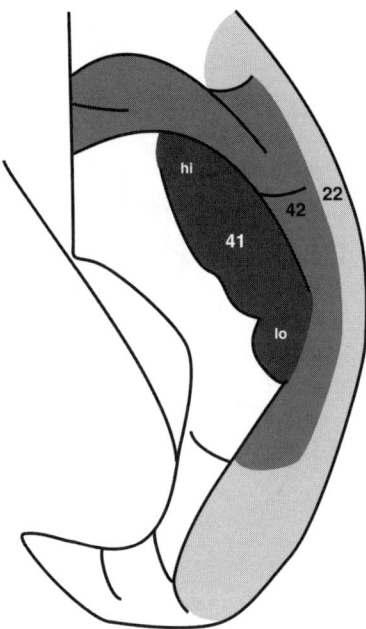

FIGURE 11.2 Auditory fields on the supratemporal plane and dorsal superior temporal gyrus of the chimpanzee. Sites within the core region (area 41, dark shading) were found to be responsive to acoustic stimulation and tonotopically organized (hi, high frequencies; lo, low frequencies). Strychnine application revealed connections between the belt region (area 42, medium shading) and a parabelt region (area 22, light shading). Locations of regions were approximated from diagrams and written descriptions. (Adapted from Bailey et al.[26])

an organizational framework linking core and belt regions across the superior temporal region and established a new foundation for the wide range of anatomical, physiological, and behavioral studies that have continued to produce refinements and extensions into the present.

II. ORGANIZATION OF PRIMATE AUDITORY CORTEX

In contrast to most other mammals, primates have a distinct temporal lobe that contains auditory sensory cortex. The fields that comprise the auditory cortex lie between the insular cortex medially and a poorly defined lateral boundary on the gyral surface ventral to the Sylvian fissure.[41] The identification of cortical regions and their subdivisions depends on profiles derived from the local architecture, interneuronal connections, and neuron response properties. Few studies have gone beyond determining basic response properties of neurons in the core and some of the belt areas. As a result, the majority of non-core areas have only been identified from anatomical studies. The precise number of areas that comprise the auditory cortex is not certain for any species, although models based on the macaque monkey have changed in relatively minor ways since Sanides.[39]

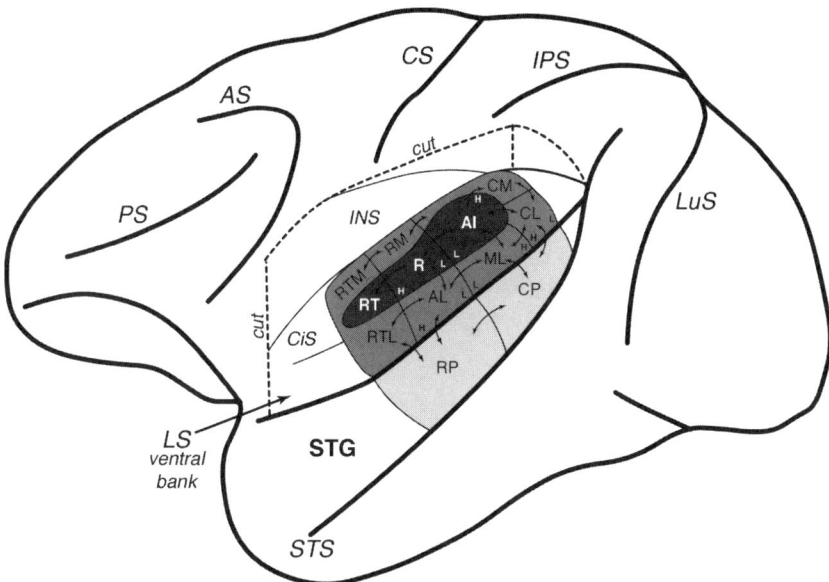

FIGURE 11.3 Schematic view of the macaque left hemisphere showing the location of auditory cortical fields and local connections. The dorsal bank of the lateral sulcus has been removed (*cut*) to expose the superior temporal plane (*LS ventral bank*). The floor and outer bank of the circular sulcus (*CiS*) have been flattened to show the medial auditory fields. The core region (dark shading) contains three subdivisions (AI, R, RT). In the belt region (light shading), seven subdivisions are proposed (CM, CL, ML, AL, RTL, RTM, RM). The parabelt region (light shading) occupies the exposed surface of the superior temporal gyrus (STG). The core fields project to surrounding belt areas (arrows). Inputs to the parabelt arise from the lateral and medial belt subdivisions. Connections between the parabelt and medial belt fields are not illustrated to improve clarity. Tonotopic gradients in the core and lateral belt fields are indicated by the letters H (high frequency) and L (low frequency). (Adapted from Hackett et al.[46])

Most recently, a working model of auditory cortical organization in non-human primates has been proposed[1,42] that attempts to account for the collective findings of the field. Using the nomenclature established in previous studies,[40,43–46] the model divides the auditory cortex into three major regions: *core*, *belt*, and *parabelt* (Figure 11.3). Each region is subdivided into smaller areas. The core region contains up to three subdivisions (AI, R, RT) that exhibit structural and functional features classically associated with "primary" auditory cortex. Encircling the core is a narrow belt of cortex containing seven subdivisions (CL, CM, RM, RTM, RTL, AL, ML). Flanking the belt ventrolaterally is the parabelt region. The parabelt may have two subdivisions (RPB, CPB) that occupy much of the posterior superior temporal gyrus (in macaque monkeys). The belt and parabelt regions have significant connections with auditory-related fields in the temporal lobe (e.g., temporal pole, superior temporal sulcus), posterior parietal cortex, and prefrontal cortex (see below). Of the three regions, the location and extent of the core are the most firmly established, as the anatomical and physiological characteristics of the core have been defined and refined for several primate species in numerous independent investigations (see Appendix A). By comparison, anatomical descriptions of the belt and parabelt regions are much more variable in the literature, and methodical recordings from some of the belt areas have only recently been acquired in macaque monkeys.[47,48]

In the remainder of this review, comparative descriptions of auditory cortical organization in non-human and human primates are presented, using the model outlined above as a guide. The structural and functional features that support this organization scheme will be described in some detail for the core, belt, and parabelt regions. In comparing species of primates, organizational features that are similar and dissimilar are highlighted and discussed.

III. CORE

As indicated in Appendix A, nearly all descriptions of the superior temporal cortex in primates have included a distinct region, commonly referred to as the *auditory core* or *primary auditory cortex*. In prosimian and simian primates, the position of the core varies systematically relative to the edge of the Sylvian fissure. In the prosimian Galago and New World monkeys, the core occupies the gyral surface of the temporal lobe and extends for a short distance onto the ventral bank of the sulcus (supratemporal plane). In Old World macaques, chimpanzees, and humans, the core is generally contained completely in the depths of the sulcus, covered by the overlying frontal and parietal opercula.[41]

In primates, the core is not a homogeneous field (Figure 11.3). Based on the recognition of nonuniformities in the architecture or connections, a variety of different schemes have been proposed,[15,16,18,37,44,46,49,50] but the findings are inconsistent. Microelectrode mapping studies in monkeys, however, have consistently subdivided the core into at least two areas (AI, R) in which the tonotopic gradients are usually found to be reversed.[40,43,44,48,49,51,52] Although the anatomical substrates of this functional pattern have not been identified, AI and R are widely recognized as core fields. A putative third subdivision, labeled RT in Figure 11.1, is the least certain member of the core. Because it has intermediate anatomical and physiological properties, RT has been included in the core by some investigators[44,46,49] and in the rostral belt region by others.[37,53,54]

A. ARCHITECTURE

The architectonic features of the core were first described by anatomists in the early part of the 20th century.[19–23] Excepting the finer details, these early descriptions of the cytoarchitecture and myeloarchitecture have scarcely been modified since their inception, owing to the conservation of basic structural features of the core in non-human primates and humans. The cytoarchitecture of the core is commonly referred to as *granulous* or *koniocellular*, so named because of the dense concentration of small cells in layers II and IV. In their comprehensive work on the cytoarchitecture of human cortex, Von Economo and Koskinas[17] introduced the general term *koniocortex* for the koniocellular ("dust-like") appearance of primary sensory cortices.

They also introduced the familiar term *rainshower formation* to describe the distribution of groups of radially aligned cells in layers II through IV of the auditory koniocortex in the temporal lobe (area TC, *supratemporalis granulosa*). Other notable features of the core include a conspicuous absence of large pyramidal cells in layer III and a relatively sparse population of pyramidal cells in layer V. With few notable exceptions, the basic cytoarchitectonic profiles of the core established in these early studies correspond well with each other and with subsequent parcellations of the superior temporal region in humans and non-human primates (see Appendix A). The exceptions relate to a "delay" in the identification of a core region in non-hominoid primates.

Although well established for the great apes and humans prior to 1930, the core was first described in the macaque monkey by Walker.[22] Brodmann[11,12] and Mauss[14] recognized auditory koniocortex in great apes and humans but did not identify a homologous region in Cercopithecine or prosimian primates.[10,13,55] The existence of the core in these primates was also not obvious to other anatomists at the time. In the marmoset, Mott et al.[56] did not specifically describe a granular region, but its location is clearly demarcated in his drawings, situated between "typical temporal cortex" (belt, parabelt) on the brain surface and the insular region deep in the Sylvian fissure. In *Tarsius*, Wollard's (1925) illustrations portray the entire superior temporal region as area 22, but in the written description he recognized the highly granular architecture of the "receptive" portion of the greater region, concluding that the auditory cortex of *Tarsius* resembled that of higher apes and humans more than the lemur (in reference to Brodmann).[57] As late as 1946, Lashley and Clark[58] concluded the following based on their studies of spider monkey (*Ateles*) and macaque (*Macaca*) cortex:

> (page 283) Except for the auditory area described below (temporal pole), we can find no areal differences in the temporal lobe which can be delimited with sufficient accuracy to have topographic value or which have characteristics of probable functional significance... (page 285) There is nowhere a change within a short distance which would serve for recognition of a boundary between areas. The fact that previous investigators have bounded their areas at the sulci indicates that they also have not been able to recognize any distinct boundaries. The sulci provide the only identifiable topographic features of the lateral surface... (page 286) In neither Macaca nor Ateles can any area on the ventral wall of the Sylvian fissure be distinguished as more granular than the lateral surface of the temporal lobe.

Architectonic studies using cell and myelin stains were both widely used until about 1930. Thereafter, the use of myelin stains in architectonic studies declined, possibly because of difficulties in obtaining reliable staining.[59] Yet, the early myeloarchitectonic descriptions of the superior temporal region agreed favorably with the cytoarchitectonic findings in prosimian and anthropoid primates, including humans.[7-9,13-16,60] The most detailed analyses of myeloarchitecture in the superior temporal region were conducted in humans and chimpanzees by Beck.[15,16] His papers included meticulous parcellations of individual specimens and summary diagrams derived from them. Although Beck's works have been criticized as difficult to interpret and unreplicable, the locations of the seven subregions abstracted from the detailed schematics approximate those depicted in both early and more recent architectonic studies (see Figure 11.1). Among these regions, Beck's *subregio temporalis transversa prima* clearly represents the core, which is described invariably in the literature as a region with an exceptionally dense matrix of small- to large-caliber fibers. The fiber density is such that the inner and outer striae of Baillarger in layers IV and Vb are difficult to resolve, whereas one or both striae are prominent in the belt and parabelt regions.[37,39,61,62]

Recent anatomical studies have made use of less traditional architectonic methods to identify auditory fields in the cortex of non-human primates and humans. Expression of the enzymes acetylcholinesterase and cytochrome oxidase is much higher in layers IIIc and IV of the core than in the belt or parabelt regions (Figure 11.4).[44,46,49,62,63] The calcium-binding protein parvalbumin is also expressed at higher levels in these laminae of the core.[46,52,54,63,64] The dense staining of the core can be instantly appreciated by inspection of Figure 11.5. The superior temporal cortex of a macaque monkey has been flattened through blunt dissection, cut parallel to the pial surface, then stained for the markers indicated. The section stained for parvalbumin (panel A) was cut through the dense layer IIIc/IV band in AI, R, and RT, clearly revealing the boundaries of the core. The other panels, stained for myelin and acetylcholinesterase, also reveal the location of the core. Although their functional significance is poorly understood at present, the co-expression of these molecules in the core region strengthens its profile as a structurally and functionally distinct assembly of areas in primates.

B. Connections

The connections of auditory cortex have been partially described for a number of primates but are the most complete for macaque monkeys. Early degeneration studies firmly established the link between the medial geniculate complex (MGC) and the superior temporal region in monkeys,[20-22,29,65] chimpanzees,[66] and humans.[30] By making selective lesions of primary or association cortex, Pandya et al.[31] revealed the intracortical connection patterns of distinct auditory regions in the macaque monkey. They presented the first clear evidence that the core region projected to adjacent belt areas but not to more distant superior temporal locations in the same hemisphere. This pattern matched the results of strychnine neuronography experiments in which macaque area 41 was found to be reciprocally connected with adjacent cortex (area 42), but not area 22, which received inputs only from area 42.[67-69]

The development of anterograde and retrograde tracing techniques for revealing interneuronal connections improved resolution, allowing for the characterization of finer topographic and areal patterns. Using these methods, a number of studies examined the cortical connections of the core region in several primates, including New World owl monkeys,[44,70] marmosets,[71,72] and Old World macaques.[49,53,54,73] These studies revealed dense connections between the core areas in both hemispheres and with adjoining areas in the medial and lateral belt areas surrounding the core (Figure 11.3). The connection patterns are also topographic, in that caudal belt fields (e.g., CM, CL, ML) have stronger connections with AI, whereas connections of the

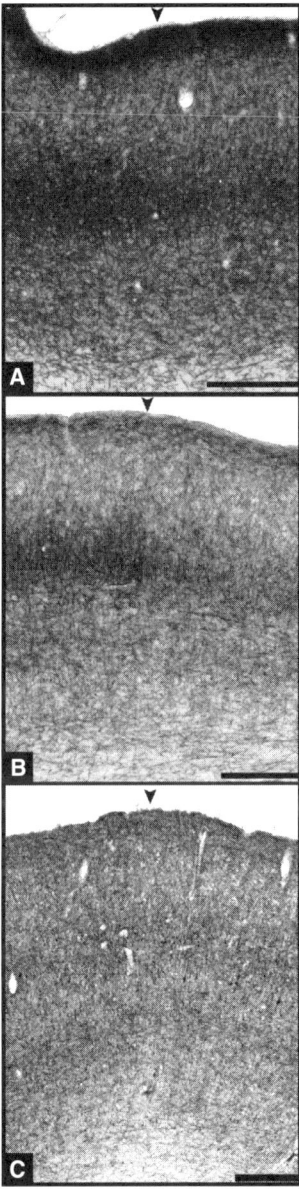

FIGURE 11.4 Acetylcholinesterase histochemistry of the auditory cortex showing borders (arrowheads) between the core and lateral belt regions. Core is to the left of the arrowhead and lateral belt is to the right. (A) Macaque monkey; (B) chimpanzee; (C) human. Scale bar = 0.5 mm. (Adapted from Hackett et al.[62])

rostral belt fields (e.g., RM, AL) tend to emphasize R. Subcortical inputs to the core have been found to arise primarily from the ventral (MGv) and magnocellular (MGm) divisions of the medial geniculate complex (MGC).[44,49,71–72,74–78] Thalamic

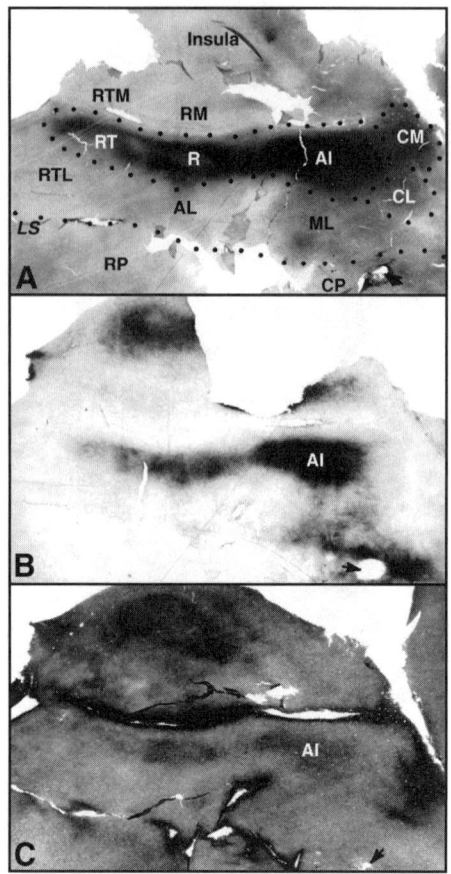

FIGURE 11.5 Architecture of macaque monkey auditory cortex. The superior temporal cortex was flattened and sectioned parallel to the pial surface at 40 μm. (A) Parvalbumin immunohistochemistry; (B) myelin stain; (C) acetylcholinesterase histochemistry. The core fields (AI, R, RT) are the most darkly stained. The caudal belt fields (ML, CL, CM) are moderately dark. Rostral belt and parabelt fields are weakly reactive. Arrows indicate location of tracer injection. (Adapted from Hackett et al.[46])

projections to the core are topographic, with more rostral portions of the core (e.g., R, RT) receiving inputs from more caudal loci in the MGC.

IV. BELT

The notion that auditory sensory cortex was organized in belts about a koniocortical core is often attributed to anatomical[79] and physiological[80] studies in cats. Yet, such systematic structural variations have long been recognized in the primate superior temporal cortex, as discussed above. With respect to the number and extent of the belt fields, however, there is significant variability between studies and species, so the identification of homologous areas outside of the core remains tentative. Von

Economo and Koskinas[17] and Beck[15,16] recognized tremendous architectonic variation throughout the human and chimpanzee temporal lobe (Figure 11.1). Excluding subdivisions, five major supratemporal regions were identified by Von Economo (TA, TB, TC, TD, and part of TG) which were later subdivided into over 30 zones and subtypes.[18] Beck[15,16] identified seven regions on the dorsal surface of the temporal lobe alone (tp, ts, tpar, ttrI, ttrII, ttrIII, tpt), encompassing about 50 subdivisions.

When minutiae such as these are set aside, however, coherence to the basic organization schemes of early investigators[7,12,17] can be recognized (Figure 11.1). Bordering the koniocortex anteromedially is a region with distinctive architecture, referred to herein as the *medial belt*. Although known by different names across studies (e.g., area 52[11,12] or proA[37,39]), the various architectonic descriptions of this region are remarkably similar. Bordering the koniocortex posterolaterally is a second belt region, referred to herein as the *lateral belt*, but better known as area 42 of Brodmann[11,12] and area TB of Von Economo and Koskinas.[17] Again, the architectonic descriptions of the lateral belt are highly consistent among investigators who recognized its presence. In Appendix A, the fields that correspond most closely to the medial and lateral belt regions are listed for humans and other primates. The existing data may not support homology in each instance, but the anatomical evidence is generally supportive.

A. Architecture

Von Bonin and Bailey[81] did not identify a distinct region (i.e., TB) surrounding the core (TC) in macaque monkeys. In their study, most of the superior temporal cortex outside of the core was classified as TA, while patches of cortex bearing the architectonic signature of TB (after Von Economo and colleagues[17,18]) were found to be confined to the superior temporal plane. Their findings echoed those of most early studies in that the narrow belt of cortex surrounding the core was usually not identified, whereas the homotypical temporal cortex (e.g., TA, area 22) was always recognized. The work of Sanides,[39] Pandya and Sanides,[37] and Merzenich and Brugge[40] combined to establish the concept of an auditory belt in primates and to further define it as being composed of multiple subdivisions. In the parcellations of Sanides[39] and Pandya and Sanides,[37] the koniocortical core region (Kam/Kalt) was separated from the insula medially by a prokoniocortical region (ProA) and bordered by parakoniocortical fields laterally (paAlt), rostrally (paAr), and caudally (paAc). These medial and lateral belt areas were surrounded by additional fields that occupied most of the remaining temporal lobe (Ts1, Ts2, Ts3, Tpt, Pro, reIt). Merzenich and Brugge[40] identified a similar array of belt fields adjacent to AI on the inferior bank of the lateral sulcus. Since then, an array of belt areas has been identified or proposed in most models of auditory cortical organization in New World and Old World primates. Von Economo and Koskinas[17] designated the auditory parakoniocortex as TB. On the bases of relative location and architectonics, TB in humans would appear to represent an expansion of the lateral belt fields in monkeys into the planum temporale.

The architectonic descriptions of the medial and lateral belt fields of non-human primates are strikingly similar to comparable fields in the human brain.

The architectonic features of belt areas are distinct in two general ways. First, compared to the core, significant reduction in granularity and cell density in the outer layers is observed, as is an increase in the size of pyramidal cells, especially in layer IIIc. Second, fiber density is greatly diminished, shifting from the astriate pattern of the core to a unistriate (layer IV band visible) or bistriate (layer IV and Vb bands visible) pattern. With respect to chemoarchitecture, the expression of acetylcholinesterase, cytochrome oxidase, and parvalbumin are significantly reduced in both lateral and medial belt regions, particularly in the layer IIIc/IV band (Figures 11.4 and 11.5). Thus, the reduction in expression is a useful marker of the border between regions and has been widely used to support patterns of connections and physiological recordings.[44,46,49,52,54,63,64] Variations in this general architectonic profile contribute to the identification of the numerous subdivisions of the belt. For example, the rostral area, paAr,[37,39] appears to correspond to the core area, RT, in the present model (Figure 11.3). Pandya and Sanides[37] commented that this area resembled koniocortex more than any other belt area because of a general parvocellularity and broad layer IV. Myelination is slightly less dense than the core (Figure 11.5B), and the pattern less typically astriate because the inner and outer bands in layers Vb and IV can be seen. Chemoarchitectonic expression density is also intermediate to core and belt areas, as shown in Figure 11.5 (see Hackett et al.,[46,62]). Note also that the caudal belt areas (CM, CL, ML) are more reactive for these markers than more rostral belt and parabelt areas.

B. Connections

As described above, tracer injections placed in the core of New World and Old World monkeys have consistently revealed dense topographic connections between subdivisions of the core and adjacent subdivisions of the belt, as illustrated in Figure 11.3.[44,49,53,54,70–72,82,83] Due to the narrow width of the belt region, and difficulties in establishing precise physiological boundaries, the literature contains few examples of either lesions or tracer injections that were confined to a single subdivision of the belt. The most precise injections appear to have been made in macaques by Jones et al.,[54] where most of the belt injections illustrated were confined to specific belt fields. Injections of the parabelt have also revealed dense topographic connections with the belt region.[46,53,54] Rostral belt areas are more strongly interconnected with the rostral divisions of the core and parabelt, whereas connections with the caudal belt areas favor AI and the caudal parabelt.[46] The belt is also known to have topographic connections with several regions in prefrontal cortex.[84–86] Thalamic inputs include the dorsal (MGd) and magnocellular (MGm) divisions of the medial geniculate complex (MGC), with few or no inputs from the ventral division (MGv).[77,87]

V. PARABELT

The parabelt region refers to that part of the superior temporal cortex adjoining to the lateral belt region ventrolaterally. In macaques, the parabelt occupies much of the caudal half of the STG but does not include area Tpt at the caudal limits of the STG or the rostral half of the temporal lobe (Figure 11.3). The ventral border of the parabelt region is not clearly defined but appears to extend across the lateral surface of the STG ventrally to a location near the upper bank of the superior temporal sulcus. Physiological maps of the region are not available, although recent studies of the lateral belt in macaques may include some sites in the parabelt.[45,88] The architectonic profile of the parabelt region in monkeys has long been associated with homotypical cortex of the superior temporal gyrus in humans and other primates, but the parabelt is likely to correspond to only part of the classically defined regions, such as TA or area 22, that cover the STG and may extend onto the ventral Sylvian surface (Figure 11.1). The homologies proposed in Appendix A should be considered especially tenuous for parabelt regions because the expansion of this region in humans, in particular, may represent the addition of areas not accounted for at present. Thus, the comparisons in Appendix A rely heavily on the author's interpretation of the architectonic data.

A. ARCHITECTURE

The cytoarchitecture of the parabelt region resembles that of the lateral belt, a fact that has likely contributed to ambiguity in the identification of the latter. As Appendix A reveals, the entire superior temporal region outside of the koniocortex (when recognized) was believed to be structurally homogeneous by some investigators, but note that areal differences have been rarely excluded in studies of the great apes or humans. One explanation may be that architectonic differentiation is simply more evident in these primates. As noted recently,[62] the relatively compact cortex of the macaque makes cytoarchitectonic identification of borders more difficult compared to the cortex of chimpanzees and humans, where distinctive features are much more obvious. Conversely, myeloarchitectonic and chemoarchitectonic transitions are more apparent in macaques, where the dense concentration of cells and fibers enhances differences (see Figure 11.4). Additionally, in New World monkeys (e.g., owl monkey, marmoset) and prosimians (e.g., Galago), we have found that cytoarchitectonic transitions in the superior temporal cortex are even less obvious than in macaques, whereas borders are fairly distinct in myelin or other histochemical stains (e.g., acetylcholinesterase, cytochrome oxidase, parvalbumin). Considering these observations, it is not altogether surprising, therefore, that the distinction between belt and parabelt regions was sometimes misidentified by early investigators.

In the cortex corresponding to the parabelt region in primates, neurons are arranged in distinct radial columns from layer VI to layer II. The term *organ pipe formation* was introduced by Von Economo and Koskinas[17] to describe this structural feature of area TA in humans. Cytoarchitectonic descriptions of the STG in macaque monkeys almost always make note of this configuration.[37,39,46,53] The cytoarchitecture of the parabelt region is distinguished from the lateral belt (TB) by the large pyramidal cells present in layer IIIc. In the parabelt, pyramidal cells in layer III are more uniform in size, often appearing to be stacked atop one another in columns (organ pipes), and the largest pyramidal cells are usually found in layer V. Myelination is of the bistriate type, but slightly less dense compared to the belt. Radial fibers can be followed from layer V through layer II in the well-spaced columns. Acetylcholinesterase, cytochrome oxidase, and parvalbumin are expressed at low levels in the parabelt, compared to the moderate levels characteristic of the lateral belt (Figures 11.4 and 11.5).[46,52,54] Layer IV is narrower, apparently contributing to the density reduction in the layer IIIc/IV band. Thus, there is a stepwise progression in the density of expression for these markers from the parabelt to the belt to the core region. These shifts in the architecture support the hierarchical pattern of interneuronal connections.

B. Connections

Pandya et al.[31] studied fiber degeneration patterns in the macaque after lesions in the superior temporal region. Lesions of the core resulted in local terminal degeneration in the belt areas surrounding the core medially and laterally, in *both* ipsilateral and contralateral hemispheres. On the exposed surface of the STG (TA), however, evidence of degeneration was minimal. Lesions confined to the caudal STG revealed fiber degeneration in cortex corresponding to the medial and lateral belt, but not within the core (TC). Similar findings were found by Pandya and Sanides[37] in one monkey with a lesion of the gyral portion of area paAlt. In subsequent studies, injections of radiolabeled amino acids placed in the parabelt region labeled many cells in the belt and medial belt fields but none in the core (KA).[53,83] Labeling in paAr, rostral to the core (RT in Figure 11.3), was weak or absent. In a recent series of experiments, multiple tracer injections were made in the parabelt region in each of five macaques.[46,89] The results showed that thalamic inputs to the parabelt include the MGd, MGm, suprageniculate, limitans, and medial pulvinar but not the MGv. Cortical inputs originate in the medial and lateral belt regions, but not the core (Figure 11.3). Rostral belt areas target the rostral parabelt, while caudal belt areas target the caudal parabelt. The medial belt area, RM, is the apparent exception to this topographical pattern, as it has diffuse connections throughout the parabelt region.

Neurons in the parabelt widely project to auditory related cortex throughout the brain. Temporal lobe projections include several regions outside of the parabelt (i.e., rostral temporal lobe and temporal pole, superior temporal sulcus, and Tpt). These areas generally receive no significant inputs from the MGC and have limited, if any, connections with the auditory belt cortex. In keeping with the rostrocaudal topography of the core and belt, the caudal parabelt has very few connections with the rostral temporal lobe. In the upper bank of the STS, however, rostral and caudal projections from the parabelt have been observed to overlap in discrete patches.[46] The rostrocaudal topography among the auditory cortical fields also extends to connections with auditory-related areas in prefrontal cortex.[84–86,90] Projections of the rostral belt and parabelt target prefrontal domains associated with the processing of nonspatial information, while the caudal belt and parabelt fields target cortex related to spatial processing.

VI. SUMMARY AND CONCLUSIONS

The model of primate auditory cortex illustrated in Figure 11.3 indicates that three major regions occupy the caudal superior temporal cortex: core, belt, parabelt. Each of these regions has more than one subdivision and represents a distinct level of cortical auditory processing.[1,91] At each successive level of processing, inputs appear to be processed in parallel by subdivisions within that region. Inputs from the primary lemniscal pathway in the brainstem target the core at a first level of cortical processing via projections from the MGv. The subdivisions of the core direct their outputs to the surrounding belt fields at a secondary stage of processing. A third stage of processing is mediated by the parabelt, which receives topographic inputs from the belt. The principal inputs to both belt and parabelt regions arise from the MGd. The outputs of the belt and parabelt target auditory-related domains in prefrontal, temporal, and posterior parietal cortex. The rostrocaudal topography maintained in the connection patterns between regions has led to speculation that there is a segregation of spatial and nonspatial processing in the auditory cortical pathways,[42,47,84,85,92] analogous to that found in the visual system.

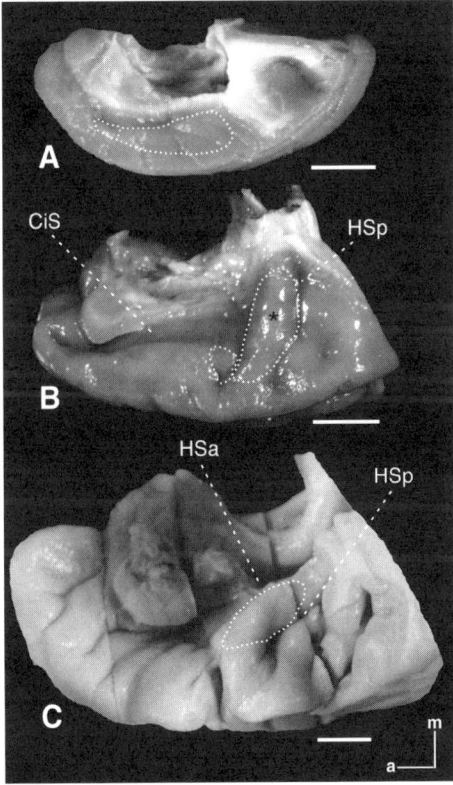

FIGURE 11.6 Dorsolateral views of the left superior temporal plane showing position of the core region. (A) macaque monkey; (B) chimpanzee; (C) human. White dashed ovoids in all panels indicate approximate boundaries of the core. Dashed straight white lines designate sulcal landmarks. CiS, circular sulcus; HSa, anterior Heschl's sulcus; HSp, posterior Heschl's sulcus; SI, sulcus intermedius; a, anterior; m, medial. Scale bars = 5 mm. (Adapted from Hackett et al.[62])

The extent to which this model may be applied to other primates, including humans, is limited both by methodological constraints and phyletic differences in brain organization. Thus, theories about potential homologies must be drafted with care. Experiments to determine connectivity and neurophysiology, as studied in experimental animals, are generally not possible in great apes or humans. In the functional imaging literature, evidence is accumulating for both serial and parallel processing among auditory regions (see Hackett et al.[62]), but it is not yet possible to relate areas of activation to anatomical constructs other than the gross anatomy, and most of the available data pertain only to humans. So, although functional comparisons between taxa are incomplete, an opportunity does remain to directly compare the underlying architectonic features that remain intact postmortem. As we have recently reported, the architectonic criteria used to define discrete auditory cortical areas of the core and belt regions in monkeys can also be used to define homologous regions in chimpanzees and humans (Figure 11.6).[62] Thus, there is reason to expect that a primate model of auditory cortical organization can be used to make accurate predictions about the nature of auditory processing in human cortex, at least for the early stages of processing. Toward this end, the identification and comparison of key architectonic features remain useful tools for bridging some of the gaps imposed by experimental and observational constraints.

ACKNOWLEDGMENTS

The author would like to acknowledge the support of the National Institutes of Health, NIDCD grants DC00249 and DC04318.

REFERENCES

1. Kaas, J.H. and Hackett, T.A., Subdivisions and levels of processing in primate auditory cortex, *Audiol. Neurootol.*, 3, 73–85, 1998.
2. Ehret, G., The auditory cortex, *J. Comp. Physiol. A*, 181, 547–557, 1997.
3. Ferrier, D., Experiments on the brain of monkeys, *Proc. Roy. Soc. London B, Biol. Sci.*, 23, 409–432, 1875.
4. Heffner, H.E., Ferrier and the study of auditory cortex, *Arch. Neurol.*, 44, 218–221, 1987
5. Broca, P., Remarques sur le siege de la faculte du langage articule, *Bull. Soc. Anthropol.*, 6, 377–393, 1865.
6. Wernicke, C., *Der aphasische Symptomenkomplex*, Cohn and Weigert, Breslau, 1874.
7. Flechsig, P., *Die Leitungsbahnen im Gehirn und Ruckenmark des Menschen, auf Grund entwicklungsgeschichtlicher Untersuchungen*, Leipzig, 1876.
8. Flechsig, P., *Anatomie des menschlichen Gehirns und Ruckenmarks auf myelogenetischer Grundlage*, Thieme, Leipzig, 1920.
9. Campbell, A.W., *Histological Studies on the Localization of Cerebral Function*, Cambridge University Press, Cambridge, U.K., 1905.
10. Brodmann, K., Beitrage zur histologischen Lokalisation der Grosshirnrinde. Dritte Mitteilung: Die Rinderfelder der niederen Affen, *J. Psych. Neurol.*, 4, 177–226, 1905.

11. Brodmann, K., Beitrage zur histologischen Localisation der Grosshirnrinde. VI. Mitteilung. Die Cortexgliederung des Menschen, *J. Psych. Neurol.*, 10, 231–246, 1908.
12. Brodmann, K., *Vergleichende Lokalisationslehre der Grosshirnrinde*, Barth, Leipzig, 1909.
13. Mauss, T., Die faserarchitektonische Gliederung der Grosshirnrinde bei den niederen Affen, *J. Psych. Neurol.*, 13, 263–325, 1908.
14. Mauss, T., Die faserarchitektonische Gliderung des Cortex Cerebri der antropomorphen Affen, *J. Psych. Neurol.*, 18, 410–467, 1911.
15. Beck, E., Die myeloarchitektonische Felderung es in der Sylvischen Furche gelegenen Teiles des menschlichen Schläfenlappens, *J. Psychol. Neurol.*, 36, s1–s21, 1928.
16. Beck, E., Der myeloarchitektonische Bau des in der Sylvischen Furche gelegenen Teiles des Schläfenlappens beim Schimpansen (*Troglodytes niger*), *J. Psychol. Neurol.*, 38, 309–428, 1929.
17. Von Economo, C. and Koskinas, G.N., *Die Cytoarchitectonik der Hirnrinde des erwachsenen menschen*, Julius Springer, Berlin, 1925.
18. Von Economo, C. and Horn, L., Über Windungsrelief, Maße und Rindenarchitektonik der Supratemporalfläche, ihre individuellen und ihre Seitenunterschiede, *Z. Neurol. Psychiatr.*, 130, 678–757, 1930.
19. Clark, W.E.L., The brain of *Microcebus murinus*, *Proc. Zool. Soc. London*, 1931, 463–486, 1931.
20. Clark, W.E.L., The thalamic connections of the temporal lobe of the brain in the monkey, *J. Anat.*, 70, 447–464, 1936.
21. Poljak, S. Origin, course, termination and internal organization of the auditory radiation, in *The Main Afferent Fiber Systems of the Cerebral Cortex in Primates*, H.M. Evans and I.M. Thompson, Eds., University of California Press, Berkeley, 1932, pp. 81–104.
22. Walker, A.E.,The projection of the medial geniculate body to the cerebral cortex in the macaque monkey, *J. Anat. (London)*, 71, 319–331, 1937.
23. Von Bonin, G., The cerebral cortex of the cebus monkey, *J. Comp. Neurol.*, 69, 181–227, 1938.
24. Ades, H.W. and Felder, R., The acoustic area of the monkey (*Macaca mulatta*), *J. Neurophysiol.*, 5, 49–59, 1942.
25. Licklider, J.C.R. and Kryter, K.D., Frequency localization in the auditory cortex of the monkey, *Fed. Proc.*, 1, 51, 1942.
26. Bailey, P., Von Bonin, G., Garol, H.W., and McCulloch, W.S., The functional organization of temporal lobe of the monkey (*Macaca mulatta*) and chimpanzee (*Pan satyrus*), *J. Neurophysiol.*, 6, 121–128, 1943.
27. Woolsey, C.N., Tonotopic organization of the auditory cortex, in *Physiology of the Auditory System*, M.B. Sachs, Ed., National Education Consultants, Baltimore, MD, 1972, pp. 271–282.
28. Nauta, W.J.H., Silver impregnation of degenerating axons, in *New Research Techniques of Neuroanatomy*, W.F. Windle, Ed., Charles C Thomas, Springfield, IL, 1957, pp. 17–26.
29. Akert, K., Woolsey, C.N., Diamond, I.T., and Neff, W.T., The cortical projection area of the posterior pole of the medial geniculate body in *Macaca mulatta*, *Anat. Rec.*, 133, 242, 1959.
30. van Buren, J.M. and Yakovlev, P.I., Connections of the temporal lobe in man, *Acta Anat.*, 39, 1–50, 1959.
31. Pandya, D.N., Hallett, M., and Mukherjee, S.K., Intra- and interhemispheric connections of the neocortical auditory system in the rhesus monkey, *Brain Res.*, 14, 49–65, 1969.

32. Pribram, K.H., Burton, S.R., and Rosenblith, W.A., Electrical responses to acoustic clicks in monkey: extent of neocortex activated, *J. Neurophysiol.*, 17, 336–344, 1954.
33. Hind, J.E. Jr., Benjamin, R.M., and Woolsey, C.N., Auditory cortex of the squirrel monkey (*Saimiri sciureus*), *Fed. Proc.*, 17, 71, 1958.
34. Katsuki, Y., Suga, N., and Kanno, Y., Neural mechanisms of the peripheral and central auditory system in monkeys, *J. Acoust. Soc. Am.*, 34, 1396–1410, 1962.
35. Gross, C.G., Schiller, P.H., Wells, C., and Gerstein, G., Single-unit activity in temporal association cortex of the monkey, *J. Neurophysiol.*, 30, 833–843, 1967.
36. Celesia, G.G. and Puletti, F., Auditory cortical areas of man, *Neurology*, 19, 211–220, 1969.
37. Pandya, D.N. and Sanides, F., Architectonic parcellation of the temporal operculum in rhesus monkey and its projection pattern, *Z. Anat. Entwickl.-Gesch.*, 139, 127–161, 1973.
38. Mesulam, M.M. and Pandya, D.N., The projections of the medial geniculate complex within the Sylvian fissure of the rhesus monkey, *Brain Res.*, 60, 315–333, 1973.
39. Sanides, F., Representation in the cerebral cortex and its areal lamination pattern, in *The Structure and Function of Nervous Tissue*, Vol. 5, G.H. Bourne, Ed., Academic Press, New York, 1972.
40. Merzenich, M.M. and Brugge, J.F., Representation of the cochlear partition on the superior temporal plane of the macaque monkey, *Brain Res.*, 50, 275–296, 1973.
41. Sanides, F., Comparative neurology of the temporal lobe in primates including man with reference to speech, *Brain Lang.*, 2, 396–419, 1975.
42. Kaas, J.H. and Hackett, T.A., Subdivisions of auditory cortex and processing streams in primates, *Proc. Natl. Acad. Sci. USA*, 97, 11793–11799, 2000.
43. Imig, T.J., Ruggero, M.A., Kitzes, L.F., Javel, E., and Brugge, J.F., Organization of auditory cortex in the owl monkey (*Aotus trivirgatus*), *J. Comp. Neurol.*, 171, 111–128, 1977.
44. Morel, A. and Kaas, J.H., Subdivisions and connections of auditory cortex in owl monkeys, *J. Comp. Neurol.*, 318, 27–63, 1992.
45. Rauschecker, J.P., Tian, B., and Hauser, M., Processing of complex sounds in the macaque nonprimary auditory cortex, *Science*, 268, 111–114, 1995.
46. Hackett, T.A., Stepniewska, I., and Kaas, J.H., Subdivisions of auditory cortex and ipsilateral cortical connections of the parabelt auditory cortex in macaque monkeys, *J. Comp. Neurol.*, 394, 475–495, 1998.
47. Rauschecker, J.P., and Tian, B., Mechanisms and streams for processing of 'what' and 'where' in auditory cortex, *Proc. Natl. Acad. Sci. USA*, 97, 11800–11806, 2000.
48. Recanzone, G.H., Spatial processing in the auditory cortex of the macaque monkey, *Proc. Natl. Acad. Sci. USA*, 97, 11829–11835, 2000.
49. Morel, A., Garraghty, P.E., and Kaas, J.H., Tonotopic organization, architectonic fields, and connections of auditory cortex in macaque monkeys, *J. Comp. Neurol.*, 335, 437–459, 1993.
50. Morosan, P., Rademacher, J., Schleicher, A., Amunts, K., Schormann, T., and Zilles, Z., Human primary auditory cortex: cytoarchitectonic subdivisions and mapping into a spatial reference system, *NeuroImage*, 13, 684–701, 2001.
51. Pfingst, B.E. and O'Connor, T.A., Characteristics of neurons in auditory cortex of monkeys performing a simple auditory task, *J. Neurophysiol.*, 45, 16–34, 1981.
52. Kosaki, H., Hashikawa, T., He, J., and Jones, E.G., Tonotopic organization of auditory cortical fields delineated by parvalbumin immunoreactivity in macaque monkeys, *J. Comp. Neurol.*, 386, 304–316, 1997.

53. Galaburda, A.M. and Pandya, D.N., The intrinsic architectonic and connectional organization of the superior temporal region of the rhesus monkey, *J. Comp. Neurol.*, 221, 169–184, 1983.
54. Jones, E.G., Dell'Anna, M.E., Molinari, M., Rausell, E., and Hashikawa, T., Subdivisions of macaque monkey auditory cortex revealed by calcium-binding protein immunoreactivity, *J. Comp. Neurol.*, 362, 153–170, 1995.
55. Brodmann, K., Beitrage zur histologischen Localisation der Grosshirnrinde: Die cytoarchitektonische Cortexgliederung der Halbaffen (Lemuriden), *J. Psych. Neurol.*, 10, 287–334, 1908.
56. Mott, F.W., Schuster, E., and Halliburton, W.D., Cortical lamination and localization in the brain of the marmoset, *Proc. Roy. Soc. London Ser. B*, 82, 124–137, 1910.
57. Wollard, H.H., The cortical lamination of *Tarsius*, *J. Anat.*, 25, 86–105, 1925.
58. Lashley, K.S. and Clark, G., The cytoarchitecture of the cerebral cortex of Ateles: a critical examination of architectonic studies, *J. Comp. Neurol.*, 85, 223–306, 1946.
59. Von Bonin, G. and Bailey, P., Pattern of the cerebral isocortex, in *Primatologia: Handbook of Primatology*, H. Hofer, A.H. Schultz, and D. Starck, Eds., Karger, New York, 1961.
60. Vogt, C. and Vogt, O., Allgemeinere Ergebnisse unserer Hirnforschung, *J. Psych. Neurol.*, 24, 279–462, 1919.
61. Hopf, P., Die myeloarchitektonic des isocortex temporalis beim menschen, *J. Hirnforsch.*, 1, 208–279, 1954.
62. Hackett, T.A., Preuss, T.M., and Kaas, J.H., Architectonic identification of the core region in auditory cortex of macaques, chimpanzees, and humans, *J. Comp. Neurol.*, 441, 197–222, 2001.
63. Rivier, F. and Clarke, S., Cytochrome oxidase, acetylcholinesterase, and NADPH-diaphorase staining in human supratemporal and insular cortex: evidence for multiple auditory areas, *NeuroImage*, 6, 288–304, 1997.
64. Nakahara, H., Yamada, S., Mizutani, T., and Murayama, S., Identification of the primary auditory field in archival human brain tissue via immunocytochemistry of parvalbumin, *Neurosci. Lett.*, 286, 29–32, 2000.
65. Hurst, E.M., Some cortical association systems related to auditory functions, *J. Comp. Neurol.*, 112, 103–119, 1959.
66. Walker, A.E., The thalamus of the chimpanzee. IV. Thalamic projections to the cerebral cortex, *J. Anat.*, 73, 37–93, 1938.
67. McCulloch, W.S., The functional organization of the cerebral cortex, *Physiol. Rev.*, 24, 390–407, 1944.
68. Ward, A.A. Jr., Peden, J.K., and Sugar, O., Cortico-cortical connections in the monkey with special reference to area 6, *J. Neurophysiol.*, 9, 455–461, 1946.
69. Sugar, O., French, J.D., and Chusid, J.G., Corticortical connections of the superior surface of the temporal operculum in the monkey (*Macaca mulatta*), *J. Neurophysiol.*, 11, 175–184, 1948.
70. Fitzpatrick, K.A., and Imig, T.J., Auditory cortico-cortical connections in the owl monkey, *J. Comp. Neurol.*, 177, 537–556, 1980.
71. Aitkin, L.J., Kudo, M., and Irvine, D.R.F., Connections of the primary auditory cortex in the common marmoset (*Callithrix jacchus jacchus*), *J. Comp. Neurol.*, 269, 235–248, 1988.
72. Luethke, L.E., Krubitzer, L.A., and Kaas, J.H., Connections of primary auditory cortex in the New World monkey, *Saguinus*, *J. Comp. Neurol.*, 285, 487–513, 1989.

73. Pandya, D.N. and Rosene, D.L., Laminar termination patterns of thalamic, callosal, and association afferents in the primary auditory area of the rhesus monkey, *Exp. Neurol.*, 119, 220–234, 1993.
74. Burton, H. and Jones, E.G., The posterior thalamic region and its cortical projection in new world and old world monkeys, *J. Comp. Neurol.*, 168, 249–302, 1976.
75. Fitzpatrick, K.A. and Imig, T.J., Projections of auditory cortex upon the thalamus and midbrain in the owl monkey, *J. Comp. Neurol.*, 177, 537–556, 1978.
76. Pandya, D.N., Rosene, D.L., and Doolittle, A.M., Corticothalmic connections of auditory-related areas of the temporal lobe in the rhesus monkey, *J. Comp. Neurol.*, 345, 447–471, 1994.
77. Molinari, M., Dell'Anna, M.E., Rausell, E., Leggio, M.G., Hashikawa, T., and Jones, E.G., Auditory thalamocortical pathways defined in monkeys by calcium-binding protein immunoreactivity, *J. Comp. Neurol.*, 362, 171–194, 1995.
78. Hashikawa, T., Molinari, E., Rausell, E., and Jones, E.G., Patchy and laminar terminations of medial geniculate axons in monkey auditory cortex, *J. Comp. Neurol.*, 362, 195–208, 1995.
79. Rose, J.E., The cellular structure of the auditory region of the cat, *J. Comp. Neurol.*, 91, 409–440, 1949.
80. Rose, J.E. and Woolsey, C.N., The relation of thalamic connections, cellular structure, and evocable electrical activity in the auditory region of the cat, *J. Comp. Neurol.*, 91, 441–466, 1949.
81. Von Bonin, G. and Bailey, P., *The Neocortex of Macaca mulatta*, University of Illinois Press, Urbana, 1947.
82. Forbes, B.F. and Moskowitz, N., Projections of auditory responsive cortex in the squirrel monkey, *Brain Res.*, 67, 239–254, 1974.
83. Cipolloni, P.B. and Pandya, D.N., Connectional analysis of the ipsilateral and contralateral afferent neurons of the superior temporal region in the rhesus monkey, *J. Comp. Neurol.*, 281, 567–585, 1989.
84. Romanski, L.M., Bates, J.F., and Goldman-Rakic, P.S., Auditory belt and parabelt projections to the prefrontal cortex in the rhesus monkey, *J. Comp. Neurol.*, 403, 141–157, 1999.
85. Romanski, L.M., Tian, B., Fritz, J., Mishkin, M., Goldman-Rakic, P.S., and Rauschecker, J.P., Dual streams of auditory afferents target multiple domains in the primate prefrontal cortex, *Nature Neurosci.*, 2, 1131–1136, 1999.
86. Romanski, L.M., Anatomy and physiology of auditory–prefontal interactions in nonhuman primates, in *Primate Audition: Ethology and Neurobiology*, A.A. Ghazanfar, Ed., CRC Press, Boca Raton, FL, 2003, chap. 14.
87. Rauschecker, J.P., Tian, B., Pons, T., and Mishkin, M., Serial and parallel processing in rhesus monkey auditory cortex, *J. Comp. Neurol.*, 382, 89–103, 1997.
88. Tian, B., Reser, D., Durham, A., Kustov, A., and Rauschecker, J.P., Functional specialization in rhesus monkey auditory cortex, *Science*, 292, 290–293, 2001.
89. Hackett, T.A., Stepniewska, I., and Kaas, J.H., Thalamocortical connections of the parabelt auditory cortex in macaque monkeys, *J. Comp. Neurol.*, 400, 271–286, 1998.
90. Hackett, T.A., Stepniewska, I., and Kaas, J.H., Prefrontal connections of the auditory parabelt cortex in macaque monkeys, *Brain Res.*, 817, 45–58, 1999.
91. Kaas, J.H., Hackett, T.A., and Tramo, M.J., Auditory processing in primate cerebral cortex, *Curr. Opin. Neurobiol.*, 9, 164–170, 1999.
92. Rauschecker, J.P., Parallel processing in the auditory cortex of primates, *Audiol. Neurootol.*, 3, 86–103, 1998.

93. Bailey, P., Von Bonin, G., and McCulloch, W.S., *The Isocortex of the Chimpanzee*, University of Illinois Press, Urbana, 1950.
94. Bailey, P. and Von Bonin, G., *The Isocortex of Man*, University of Illinois Press, Urbana, 1951.
95. Von Bonin, G., *The Cortex of Galago: Its Relations to the Pattern of the Primate Cortex*, University of Illinois Press, Urbana, 1945.
96. Braak, H., The pigment architecture of the human temporal lobe, *Anat. Embryol.*, 154, 43–240, 1978.
97. Brugge, J.F., Auditory cortical areas in primates, in *Cortical Sensory Organization*, Vol. 3, C.N. Woolsey, Ed., Humana Press, Clifton, NJ, 1982, pp. 59–70.
98. Galaburda, A.M. and Sanides, F., Cytoarchitectonic organization of the human auditory cortex, *J. Comp. Neurol.*, 190, 597–610, 1980.
99. Ong, W.Y. and Garey, L.J., Distribution of GABA and neuropeptides in the human cerebral cortex. A light and electron microscopic study, *Anat. Embryol. (Berlin)*, 183, 397–413, 1991.
100. Peden, J.K. and Von Bonin, G., The neocortex of hapale, *J. Comp. Neurol.*, 86, 37–64, 1947.
101. Rademacher, J., Caviness, V., Steinmetz, H., and Galaburda, A., Topographical variation of the human primary cortices: implications for neuroimaging, brain mapping and neurobiology, *Cereb. Cortex*, 3, 313–329, 1993.
102. Ravizza, R. and Diamond, I.T., Role of auditory cortex in sound localization: a comparative ablation study of hedgehog and bush-baby, *Fed. Proc.*, 33, 1915–1919, 1974.
103. Sanides, F. and Krishnamurti, A., Cytoarchitectonic subdivisions of sensorimotor and prefrontal regions and of bordering insular and limbic fields in slow loris (*Nycticebus coucang coucang*), *J. Hirnforsch.*, 9, 225–252, 1967.
104. Smith, G.E., A new topographical survey of the human cerebral cortex, being an account of the distribution of the anatomically distinct cortical areas and their relationship to the cerebral sulci, *J. Anat. Physiol.*, 41, 237–254, 1907.
105. Smith, G.E. and Moskowitz, N., Ultrastructure of layer IV of the primary auditory cortex of the squirrel monkey, *Neuroscience*, 4, 349–359, 1979.
106. Zilles, K., Rehkamper, G., Stephan, H., and Schleicher, A., A quantitative approach to cytoarchitectonics. IV. The areal pattern of the cortex of *Galago demidovii*, *Anat. Embryol.*, 157, 81–103, 1979.
107. Zilles, K., Rehkamper, G., and Schleicher, A., A quantitative approach to cytoarchitectonics. IV. The areal pattern of the cortex of *Microcebus murinus*, *Anat. Embryol.*, 157, 269–289, 1979.

Appendix A

Proposed Homologous Auditory Cortical Regions in Non-human Primates and Humans

For each study cited, the species is listed, along with the designation (name) given to the region and the method(s) used to identify the region.

Study	Species	Core	Medial Belt	Lateral Belt	Parabelt	Method[a]
Aitkin et al. (1988)[71]	Marmoset	AI	AI	ND	ND	E, T, C
Bailey et al. (1950)[93]	Chimpanzee	TC	TC	TB	TA	C
Bailey and Von Bonin (1951)[94]	Human	TC, Koniosus supratemporalis	TB, 42	TB, 42	TA, 22	C
Beck (1928, 1929)[15,16]	Human, chimpanzee	Ttrli/e	Tsm	TtrlI	ND	M
Von Bonin (1938)[23]	Cebus	Supratemporalis granulosa	ND	TB	TA	C
Von Bonin (1945)[95]	Galago	41	ND	ND	NDs	C
Von Bonin and Bailey (1947)[81]	Macaque	TC	TB[b]	TB[b]	TA	C
Braak (1978)[96]	Human	Temporalis granulosa	Temporalis progranulosa	Temporalis paragranulosa	Temoralis magnopyramidalis	L, C
Brodmann (1905)[10]	Cercopithecus	22[c]	22[c]	22[c]	22[c]	C
Brodmann (1908)[55]	Lemur	22[c]	22[c]	22[c]	22[c]	C
Brodmann (1908, 1909)[11,12]	Human	41	52	42	22	C
Brugge (1982)[97]	Galago crassicaudatus	AI	CM	PL	ND	E (?)
Burton and Jones (1976)[74]	Squirrel and macaque monkeys	AI	Pi	T1, Pa	T2	C, T
Campbell (1905)[9]	Human, chimpanzee, orangutan	Audito-sensory	ND	ND	Audito-psychic	C, M
Von Economo and Koskinas (1925)[17]	Human	TC, supratemporalis granulosa	TB, TD	TB	TA$_1$, TA$_2$ (?)	C
Von Economo and Horn (1930)[18]	Human	TC, supratemporalis granulosa	TD, (TA$_2$a)	TBma	TA$_1$, TA$_2$ (?)	C
Flechsig (1876)[7]	Human	7	14 (?)	14 (?)	14	M
Flechsig (1920)[8]	Human	10	18 (?)	19 (?)	19 (?)	M

Appendix A

Reference	Species					
Fitzpatrick and Imig (1980)[70]	Owl monkey	AI, R	CM	AL, PL	ND	C, E, T
Forbes and Moskowitz (1974)[82]	Squirrel monkey	Primary auditory cortex	ND (STP)	ND (STG)	ND (STG)	C, Lx
Galaburda and Pandya (1983)[53]	Macaque	KA	ProA	PaAlt	Ts2, Ts3 (?)	C, T
Galaburda and Sanides (1980)[58]	Human	KAm, KAlt	ProA, PaAc/d	PaAi	PaAe	C, M
Hackett et al. (1998)[46]	Macaque	Core (AI, R, RT)	RM, RTM, CM	RTL, AL, ML, CL	RPB, CPB	C, M, A, CO, PV, T
Hackett et al. (2001)[62]	Human, chimpanzee, macaque	Core	Medial belt	Lateral belt	ND	C, M, A
Hopf (1954)[61]	Human	Ttr1	Tsep	Ttr2	Tpart, Tmag,	M
Imig et al. (1977)[43]	Owl monkey	AI, R	CM	AL, PL	ND	C, E
Jones et al. (1995)[54]	Macaque	AI, RL	A-m, P-m	L, A-m, P-m	ND	C, PV, T
Kosaki et al. (1997)[52]	Macaque	AI, R	M, A-m, P-m	L	ND	C, PV, E
Lashley and Clark (1946)[58]	Spider monkey, macaque	ND	ND	ND	ND	C, M
Luethke et al. (1989)[72]	Marmoset	AI, R	Lateral	Medial	ND	M, T, E
Mauss (1908)[13]	Macaque, Cercopithecus	22ᶜ	22ᶜ	22ᶜ	22ᶜ	M
Mauss (1911)[14]	Orangutan, gibbon	40, temporalis transversa sive profunda				M
Merzenich and Brugge (1973)[40]	Macaque	AI	a, CM	L	ND	C, E
Morel and Kaas (1992)[44]	Owl monkey	AI, R, RT	CM, RM, MRT	PL, AL, LRT	Parabelt	C, M, A, CO, T
Morel et al. (1993)[49]	Macaque	AI, R, RT	CM, RM	PL, AL	Parabelt	C, M, A, CO, T

(continued)

Study	Species	Core	Medial Belt	Lateral Belt	Parabelt	Method[a]
Morosan et al. (2001)[50]	Human	Te1.0	TI1	Te2	Te3	C
Mott et al. (1910)[56]	Marmoset	Temporal area, atypical	?	Temporal area, typical	Temporal area, typical	C
Nakahara et al. (2000)[64]	Human	Zone 1	Zone 2	Zone 2	ND	PV, C
Ong and Garey (1990)[99]	Human	41	ND	42	22	C, G, M
Pandya et al. (1969)[31]	Macaque	TC	ND	TB	TA	C, Lx
Pandya and Sanides (1973)[37]	Macaque	Kam, KaIt	ProA	PaAlt	PaAlt, Ts3	C, M
Peden and Von Bonin (1947)[100]	Hapale (marmoset)	TC	ND	TA	TA	C, M
Rademacher et al. (1993)[101]	Human	41	ND	ND	ND	C
Ravizza and Diamond (1974)[102]	Galago senegalensis	AI	ND	EP	EP	C
Rivier and Clarke (1997)[63]	Human	AI	MA	LA, PA	PA, STA	A, CO, NADPH
Sanides and Krishnamurti (1967)[103]	Slow loris	Ka, koniocortex	Prokoniocortex (AII)	Parakoniocortex	ND	C
Smith (1907)[104]	Human	No. 27	No. 21, postcentral insular	No. 26, temporalis superior	No. 26, temporalis superior	G
Smith and Moskowitz (1979)[105]	Squirrel monkey	Primary	Primary	ND	ND	C
Vogt and Vogt (1919)[60]	Human, macaque	41, temporalis transversa interna	ND	42, temporalis transversa externa	22aB	M, C

	Macaque	Koniocortex				
Walker (1937)[22]			ND	22	22	C, M
Wollard (1925)[57]	Tarsius	22[c]	22[c]	22[c]	22[c]	C
Zilles et al. (1979)[106]	Galago demidovii	Te1	Te2.1	Te2.2	(?)	C
Zilles et al. (1979)[107]	Microcebus murinus	Te1	Te2.1	Te2.2	(?)	C

[a] A, acetylcholinesterase; C, cytoarchitecture; CO, cytochrome oxidase; E, electrophysiology; G, Golgi; L, lipofuscin (pigment architecture); Lx, tissue lesions; M, myeloarchitecture; NADPH, NADPH-diaphorase; PV, parvalbumin.

[b] Not definitive as a belt surrounding TC, but found in patches.

[c] Area 22 was used to designate all cortex on lower bank of Sylvian fissure and surface of the superior temporal gyrus. No regional variations are identified.

12 Auditory Communication and Central Auditory Mechanisms in the Squirrel Monkey: Past and Present

John D. Newman

CONTENTS

I. Introduction 228
II. Vocal Communication in the Squirrel Monkey 228
   A. Introduction 228
   B. Review of Squirrel Monkey Vocal Behavior 228
      1. Isolation Peep 229
      2. Chuck Call 230
      3. Play Peep 231
      4. Twitter 231
      5. Caregiver Call 231
      6. Purr and Err 231
III. Central Auditory Mechanisms 232
   A. Introduction 232
   B. Early Studies of Squirrel Monkey Auditory Cortex 232
   C. Psychoacoustics and Central Auditory Mechanisms 233
   D. Neuroanatomy and Neurophysiology of the Auditory Cortex 233
   E. Auditory Cortical Mechanisms Relevant to Vocal Communication: A Historical Perspective 233
      1. The NIH Studies 234
      2. Studies in Israel 237
      3. Studies in Germany 237
   F. Studies of Auditory Structures below the Auditory Cortex 239
      1. Medial Geniculate Nucleus 239

        2. Inferior Colliculus 239
    G. Auditory Responses within the Frontal Lobe....................................240
 IV. Conclusions...240
References..241

I. INTRODUCTION

The squirrel monkey (*Saimiri*) is a small, cat-sized primate native to tropical forests of South and Central America (see Baldwin and Baldwin[1]). It is one of the more common New World primates over much of its range, is highly social, adapts well to captivity, and has been the subject of research on its vocal behavior and associated auditory mechanisms for nearly 40 years. How this primate came to be the focus of so much research on auditory communication is something of a historical accident, but its selection has proven fortuitous. This chapter will present an overview of our current understanding of its vocal behavior, of its central auditory mechanisms from the perspective of their relevance to vocal communication, and of its general auditory mechanisms.

More is known about the central mechanisms mediating vocal communication in the squirrel monkey than any other New World primate. General overviews of the neuroethology of primate vocal communication[2] and specifically for the squirrel monkey[3,4] can be found elsewhere. Behavioral analyses of primate vocal communication can be found in Seyfarth and Cheney.[5]

II. VOCAL COMMUNICATION IN THE SQUIRREL MONKEY

A. INTRODUCTION

Squirrel monkeys are vocal animals. This may be an adaptation in primates that forage for food by moving through the forest canopy and that live in large groups. Given the low visibility in forest environments, both are situations that make it desirable for troop members to maintain contact with each other using vocalizations.

B. REVIEW OF SQUIRREL MONKEY VOCAL BEHAVIOR

A great deal has been published about the vocal behavior of squirrel monkeys, although much of this is based on captive groups.[6-9] Early on in this line of research, a question of considerable significance was whether or not squirrel monkeys (or any non-human primate) learned to produce adult forms of their species-specific vocalizations through individual experience (as is the case with song in songbirds). Isolation rearing failed to demonstrate a significant difference in vocal behavior compared to normally reared infants.[10] Deafening infants[10] or adult[11] squirrel monkeys yielded similar results. Evidence for a major role of genetic programming in producing one vocalization, the isolation peep, came from an analysis of hybrid calls.[12] On the other hand, some evidence that individual experience is important for responding appropriately came from a study of alarm peep responses during development.[13] Another study, however, using infants raised only with a surrogate, indicated that infants with this restricted rearing showed the appropriate response to alarm peep playbacks.[14] Two subtypes of vocalization, the isolation peep and the chuck, have been the subjects of the most intensive study. For both, strong evidence indicates that individual experience plays a role in determining the selectivity of responders.

1. Isolation Peep

In the original description of squirrel monkey vocalizations,[6] the *isolation peep* was described as a high-pitched (9 to 12 kHz), stereotyped, loud call of long duration, and emitted when an individual lost contact or was separated by a long distance from the group. It was also reported that group members, upon hearing this sound, would answer in the same manner, resulting in an exchange of isolation peeps. A subsequent paper[15] reported that the isolation peep acoustic structure varied depending on whether the vocalizer was the *Gothic-arch* or *Roman-arch* variety of squirrel monkey. The call of the Roman-arch variety ascends in frequency at the end of the call, while that of the Gothic type descends in frequency. The Gothic-arch/Roman-arch dichotomy was originally formulated by MacLean,[16] who, in addition to describing the facial feature differences that are diagnostic, also described characteristic differences in genital display behavior in the males of the two types. Subsequently, Hershkovitz[17] formalized this Roman-arch/Gothic-arch distinction by creating the taxa *Saimiri boliviensis boliviensis*, *S. boliviensis peruviensis* (the Roman-arch forms), *S. sciureus* (with four subspecies), *S. ustus*, and *S. oerstedi* (the Gothic-arch forms). The acoustic differences between the Roman and Gothic isolation peeps are attended to by listeners.[18] Adults of the Roman-arch and Gothic-arch subtypes responded (as measured by increased activity, decreased huddling, and approaches to the speaker) only when recordings of isolation peeps from infants of their own subtype were played back. Whether adult Roman and Gothic peeps likewise produce differential responding has yet to be studied in detail, but results of a preliminary study presented here suggest that this is the case (see Table 12.1).

Listeners also attend to other isolation peep acoustic attributes, besides those that differentiate Roman from Gothic calls. Individual differences in the acoustic structure of isolation peeps have been described for both adults[19] and infants[20] of both Roman and Gothic squirrel monkeys. Mothers can discriminate the isolation peeps of their infants,[21,22] thereby providing an efficient means of reunion between mothers and infants. A functional role for individual differences in adult calls has yet to be demonstrated, but adult squirrel monkeys can readily discriminate the acoustic differences that characterize individuals.[23] Squirrel monkeys also appear to be able to judge the distance between themselves and distant vocalizing monkeys and vary their isolation peep structure accordingly. By systematically varying the distance between a separated infant (held captive in a wire-mesh cage) and its natal group, it was found that both the infant and the group members answering the infant increased the duration of their isolation peeps with increasing distance between infant and group.[24]

TABLE 12.1
Isolation Peep Answers to Isolation Peep Playbacks

	Subject				
Stimulus	776 Roman	302 Gothic	303 Gothic	629 Gothic	986 Gothic
Gothic 1	0	3	11	0	20
Gothic 2	0	0	9	0	23
Gothic 3	0	51	25	18	15
Gothic 4	0	32	8	5	15
Roman 1	38	21	4	8	10

Note: Subjects were alone in a wire-mesh cage in a sound-attenuating room. Playback stimuli were presented through a loudspeaker 1 m from the subject and presented at approximately 60 dB SPL. Subjects were left alone for 15 min prior to testing, during which time spontaneous isolation peeps were given. Playbacks were initiated when spontaneous calling had ended. For each trial, one repetition of each of the five stimuli (in the same order as in the table) was presented, separated by 30 sec. A total of 16 trials were conducted for each subject.

2. Chuck Call

Winter et al.[6] originally described the *chuck* (p. 368) as follows:

> A twit element or a short twitter containing only one part of the up- or down-ward sweep of the fundamental frequency, but with a steeper descending slope, is designated as a chuck sound. Single and double forms have been registered. Unlike most other calls, a chuck is not likely to be repeated. Instead, it tends to appear with churrs, arrs, and less often with trills. Huddling animals, when becoming restless or trying to change the huddling order, are prone to use this call. It can, however, be registered during other situations. This call presumably has a significance in maintaining a given distance between the animals of a group.

Further studies have shown that the chuck is much more interesting than this initial description would suggest. Careful analysis of the situations in which chucks were made demonstrated that this call is produced mainly by adult females and is given preferentially during affiliative interactions (huddling or sitting near) between female "friends" or as chuck exchanges between these same individuals.[25,26] The calls that make up these exchanges (typically one chuck from each of two participants) often occur within 2 seconds or less.[27] We described the chuck as consisting of a sequence of acoustic elements: *flag*, *mast*, and *cackle*. Detailed measurement of these elements from the chucks of identified vocalizers indicated a high degree of individuality in chuck structure.[28] That the monkeys used the individual differences in acoustic structure was shown in a series of playback experiments. Females make chuck answers more frequently to the chucks of familiar group members than to the chucks of unfamiliar individuals.[29,30] Analyses of the acoustic structure of the chucks from a given female suggest that the calls with which she initiates exchanges (dubbed her *question chucks*) and the calls that she uses to respond to another female (her *answer chucks*) differ in small but statistically significant values. Most notably, the peak frequency of the flag and mast differ, a finding that is supported by the fact that listeners preferentially respond to question chucks over answer chucks from a given familiar female.[31–33] Some evidence exists for similarity in the acoustic structure of the chucks of adult females and their daughters, and that the adults respond more to the chucks of their daughters (even when separated for 1 year or more) than to the chucks of strangers.[34] Of particular interest is evidence that the process by which a female comes to distinguish between the chucks of familiar group members and the chucks of strangers is a gradual one and is not clearly established until a female is more than 2 years old.[35]

3. Play Peep

Squirrel monkeys have a prominent vocalization produced during play bouts known as the *play peep*, which is characterized by a rapid and sudden increase in frequency. It occurs mainly during play bouts between juveniles but also during genital displays by young animals.[6] Loud vocalizations during play are unusual in other species, not surprising given that the playing animals are generally oblivious to their surroundings and the vocalizations conceivably could attract predators. However, squirrel monkey play calls may be monitored by the parents of the play partners, who modify their behavior both by increasing their vigilance for possible predators and also by standing ready to intervene when the playing becomes too rough.[36]

4. Twitter

Twitters occur during reunions between separated individuals (as a greeting) and during feeding.[6] Twitters (along with chucks) also are the most frequent vocalization during the period after the lights are turned off in the holding rooms of captive squirrel monkey groups and possibly function to reinforce social bonds in the absence of visual information.[37] The structure of twitter calls is rather complex, consisting of sequences of several acoustically distinct syllables.[38] For a given individual, the particular combination of syllables in its twitters varies, depending on whether it was actively feeding or visually separated from a familiar companion.

5. Caregiver Call

Caregiver calls, a vocalization not described in the original paper of Winter et al.,[6] are produced by adult females during interactions with their, or another female's, infant. The calls have a prominent harmonic structure but vary in acoustic details according to the specific context (dorsal contact, nursing, retrieval, or inspection).[39] The calls appear to be used to gain an infant's attention. During periods when infants are old enough to move about independently, mothers and infants engage in bouts of antiphonal calling, the mothers making caregiver calls and the infants making peeps.[40]

6. Purr and Err

Winter et al.[6] described *errs* as noisy, pulsatile calls given during periods of mild annoyance and *purrs* as longer pulsatile calls given by both mother and infant during nursing bouts. We observed an increase in errs and purrs by adult females in the presence of an adult male, accompanied by the females approaching the male and soliciting mounting.[41] We concluded that these vocalizations were used to signal the receptivity of the females and possibly to reduce aggression by a male during the close approach by the female.

III. CENTRAL AUDITORY MECHANISMS

A. INTRODUCTION

Knowledge of the central auditory mechanisms in squirrel monkeys is of particular interest and importance, given the well-studied and elaborate vocal communication system in these primates. However, traditional methods for studying central auditory mechanisms have failed to reveal anything remarkable about the squirrel monkey auditory system that might give clues as to an elaborate central processing system comparable to the elaborate system of auditory communication sounds. This is certainly true of physiological properties of the cochlea[42] and the auditory nerve.[43-49]

B. EARLY STUDIES OF SQUIRREL MONKEY AUDITORY CORTEX

It is a historical anomaly that the temporal lobe auditory cortex (AC) of the squirrel monkey has received less detailed study than another New World monkey, the owl monkey (*Aotus*), with a less elaborate vocal repertoire and with no attempts to study its auditory cortex using species-specific vocalizations. However, it seems probable that the basic organization of the owl monkey auditory cortex closely resembles that of the squirrel monkey. Systematic microelectrode mapping of the superior temporal gyrus in the anesthetized owl monkey found the presence of multiple frequency representations, whose boundaries closely matched anatomical subdivisions.[50] Five subdivisions were recognized: the primary auditory cortex (AI), a rostral field, and a surrounding belt zone of three fields containing units that are less responsive to artificial sounds. An additional area of auditory responsiveness, the rostromedial field, is located between the rostral field and the insula. The insula itself also contains units responsive to sounds. Projections from owl monkey AI and the rostral field have been described, both within the cortex[51] and to the thalamus and midbrain.[52]

The ultrastructure of layer IV in the temporal lobe cortex of the squirrel monkey has been described and identified as primary auditory cortex on the basis of columns of granule cells situated at a right angle to the pial surface.[53] Close soma-to-soma connections of the granule cells suggested a kind of electrotonic coupling, known in other neural systems to be a specialization for synchronous firing of groups of neurons. No functional correlate of this anatomical feature has ever been described. Evoked potential recordings in squirrel monkeys have identified the auditory-responsive cortex as the superior temporal gyrus, the overlying frontoparietal operculum, and the insula.[54] Cells within the supratemporal plane and the dorsolateral superior temporal gyrus project to auditory nuclei (medial geniculate nucleus and inferior colliculus).[55] A tracer study mapped the cortico-cortical projections from the superior temporal gyrus and determined that the superior temporal plane projects only to adjacent superior temporal gyrus and to the contralateral homologous areas.[56] The superior temporal gyrus projects to the insula, the frontoparietal operculum, and parts of the frontal lobe, in addition to other areas within the superior temporal gyrus.[56]

C. Psychoacoustics and Central Auditory Mechanisms

The basic psychoacoustic parameters for squirrel monkeys have been worked out. Squirrel monkeys can detect pure-tone stimuli over a wide frequency range, from around 100 Hz to above 40 kHz.[57] Frequency discrimination is maximum in the range of highest auditory sensitivity (4–8 kHz), approximating 40 Hz, but deteriorates to around 80 Hz at frequencies of 500 Hz and below.[58] Brainstem auditory-evoked potentials have been studied in squirrel monkeys (e.g., see Pineda et al.[59]) but have not been employed to study species-specific vocalizations, as they have been in chimpanzees.[60]

D. Neuroanatomy and Neurophysiology of the Auditory Cortex

Within the superior temporal gyrus (STG), several cytoarchitectural fields have been identified and their thalamic projections determined. Within the medial geniculate nucleus (MGN) — the thalamic component of the classical central auditory pathway — are a ventral division that projects to field AI in the STG, an anterodorsal division that projects to two fields on the STG, and a posterodorsal component that projects to areas lying anterior and medial to AI.[61] The magnocellular MGN sends a diffuse projection to all auditory fields within the STG. The STG contains first and second temporal fields (T1 and T2), a rostrolateral auditory field (RL), and a third temporal area (T3). Within the medial supratemporal plane is a parainsular field (Pi). The cytoarchitectural anatomy of the STG and adjacent insula have also been described, as well as details on laminar differences between these areas.[62]

Until recently, the functional organization of the squirrel monkey auditory cortex had not been studied in any detail; however, the distribution of characteristic frequencies (CFs) on the exposed STG has recently been mapped.[63] A systematic shift from low CFs in the caudoventral quadrant to high frequencies in the rostrodorsal quadrant was found.

E. Auditory Cortical Mechanisms Relevant to Vocal Communication: A Historical Perspective

The squirrel monkey has been the subject of studies devoted to central auditory mechanisms since the 1960s. That is about the same time frame for research devoted to auditory communication in this primate. The first paper on central auditory mechanisms with relevance to auditory communication was not published until 1970; investigations devoted to this topic and those devoted to more traditional auditory neuroscience questions continued with little interaction over the ensuing years. As it happens, much of the work on central auditory mechanisms relating to auditory communication has come from the same group of investigators. The progenitor of the research on this topic should properly be identified as Prof. Detlev Ploog, who conducted the original studies on auditory communication. A researcher in the Ploog group, Peter Winter, came to the National Institutes of Health (NIH) in the late 1960s to work with Phillip Nelson, who was involved in early studies of the mechanisms of inhibition in central auditory pathways. Winter was joined initially by Harris Funkenstein, a clinical neurologist, and, from 1969 to 1970, by Zvi Wollberg and myself. Winter returned to Germany, only to be killed in 1972 in a tragic skiing accident. Wollberg returned to Israel and continued working on squirrel monkey central auditory mechanisms in his own laboratory, while I continued this line of research at the NIH, together with David Symmes. More recently, Uwe Jürgens, another investigator from the Ploog group in Munich, became head of the Department of Neurobiology at the German Primate Center in Göttingen and pursued his own research on auditory and vocal production mechanisms in the squirrel monkey. From 1974 to 1975, I conducted auditory physiology research at the Munich lab while training a neuroanatomist, Peter Müller-Preuss, in auditory neurophysiology techniques. After my departure, Müller-Preuss continued research in auditory neurophysiology at the Munich lab. What follows in the remainder of this chapter is an overview of the research that I have been involved with, along with a review of the studies of central auditory pathways and auditory mechanisms in the squirrel monkey performed by other researchers.

1. The NIH Studies

The initial study[64] was pioneering in at least two respects. First, it involved recording single units in the auditory cortex of an awake primate, a novelty at that time as previous studies were done in anesthetized animals. Second, in addition to the standard tone bursts, white noise bursts, and clicks, tape-recorded species-specific vocalizations were used as acoustic stimuli. While more units responded to pure tones (71%) than to vocalizations (41%), this was the first report of units responding to vocalizations but not to any artificial acoustic stimulus (albeit only 2 to 3% of those tested). A subsequent paper[65] reported the same findings, as well as the first report of cortical units responding to discrete components of more complex vocalizations. The paper concluded (p. 314) "…that cortical cells in the squirrel monkey are specialized for the detection of certain features of the acoustic environment, features that are relevant to the recognition of monkey calls and possibly of other biologically important sounds as well." Two papers by Funkenstein and Winter, published after Winter's death, summarized the studies that these two investigators conducted at the NIH. One paper[66] reported the responses of units in the superior temporal gyrus of awake squirrel monkeys to all acoustic stimuli and presented the basic methods used to prepare the monkey for testing, for stimulus presentation, data analysis, and histology. Unlike previous studies of auditory cortical neurons in anesthetized animals, the responses of neurons in the awake monkey — even to pure tones — were quite complex. In addition to tone bursts of a single frequency, frequency-modulated (FM) tones were presented. About 10% of the units tested responded to FM stimuli but not to steady tones. Units that exhibited a combination of excitatory and inhibitory responses to tones of different frequencies also exhibited more complex responses to FM stimuli. This paper also reported the first detailed histological examination of the extent of auditory cortex, based on single-unit responses, in the squirrel monkey. No units in the overlying parietal lobe responded to sounds, but as soon as the microelectrode penetrated the (buried) supratemporal plane acoustic responses were found. A few acoustically responsive units in the insula were found. Some evidence for tonotopic organization was discerned, with a high-to-low frequency response progressively moving from the rear to the rostral portion of the area examined on the superior temporal gyrus and underlying the supratemporal plane. The paper concludes (p. 487): "If the auditory cortex is specialized in any way for species-specific communication, then that specificity will have to be sought at some level other than that of standard acoustic stimuli." A second paper by Winter and Funkenstein[67] examined the effects of species-specific vocalizations on the discharge patterns of auditory cortical cells in awake squirrel monkeys. Twenty-four calls, judged to be typical and covering the entire vocal repertoire, were used. In that paper, the acoustic structure of the stimuli was illustrated by sound spectrograms, which showed the time-varying frequency changes and other acoustic details of the vocal stimuli. In that initial study, calls were presented three times in succession (each separated by 3 sec); the small number of trials was necessitated by the desire to present as many different call types to each unit as possible. In some cases, when a limited number of calls were found to be particularly effective, a larger number of stimulus repetitions were employed. This made it possible to discern the precise timing of a unit's response relative to the time-course of the vocalization and suggested that the neurons were sensitive to limited parts of a call. This study reported that 56% of the cells responding to vocalizations did so to no more than two acoustically similar calls, suggesting that certain common acoustic features were responsible for activating the unit.

Zvi Wollberg and I modified some of the methods used by Winter and Funkenstein, chiefly by using a different sedative agent (halothane–air mix instead of pentobarbital) prior to harnessing the monkey subject to the recording apparatus. This resulted in an alert monkey shortly after placing it in the restraining chair, and experiments could start soon thereafter. Furthermore, a stimulus tape recording was used, instead of the "endless loop" arrangement used earlier. A trigger pulse on a second channel of the tape 200 msec prior to the onset of the vocal stimulus served to trigger the neurophysiology data collection apparatus. Each repetition of a given call was separated from the preceding one by 4.5 sec. To determine the response selectivity of individual units, 12 different vocalizations, representing 6 major call groups, were used. In addition, each vocal stimulus was repeated 24 times, which generally facilitated deciding whether a unit was responding to a particular stimulus or not. Eight of the vocalizations were taken from the same stimulus tape recordings used by Winter and Funkenstein, while the rest were from recordings made from monkeys in our colony prior to the start of the experiments. The results of these studies[68,69] can be summarized as follows. Response selectivity was generally related to spontaneous firing rate (i.e., the higher the spontaneous rate, the more stimulus calls elicited a response); different units showed a wide range of response patterns to the same set of vocalizations; in most cases, the frequency range over which a cell responded to steady tone bursts did not serve to readily explain the responsiveness of that cell to vocalizations; and the isolation peep was the most effective stimulus. In addition, a larger percentage of units responding to vocalizations and a larger percentage responding to most call types were found in these studies compared with the Winter and Funkenstein results. These latter two findings were interpreted as being due to the greater number of presentations/stimulus, making it easier to determine that a particular vocalization was effective. This work also clearly confirmed that auditory cortex units typically were responding to discrete parts of a vocal stimulus, because electronic deletion of limited parts of a call caused a response to diminish or disappear entirely; conversely, presenting only a limited part of a call continued to produce a response that was equivalent in magnitude to the response to the whole call.[68] A follow-up study[70] addressed the issue of response selectivity to acoustic variants of the isolation peep. Although 76% of the tested units responded to all or most of the stimuli, the remaining 24% showed selective responsiveness to different subgroups of isolation peeps, leading to the conclusion that the isolation peeps were not being categorized on the basis of simple acoustic features. A behavioral study using the same stimuli demonstrated that squirrel monkeys could discriminate between isolation peeps based on small acoustic differences.[23]

Subsequent experiments at the NIH were done in collaboration with David Symmes and Gary Alexander. Based on reports that stimulation of the midbrain reticular formation (MRF) enhances perceptual performance, Symmes and I examined the effect of MRF stimulation on the responsiveness of units in the auditory cortex to vocalization stimuli.[71] Epidural electrodes in this experiment monitored cortical arousal. MRF stimulation reliably resulted in cortical arousal, as well as signs of behavioral arousal such as wider opening of the eyes, some body movement, and vocalization. Vocal stimuli were presented with and without concurrent MRF stimulation. This study resulted in several new findings: response selectivity of auditory cortex units to a battery of vocal stimuli is stable over times of 20 min to 1 hr; response selectivity is independent of short-term changes in cortical arousal; and MRF stimulation can result in increased consistency or strength of a unit's response to effective stimuli but has little effect on overall response selectivity to the full stimulus repertoire. A subsequent study[72] examined the role of connections from the dorsolateral prefrontal cortex (PFC) on unit activity in the temporal lobe auditory association cortex. Electrical stimulation of the PFC had a general inhibitory effect on the responses of auditory cortex units to vocalizations. PFC stimulation also had a profound inhibitory effect on the spontaneous activity of auditory cortex units, often lasting beyond the end of the electrical stimulation. This study led to the conclusion that the connections of the PFC to auditory association cortex are mainly inhibitory in nature, and that these effects impinge directly on sensory function of the auditory association cortex.

Two studies by our NIH group gave evidence of the powerful interaction between excitatory and inhibitory acoustic inputs on auditory cortex unit responses. In one, portions of vocal stimuli were deleted, and the response to the remaining stimulus components were compared to the response to the entire intact vocalization.[73] No units were found in which the response to the intact call was stronger than to effective vocal fragments. On the contrary, responses were often stronger to the fragment than to the intact call, suggesting an inhibitory effect of antecedent parts of the call. Another study, using combinations of steady tone bursts,[74] demonstrated the inhibitory effects of tonal frequencies well outside the range of tone frequencies that excite a neuron. Several reviews of our work have been published.[75-78]

2. Studies in Israel

Zvi Wollberg and co-workers have published several important papers on single-unit response properties in the squirrel monkey auditory cortex. One paper dealt with the basic tuning properties of auditory cortex cells.[79] This is one of the few studies to vary both stimulus frequency and intensity when determining the responses to pure tones. Separate excitatory and inhibitory receptive fields were described for some units. The authors found a wide range of tuning sharpness (Q10 value) and threshold intensity. The other papers from this group have focused on evidence for or against the existence of *call detector* neurons in the squirrel monkey auditory cortex. One study[80] found a considerable degree of lability in unit responses, concentrating recording efforts in the auditory association cortex of the superior temporal gyrus. The authors concluded that the lability in response strength and selectivity observed in their study made it unlikely that individual neurons can function reliably as call detector neurons; rather, an ensemble of neural elements must underlie discrimination of specific calls.

Two further studies focused on whether neurons provide evidence for discriminating between natural vocalizations and the same frequency information provided by playing the calls backwards.[81,82] Evidence obtained from these studies indicated that neurons in both primary and association auditory cortex are sensitive to discrete acoustic elements of more complex vocalizations ("transients" in these authors' terminology), and that responsiveness across a group of vocal stimuli reflects the presence of shared acoustic transients in all of the stimuli. Another study using similar methods[83] concluded that neurons are sensitive to certain transient components (*peaks*) and that the *peak tracing* (or *peak tracking*) responses of units in the auditory cortex and medial geniculate body were similar. The peak tracing phenomenon showed a sensitivity to intensity, in that peak tracing was more evident at some intensities (not necessarily the loudest) than others.

3. Studies in Germany

Two of the most important studies relating to auditory mechanisms in the squirrel monkey have come out of the Munich laboratory. In one (one of only two studies evaluating the effect of ablating all or part of the superior temporal gyrus of squirrel monkeys on auditory function), squirrel monkeys were trained to discriminate between species-specific vocalizations and other natural sounds using a go/no-go paradigm and exposure to a conspecific as a reward.[84] Upon reaching criterion performance, subjects received bilateral ablations, of varying size, of the superior temporal gyrus. Small lesions did not interfere with the discrimination task. Lesions destroying about 75% of the STG led to a loss of discriminative ability, but the animals were able to relearn the discrimination at above-chance levels. After nearly total ablation of the auditory cortex, monkeys could not relearn the discrimination, although the monkeys could detect the presentation of a vocalization. Unilateral ablations of either side were without significant effect. The authors concluded that the auditory cortex is important for recognizing complex sounds based on their entire structure (Gestalt), although it must be said that the existence of a feature-detection mechanism for call recognition was not tested. The second study is of a completely different nature, in this case testing the functional significance of a pathway existing between the anterior cingulate gyrus (along the medial wall of the frontal lobe) and the auditory association cortex of the superior temporal gyrus. Electrical stimulation of the cingulate region results in phonation in squirrel monkeys. Also, stimulation (at an intensity below that required to evoke vocalization) inhibits spontaneous activity and auditory responses of neurons in auditory association cortex.[85] The first author of that study, Müller-Preuss, went on to test the hypothesis that this pathway provides feedback information from a vocalization-generating structure (the anterior cingulate gyrus) to a vocalization-perceiving structure (the auditory association cortex) that results in an inhibition of auditory cortex neurons when the monkey vocalizes. This inhibitory pathway would enable the auditory cortex to distinguish between self-generated vocalizations and vocalizations produced by other monkeys. Müller-Preuss demonstrated that about half of the neurons in the auditory cortex failed to respond to self-generated vocalizations but did respond to the same (tape-recorded) vocalization when it was presented as an auditory stimulus.[86,87] This response selectivity was not found in neurons of the inferior colliculus; neurons in the medial geniculate body were also inhibited by self-produced vocalizations, but at a lower incidence than was the case for auditory cortex neurons. Other investigations in the Munich lab addressed the issue of whether the response properties of neurons in the auditory cortex were consistent with the call detector neuron hypothesis.[88,89] These studies attempted to test individual neurons over periods exceeding one hour, with the goal of assessing any variability in response selectivity to species-specific vocalizations. Neurons studied along penetrations made within 3 mm of the Sylvian fissure were considered to be located within the primary auditory cortex, while neurons recorded from the penetrations further down in the superior temporal gyrus were considered to be within the secondary auditory cortex. In neither the primary nor secondary fields were neurons found that had invariant selective responses to vocalizations when studied over longer time periods. The authors concluded that evidence is lacking to support the view that responses to vocalizations are more stable over time than responses of the same neurons to artificial stimuli, and that no evidence suggests that the secondary auditory cortex is the next step from the primary cortex in a rigid hierarchical system for detecting species-specific vocalizations. Müller-Preuss and co-workers also did a series of studies examining the coding of intensity peaks of species-specific vocalizations in the auditory cortex. A comparison of unit responses in the inferior colliculus (IC), MGN, and AC indicated that the IC and MGN responded reliably to multiple amplitude peaks, but AC neurons could only follow amplitude peaks at lower rates. Bieser and Müller-Preuss[90] found regional differences in the auditory cortex and adjacent insula with respect to following the amplitude envelope of species-specific vocalizations. This study also identified two additional functional subdivisions of the auditory cortex: the first temporal field (T1) and the parainsular auditory field (Pi). The primary auditory cortex showed the best temporal resolution, while neurons in the insula showed poor following of amplitude-modulated (AM) components. Bieser[91] studied unit responses in the primary auditory cortex and rostral field to the twitter call, a vocalization with both prominent AM and FM. A higher percentage of neurons in the rostral field were able to encode all FM elements within a twitter.

F. STUDIES OF AUDITORY STRUCTURES BELOW THE AUDITORY CORTEX

1. Medial Geniculate Nucleus

The cyto- and myelo-architecture of the MGN in squirrel monkeys has been investigated by Jordan,[92] who distinguished three nuclear subdivisions: the MGNa, MGNb and MBNc. The MGNa is located in a ventrocaudal position and is distinguished by a striking poverty of myelinated nerve fibers. The cells in this subdivision are the smallest of the entire nucleus. MGNb occupies most of the medial extent of the MGN. The neurons are irregularly arranged and often form small clusters; larger multipolar neurons are found rostromedially. These are the largest neurons of the MGN, contain abundant Nissl bodies, and are embedded in a dense meshwork of myelinated nerve fibers. MGNc forms the lateral aspect of the nucleus and is characterized by densely packed, medium-sized neurons. In the anesthetized squirrel monkey, the following divisions have been identified:[93] (1) a tonotopically organized, small-cell division; (2) a large-cell dorsal division with cells having definable best frequencies; and (3) a second ventral, large-cell division with cells responding to tones and with best frequencies organized into several high-to-low subdivisions.

Using Jordan's[92] terminology, more recent physiological studies of the awake squirrel monkey MGN have found that neurons within MGNb had the highest rates of spontaneous activity, but cells in all three divisions had similar response properties.[94] Tonotopic organization was also clearest in MGNb. In regard to vocalizations, no significant differences were found between the three MGN divisions in terms of neuronal selectivity to vocalizations, although cells within MGNb had similar, simple response patterns to the suite of stimuli, whereas neurons in MGNa and MGNc tended to have more complex response patterns.[95] Burton and Jones[61] arrived at a different organization scheme and terminology than that of Jordan, and one study compared the response properties in the MGN using this organizational scheme.[96] This study found the ventral and anterodorsal divisions to be functionally indistinguishable, but the magnocellular division had some neurons that were nonresponsive to acoustic stimuli and a smaller proportion of cells with definable best frequencies. Overall, responses of neurons in the ventral and anterodorsal regions were vigorous to all vocal stimuli and equally strong to pure tones. Note, however, that one study examining the processing of amplitude-modulated sounds found numerous MGN neurons that had a preferred modulation rate, which resembled the amplitude-modulation rate of some species-specific vocalizations.[97]

2. Inferior Colliculus

As characterized by FitzPatrick,[98] the IC is made up of a large central nucleus and two smaller, bordering nuclei: the external and pericentral nuclei. A single tonotopic representation of audible frequencies is present in the central nucleus. The octave band from 8 to 15 kHz is represented by the greatest amount of collicular tissue. Neurons in the pericentral nucleus respond to pure tones, but they tend to respond to a wide frequency range, even at low sound intensities. Neurons of the external nucleus exhibited a generally low- to high-frequency progression, but this was very irregular, and all best frequencies were below 4 kHz. Profiles generated from click-evoked field potentials have been used to provide a physiological marker for the three main IC subdivisions.[99] A study examining responses to amplitude modulation found that all neurons tested exhibited a preferred rate of modulation.[100] Encoding of amplitude-modulated sounds occurs to a greater extent through phase-locking of discharges than through changes in spike number. Another study examined the responses of IC neurons to pure tones and vocal stimuli in awake monkeys.[101] Of the tested units, 90% responded to at least five of the eight vocal stimuli, and no units responded only to vocalizations without also responding to artificial stimuli (tones, clicks, white noise). In the superior colliculus of awake squirrel monkeys, most neurons have simple responses to all vocal stimuli, although a few neurons, all of which are broadly tuned to pure tone stimuli, have more complex responses.[102]

G. Auditory Responses within the Frontal Lobe

In chloralose-anesthetized squirrel monkeys, the first study of auditory responses of single units in the dorsolateral surface of the frontal lobe found that about 25% of the units responded to simple click stimuli.[103] A subsequent study using halothane-anesthetized subjects found that only 10% of the units in the same general part of the frontal cortex responded to clicks.[104] In awake monkeys, and using a wider variety of acoustic stimuli including species-specific vocalizations, Newman and Lindsley[105] found that about 20% of the units responded to sounds, primarily in the vicinity of the principal sulcus. Units generally responded to tones over a wide frequency range; correspondingly, the most effective vocalizations were "noisy" (containing energy over a wide frequency range). In more caudal areas in the dorsolateral prefrontal cortex, tonal and noisy vocal stimuli are equally effective in eliciting unit responses.[106]

IV. CONCLUSIONS

It seems clear that there is still a great deal to discover with respect to how the central auditory system provides the behaving animal with the information needed to discriminate between acoustically diverse vocalizations, as each of these calls elicits a distinctive behavioral response profile. The evidence presented in this review indicates that neural elements within the temporal lobe fields identified as part of the auditory cortex do not individually possess specialized coding properties that would facilitate behavioral responses to different vocalizations. Two possibilities that require further study are that groups of neurons work in concert to provide this capability or that other, poorly studied cortical structures provide this capability. Current methods for conducting auditory physiology studies also are generally not conducive to allowing the experimental animals to behaviorally register a response to the stimulus. Thus, future studies might profitably combine behavioral with physiological studies, first by characterizing the behavioral profiles to species-specific vocalizations, then by exploring higher cortical areas for neurons with response profiles predicted by the behavioral responses. There is also a great need for studies of other non-human primate species.

The acoustic components that activate cortical units need to be examined in greater detail, from both a behavioral and neurophysiological perspective, in order to determine whether neurons in the STG exhibit coding properties that are predicted from behavioral responses. The selectivity of neurons in the STG to variants of particular vocalizations, especially the isolation peep, should be examined in the brains of individuals with different genetic and experiential backgrounds to determine the extent to which neurons in the STG exhibit coding properties that are predicted based on the genetic and experiential backgrounds of each subject. These are more sophisticated approaches than have been used in previous studies of non-human primates, but would appear to be necessary to advance our understanding of the perceptual processes underlying auditory communication. It would seem safe to conclude that the squirrel monkey will remain an important primate for studies of auditory communication and auditory physiology into the foreseeable future.

REFERENCES

1. Baldwin, J.D. and Baldwin, J.I., The squirrel monkeys, genus *Saimiri*, in *Ecology and Behavior of Neotropical Primates*, Vol. 1, A.F. Coimbra-Filho and R.A. Mittermeier, Eds., Academia Brasileira de Ciencias, Rio de Janeiro, 1981, p. 277.
2. Ghazanfar, A.A. and Hauser, M.D., The neuroethology of primate vocal communication: substrates for the evolution of speech, *Trends Cogn. Sci.*, 3, 377, 1999.
3. Ploog, D., Neurobiology of primate audio-vocal behavior, *Brain Res. Rev.*, 3, 35, 1981.
4. Ploog, D., Hupfer, K., Jürgens, U., and Newman, J.D., Neuroethological studies of vocalization in squirrel monkeys with special reference to genetic differences of calling in two subspecies, in *Growth and Development of the Brain*, M.A.B. Brazier, Ed., Raven Press, New York, 1975, p. 254.
5. Seyfarth, R.M. and Cheney, D.L., Behavioral mechanisms underlying vocal communication in nonhuman primates, *Animal Learning Behav.*, 25, 249, 1997.
6. Winter, P., Ploog, D., and Latta, J., Vocal repertoire of the squirrel monkey (*Saimiri sciureus*), its analysis and significance, *Exp. Brain Res.*, 1, 359, 1966.
7. Newman, J.D., Squirrel monkey communication, in *Handbook of Squirrel Monkey Research*, L.A. Rosenblum and C.L. Coe, Eds., Plenum, New York, 1985, p. 99.
8. Jürgens, U., Vocalization as an emotional indicator: a neuroethological study in the squirrel monkey, *Behaviour*, 49, 88, 1979.
9. Winter, P., Social communication in the squirrel monkey, in *The Squirrel Monkey*, L.A. Rosenblum and R.W. Cooper, Eds., Academic Press, New York, 1968, p. 235.
10. Winter, P., Handley, P., Ploog, D., and Schott, D., Ontogeny of squirrel monkey calls under normal conditions and under acoustic isolation, *Behaviour*, 47, 230, 1973.
11. Talmage-Riggs, G., Winter, P., Ploog, D., and Mayer, W., Effect of deafening on the vocal behavior of the squirrel monkey (*Saimiri sciureus*), *Folia Primatol.*, 17, 404, 1972.
12. Newman, J.D. and Symmes, D., Inheritance and experience in the acquisition of primate acoustic behavior, in *Primate Communication*, C.T. Snowdon, C.H. Brown, and M.R. Petersen, Eds., Cambridge University, Cambridge, U.K., 1982, p. 259.

13. McCowan, B., Franceschini, N.V., and Vicino, G.A., Age differences and developmental trends in alarm peep responses by squirrel monkeys (*Saimiri sciureus*), *Am. J. Primatol.*, 53, 19, 2001.
14. Hopf, S., Herzog, M., and Ploog, D., Development of attachment and exploratory behavior in infant squirrel monkeys under controlled rearing conditions, *Int. J. Behav. Dev.*, 8, 55, 1985.
15. Winter, P., Dialects in squirrel monkeys: vocalization of the Roman arch type, *Folia Primatol.*, 10, 216, 1969.
16. MacLean, P.D., Mirror display in the squirrel monkey, *Saimiri sciureus*, *Science*, 146, 950, 1964.
17. Hershkovitz, P., Taxonomy of squirrel monkeys genus *Saimiri* (Cebidae: Platyrrhini): a preliminary report with description of a hitherto unnamed form, *Am. J. Primatol.*, 6, 257, 1984.
18. Snowdon, C.T., Coe, C.L., and Hodun, A., Population recognition of infant isolation peeps in the squirrel monkey, *Animal Behav.*, 33, 1145, 1985.
19. Symmes, D., Newman, J.D., Talmage-Riggs, G., and Lieblich, A.K., Individuality and stability of isolation peeps in squirrel monkeys, *Animal Behav.*, 27, 1142, 1979.
20. Lieblich, A.K., Symmes, D., Newman, J.D., and Shapiro, M., Development of the isolation peep in laboratory-bred squirrel monkeys, *Animal Behav.*, 28, 1, 1980.
21. Kaplan, J.N., Winship-Ball, A., and Sim, L., Maternal discrimination of infant vocalizations in squirrel monkeys, *Primates*, 19, 187, 1978.
22. Symmes, D. and Biben, M., Maternal recognition of individual infant squirrel monkeys from isolation call playbacks, *Am. J. Primatol.*, 9, 39, 1985.
23. Symmes, D. and Newman, J.D., Discrimination of isolation peep variants by squirrel monkeys, *Exp. Brain Res.*, 19, 365, 1974.
24. Masataka, N. and Symmes, D., Effect of separation distance on isolation call structure in squirrel monkeys (*Saimiri sciureus*), *Am. J. Primatol.*, 10, 271, 1986.
25. Smith, H.J., Newman, J.D., and Symmes, D., Vocal concomitants of affiliative behavior in squirrel monkeys (*Saimiri sciureus*), in *Primate Communication*, C.T. Snowdon, C.H. Brown, and M.R. Petersen, Eds., Cambridge University, Cambridge, U.K., 1982, p. 30.
26. Newman, J.D. and Bernhards, D.E., The affiliative vocal subsystem of squirrel monkeys, *Ann. N.Y. Acad. Sci.*, 807, 546, 1997.
27. Masataka, N. and Biben, M., Temporal rules regulating affiliative vocal exchanges of squirrel monkeys, *Behaviour*, 101, 311, 1987.
28. Smith, H.J., Newman, J.D., Hoffman, H.J., and Fetterly, K., Statistical discrimination among vocalizations of individual squirrel monkeys (*Saimiri sciureus*), *Folia Primatol.*, 37, 267, 1982.
29. Biben, M., Symmes, D., and Masataka, N., Temporal and structural analysis of affiliative vocal exchanges in squirrel monkeys (*Saimiri sciureus*), *Behaviour*, 98, 259, 1986.
30. Biben, M. and Symmes, D., Playback studies of affiliative vocalizing in captive squirrel monkeys: familiarity as a cue to response, *Behaviour*, 117, 1, 1991.
31. Symmes, D. and Biben, M., Conversational vocal exchanges in squirrel monkeys, in *Primate Vocal Communication*, D. Todt, P. Goedeking, and D. Symmes, Eds., Springer–Verlag, Berlin, 1988, p. 124.
32. Biben, M., Recognition of order effects in squirrel monkey antiphonal call sequences, *Am. J. Primatol.*, 29, 109, 1993.
33. Biben, M., Playback studies of social communication in the squirrel monkey (*Saimiri sciureus*), in *Current Primatology*, Vol. II, *Social Development, Learning and Behaviour*, J.J. Roeder, B. Thierry, J.R. Anderson, and N. Herrenschmidt, Eds., University Louis Pasteur, Strasbourg, 1994, p. 207.

34. Biben, M. and Bernhards, D., Naïve recognition of chuck calls in squirrel monkeys (*Saimiri sciureus macrodon*), *Language Commun.*, 14, 167, 1994.
35. McCowan, B. and Newman, J.D., The role of learning in chuck call recognition by squirrel monkeys (*Saimiri sciureus*), *Behaviour*, 137, 279, 2000.
36. Biben, M. and Symmes, D., Play vocalizations of squirrel monkeys (*Saimiri sciureus*), *Folia Primatol.*, 46, 173, 1986.
37. Symmes, D. and Goedeking, P., Nocturnal vocalizations by squirrel monkeys (*Saimiri sciureus*), *Folia Primatol.*, 51, 143, 1988.
38. Newman, J.D., Lieblich, A.K., Talmage-Riggs, G., and Symmes, D., Syllable classification and sequencing in twitter calls of squirrel monkeys (*Saimiri sciureus*), *Z. Tierpsychol.*, 47, 77, 1978.
39. Biben, M., Allomaternal vocal behavior in squirrel monkeys, *Dev. Psychobiol.*, 25, 79, 1992.
40. Biben, M., Symmes, D., and Bernhards, D., Contour variables in vocal communication between squirrel monkey mothers and infants, *Dev. Psychobiol.*, 22, 617, 1988.
41. Smith, H.J., Newman, J.D., Bernhards, D.E., and Symmes, D., Effects of reproductive state on vocalizations in squirrel monkeys (*Saimiri sciureus*), *Folia Primatol.*, 40, 233, 1983.
42. Fernandez, C., Butler, R., Konishi, T., and Honrubia, V., Cochlear potentials in the rhesus and squirrel monkey, *J. Acoust. Soc. Am.*, 34, 1411, 1962.
43. Rose, J.E., Brugge, J.F., Anderson, D.J., and Hind, J.E., Phase-locked response to low-frequency tones in single auditory nerve fibers of the squirrel monkey, *J. Neurophysiol.*, 30, 769, 1967.
44. Hind, J.E., Anderson, D.J., Brugge, J.F., and Rose, J.E., Coding of information pertaining to paired low-frequency tones in single auditory nerve fibers of the squirrel monkey, *J. Neurophysiol.*, 30, 794, 1967.
45. Rose, J.E., Brugge, J.F., Anderson, D.J., and Hind, J.E., Patterns of activity in single auditory nerve fibres of the squirrel monkey, in *Hearing Mechanisms in Vertebrates*, A.V.S. de Reuck and J. Knight, Eds., Little, Brown, Boston, 1968, p. 144.
46. Ruggero, M.A., Response to noise of auditory nerve fibers in the squirrel monkey, *J. Neurophysiol.*, 36, 569, 1973.
47. Brugge, J.F., Anderson, D.J., Hind, J.E., and Rose, J.E., Time structure of discharges in single auditory nerve fibers of the squirrel monkey in response to complex periodic sounds, *J. Neurophysiol.*, 32, 386, 1969.
48. Rose, J.E., Brugge, J.F., Anderson, D.J., and Hind, J.E., Some possible neural correlates of combination tones, *J. Neurophysiol.*, 32, 402, 1969.
49. Rose, J.E., Hind, J.E., Anderson, D.J., and Brugge, J.F., Some effects of stimulus intensity on response of auditory nerve fibers in the squirrel monkey, *J. Neurophysiol.*, 34, 685, 1971.
50. Imig, T.J., Ruggero, M.A., Kitzes, L.M., Javel, E., and Brugge, J.F., Organization of auditory cortex in the owl monkey (*Aotus trivirgatus*), *J. Comp. Neurol.*, 171, 111, 1977.
51. Fitzpatrick, K.A. and Imig, T.J., Auditory cortico-cortical connections in the owl monkey, *J. Comp. Neurol.*, 192, 589, 1980.
52. Fitzpatrick, K.A. and Imig, J., Projections of auditory cortex upon the thalamus and midbrain in the owl monkey, *J. Comp. Neurol.*, 177, 537, 1978.
53. Smith, D.L. and Moskowitz, N., Ultrastructure of layer IV of the primary auditory cortex of the squirrel monkey, *Neuroscience*, 4, 349, 1979.
54. Massopust, L.C., Wolin, L.R., and Kadoya, S., Evoked responses in the auditory cortex of the squirrel monkey, *Exp. Neurol.*, 21, 35, 1968.

55. Forbes, B.F. and Moskowitz, N., Projections of auditory responsive cortex in the squirrel monkey, *Brain Res.*, 67, 239, 1974.
56. Forbes, B.F. and Moskowitz, N., Cortico-cortical connections of the superior temporal gyrus in the squirrel monkey, *Brain Res.*, 136, 547, 1977.
57. Beecher, M.D., Pure-tone thresholds of the squirrel monkey (*Saimiri sciureus*), *J. Acoust. Soc. Am.*, 55, 196, 1974.
58. Wiernicke, A., Haüsler, U., and Jürgens, U., Auditory frequency discrimination in the squirrel monkey, *J. Comp. Physiol. A*, 187, 189, 2001.
59. Pineda, J.A., Holmes, T.C., Swick, D., and Foote, S.L., Brain-stem auditory evoked potentials in squirrel monkey (*Saimiri sciureus*), *Electroencephalogr. Clin. Neurophysiol.*, 73, 532, 1989.
60. Berntson, G.G., Boysen, S.T., and Torello, M.W., Vocal perception: brain event-related potentials in a chimpanzee, *Dev. Psychobiol.*, 26, 305, 1993.
61. Burton, H. and Jones, E.G., The posterior thalamic region and its cortical projection in New World and Old World monkeys, *J. Comp. Neurol.*, 168, 249, 1976.
62. Jones, E.G. and Burton, H., Areal differences in the laminar distribution of thalamic afferents in cortical fields of the insular, parietal and temporal regions of primates, *J. Comp. Neurol.*, 168, 197, 1976.
63. Cheung, S.W., Bedenbaugh, P.H., Nagarajan, S.S., and Scheiner, C.E., Functional organization of squirrel monkey primary auditory cortex: responses to pure tones, *J. Neurophysiol.*, 85, 1732, 2001.
64. Winter, P. and Funkenstein, H.H., The auditory cortex of the squirrel monkey: neuronal discharge patterns to auditory stimuli, *Proc. 3rd Int. Congr. Primatol.*, 2, 24, 1971.
65. Funkenstein, H.H., Nelson, P.G., Winter, P., Wollberg, Z., and Newman, J.D., Unit responses in auditory cortex of awake squirrel monkeys to vocal stimulation, in *Physiology of the Auditory System: A Workshop*, M. Sachs, Ed., National Education Consultants, Baltimore, MD, 1972, p. 307.
66. Funkenstein, H.H. and Winter, P., Responses to acoustic stimuli of units in the auditory cortex of awake squirrel monkeys, *Exp. Brain Res.*, 18, 464, 1973.
67. Winter, P. and Funkenstein, H.H., The effect of species-specific vocalization on the discharge of auditory cortical cells in the awake squirrel monkey (*Saimiri sciureus*), *Exp. Brain Res.*, 18, 489, 1973.
68. Wollberg, Z. and Newman, J.D., Auditory cortex of squirrel monkey: response patterns of single cells to species-specific vocalizations, *Science*, 175, 212, 1972.
69. Newman, J.D. and Wollberg, Z., Multiple coding of species-specific vocalizations in the auditory cortex of squirrel monkeys, *Brain Res.*, 54, 287, 1973.
70. Newman, J.D. and Wollberg, Z., Responses of single neurons in the auditory cortex of squirrel monkeys to variants of a single call type, *Exp. Neurol.*, 40, 821, 1973.
71. Newman, J.D. and Symmes, D., Arousal effects on unit responsiveness to vocalizations in squirrel monkey auditory cortex, *Brain Res.*, 78, 125, 1974.
72. Alexander, G.E., Newman, J.D., and Symmes, D., Convergence of prefrontal and acoustic inputs upon neurons in the superior temporal gyrus of the awake squirrel monkey, *Brain Res.*, 116, 334, 1976.
73. Newman, J.D. and Symmes, D., Feature detection by single units in squirrel monkey auditory cortex, *Exp. Brain Res. Suppl.*, 2, 140, 1979.
74. Shamma, S.A. and Symmes, D., Patterns of inhibition in auditory cortical cells in awake squirrel monkeys, *Hearing Res.*, 19, 1, 1985.
75. Newman, J.D., Detection of biologically significant sounds by single neurons, in *Recent Advances in Primatology*, Vol. 1, D.J. Chivers and J. Herbert, Eds., Academic Press, London, 1978, p. 755.

76. Newman, J.D., Perception of sounds used in species-specific communication: the auditory cortex and beyond, *J. Med. Primatol.*, 7, 98, 1978.
77. Newman, J.D., Central nervous system processing of sounds in primates, in *Neurobiology of Social Communication in Primates*, H. Steklis and M. Raleigh, Eds., Academic Press, New York, 1979, p. 69.
78. Newman, J.D., Primate hearing mechanisms, in *Comparative Primate Biology*, Vol. 4, *Neurosciences*, H.D. Steklis and J. Erwin, Eds., Alan R. Liss, New York, 1988, p. 469.
79. Pelleg-Toiba, R. and Wollberg, Z., Tuning properties of auditory cortex cells in the awake squirrel monkey, *Exp. Brain Res.*, 74, 353, 1989.
80. Glass, I. and Wollberg, Z., Lability in the responses of cells in the auditory cortex of squirrel monkeys to species-specific vocalizations, *Exp. Brain Res.*, 34, 489, 1979.
81. Glass, I. and Wollberg, Z., Auditory cortex responses to sequences of normal and reversed squirrel monkey vocalizations, *Brain Behav. Evol.*, 22, 13, 1983.
82. Glass, I. and Wollberg, Z., Responses of cells in the auditory cortex of awake squirrel monkeys to normal and reversed species-specific vocalizations, *Hearing Res.*, 9, 27, 1983.
83. Pelleg-Toiba, R. and Wollberg, Z., Discrimination of communication calls in the squirrel monkey: 'call detectors' or 'cell ensembles'?, *J. Basic Clin. Physiol. Pharmacol.*, 2, 257, 1991.
84. Hupfer, K., Jürgens, U., and Ploog, D., The effect of superior temporal lesions on the recognition of species-specific calls in the squirrel monkey, *Exp. Brain Res.*, 30, 75, 1977.
85. Müller-Preuss, P., Newman, J.D., and Jürgens, U., Anatomical and physiological evidence for a relationship between the 'cingular' vocalization area and the auditory cortex in the squirrel monkey, *Brain Res.*, 202, 307, 1980.
86. Müller-Preuss, P., Acoustic properties of central auditory pathway neurons during phonation in the squirrel monkey, in *Neuronal Mechanisms of Hearing*, J. Syka, J. and L. Aitkin, Eds., Plenum, New York, 1981, p. 311.
87. Müller-Preuss, P. and Ploog, D., Inhibition of auditory cortical neurons during phonation, *Brain Res.*, 215, 61, 1981.
88. Manley, J.A. and Müller-Preuss, P., Response variability of auditory cortex cells in the squirrel monkey to constant acoustic stimuli, *Exp. Brain Res.*, 32, 171, 1978.
89. Manley, J.A. and Mueller-Preuss, P., Response variability in the mammalian auditory cortex: an objection to feature detection?, *Fed. Proc.*, 37, 2355, 1978.
90. Bieser, A. and Müller-Preuss, P., Auditory responsive cortex in the squirrel monkey: neural responses to amplitude-modulated sounds, *Exp. Brain Res.*, 108, 273, 1996.
91. Bieser, A., Processing of twitter-call fundamental frequencies in insula and auditory cortex of squirrel monkeys, *Exp. Brain Res.*, 122, 139, 1998.
92. Jordan, H., The structure of the medial geniculate nucleus (MGN): a cyto- and myeloarchitectonic study in the squirrel monkey, *J. Comp. Neurol.*, 148, 469, 1973.
93. Gross, N.B., Lifschitz, W.S., and Anderson, D.J., The tonotopic organization of the auditory thalamus of the squirrel monkey (*Saimiri sciureus*), *Brain Res.*, 65, 323, 1974.
94. Allon, N., Yeshurun, Y., and Wollberg, Z., Responses of single cells in the medial geniculate body of awake squirrel monkeys, *Exp. Brain Res.*, 41, 222, 1981.
95. Allon, N. and Yeshurun, Y., Functional organization of the medial geniculate body's subdivisions of the awake squirrel monkey, *Brain Res.*, 360, 75, 1985.
96. Symmes, D., Alexander, G.E., and Newman, J.D., Neural processing of vocalizations and artificial stimuli in the medial geniculate body of squirrel monkey, *Hearing Res.*, 3, 133, 1980.

97. Preuss, A. and Müller-Preuss, P., Processing of amplitude modulated sounds in the medial geniculate body of squirrel monkeys, *Exp. Brain Res.*, 79, 207, 1990.
98. Fitzpatrick, K.A., Cellular architecture and topographic organization of the inferior colliculus of the squirrel monkey, *J. Comp. Neurol.*, 164, 185, 1975.
99. Müller-Preuss, P. and Mitzdorf, U., Functional anatomy of the inferior colliculus and the auditory cortex: current source density analyses of click-evoked potentials, *Hearing Res.*, 16, 133, 1984.
100. Müller-Preuss, P., Flachskamm, C., and Bieser, A., Neural encoding of amplitude modulation within the auditory midbrain of squirrel monkeys, *Hearing Res.*, 80, 197, 1994.
101. Manley, J.A. and Müller-Preuss, P., A comparison of the responses evoked by artificial stimuli and vocalizations in the inferior colliculus of squirrel monkeys, in *Neuronal Mechanisms of Hearing*, J. Syka and L. Aitkin, Eds., Plenum, New York, 1981, p. 307.
102. Allon, N. and Wollberg, Z., Responses of cells in the superior colliculus of the squirrel monkey to auditory stimuli, *Brain Res.*, 159, 321, 1978.
103. Nelson, C.N. and Bignall, K.E., Interactions of sensory and nonspecific thalamic inputs to cortical polysensory units in the squirrel monkey, *Exp. Neurol.*, 40, 189, 1973.
104. Schechter, P.B. and Murphy, E.H., Response characteristics of single cells in squirrel monkey frontal cortex, *Brain Res.*, 96, 66, 1975.
105. Newman, J.D. and Lindsley, D., Single unit analysis of auditory processing in squirrel monkey frontal cortex, *Exp. Brain Res.*, 25, 169, 1976.
106. Wollberg, Z. and Sela, J., Frontal cortex of the awake squirrel monkey: responses of single cells to visual and auditory stimuli, *Brain Res.*, 198, 216, 1980.

13 Cortical Mechanisms of Sound Localization and Plasticity in Primates

Gregg H. Recanzone

CONTENTS

I. Introduction 247
II. Sound Localization Cues and Perception 248
III. The Primate Auditory Cortex 248
IV. Lesion Studies 249
V. Electrophysiological Studies 250
VI. Plasticity of Sound Localization 252
    A. Auditory Cortical Plasticity 252
    B. Cortical Plasticity and Sound Localization 254
VII. Summary 254
References 255

I. INTRODUCTION

Sound localization has been studied extensively for many decades at the perceptual level, particularly in humans[1-3] but also in non-human primates[4-6] and a variety of other species (e.g., see References 7 to 9). However, at the neurophysiological level, particularly in the cerebral cortex, relatively few studies have been conducted in either monkeys or humans. The non-human primate, particularly the macaque monkey, has several advantages as an animal model of human auditory processing in studies correlating neurophysiological responses to behavior and perception. This is particularly true for studies on auditory cortical function because (1) this species has perceptual abilities similar to humans in a variety of auditory discrimination tasks that are dependent on normal cortical activity (for example, sound localization); (2) it is possible to routinely record single-cell responses in awake monkeys actively participating in discrimination experiments; and, therefore, (3) it is possible to directly relate neuronal activity at the level of the single neuron to the auditory perceptions of the animal. This chapter will concentrate on comparing both the perceptual abilities and the neurophysiological correlates of sound localization in non-human primates to those in humans, with some references to other species when considering areas where little is currently known in primates. It will also explore the more general phenomenon of plasticity of the auditory cortex with reference to how changes in the neuronal representations of acoustic stimuli can be related to changes in perception.

II. SOUND LOCALIZATION CUES AND PERCEPTION

Because the auditory receptors, the hair cells within the cochlea, contain no spatial information individually, the spatial location of acoustic stimuli must be computed by the nervous system. Three principal cues are used in this computation.[2] The first two are based on differences in the sound waves that reach the two ears, known as *binaural cues*. Interaural *intensity* differences result when a sound source is laterally placed and the shadowing of the sound waves by the head and torso decreases the intensity of the sound at the far ear with respect to the intensity at the near ear. This cue is most effective for the higher frequency components of the sound, generally above 4 kHz. The time of arrival and the phase of the sounds will also differ between the two ears as a function of the laterality of the stimulus. This cue, based on interaural *time* differences, is most effective for the low-frequency components, generally below 1 kHz. These two interaural cues are used for determining the sound location along the horizontal plane (azimuth) but are essentially ineffective for stimuli that vary in elevation. Elevation cues are believed to arise from the differences in the filtering of the acoustic stimulus by the pinnae, head, and torso.[10,11] These filtering properties generate spectral peaks and notches, referred to as the head-related transfer function (HRTF), with the frequency of these peaks and particularly the notches varying systematically as a function of stimulus elevation. Thus, all three localization cues are available only for stimuli with a broad spectral bandwidth.

A comparison of studies across different paradigms and acoustic stimuli indicate that the sound localization ability of humans and macaque monkeys are similar. For example, monkeys and humans localize broadband noise stimuli best and band-passed stimuli less well, and exhibit the poorest localization performance for tonal stimuli (e.g., see Recanzone et al.[6,12]). Monkeys and humans also need the high-frequency spectral cues to localize sounds in elevation.[4-6,12] These observations indicate that monkeys and humans take advantage of the same localization cues and that the two species presumably have similar neuronal mechanisms subserving their sound localization ability.

III. THE PRIMATE AUDITORY CORTEX

While the sound localization ability of primates and other species has been well documented, the neuronal mechanisms of sound location perception are poorly understood. Information from both ears is processed early in the ascending auditory pathway, and several brainstem and midbrain structures, particularly the superior olivary complex, are believed to be integral in processing binaural cues (for a review, see Phillips and Brugge[13]). However, lesion studies indicate that the auditory divisions of the cerebral cortex are critical for the perception of acoustic space. The auditory cortex of the primate is composed of multiple cortical fields organized into a "core" region including the primary auditory cortex (AI), a surrounding "belt" region including the caudomedial field (CM) and caudolateral field (CL), and a more lateral "parabelt" region. This delineation of cortical fields is based on histochemical and cytoarchitectonic analysis, the cortico-cortical and thalamo-cortical anatomical connectivity, reversals in the frequency response properties to tones and band-passed noise, and differences in physiological responses such as frequency tuning, intensity tuning, and latency.[14-25]

It is currently unclear how spatial information is processed throughout the auditory cortex, although neurons located caudally and laterally in the belt areas (e.g., area CL) have sharper spatial tuning than neurons located more rostrally.[25] This is consistent with the notion that auditory cortical fields are better tuned for spatial processing in the more caudal regions of auditory cortex (i.e., CL and CM) than neurons in the more rostral regions (see Woods et al.[26]) which can potentially better encode spectral or temporal features of sound stimuli (see Rauschecker[27] and Kass et al.[28]). This is similar to the dorsal "what" and ventral "where" processing proposed in the visual cortex of both monkeys[29] and humans.[30]

IV. LESION STUDIES

The results from lesion studies in primates present strong evidence that the auditory cortex plays a pivotal role in sound localization; however, the effects of lesions restricted to one or even a few of the primate auditory cortical areas have yet to be tested. Large bilateral lesions that likely incorporated all auditory cortical areas resulted in sound localization deficits in macaque monkeys.[31] An earlier study in the squirrel monkey also showed sound localization deficits following unilateral lesions.[32] In this study, the lesions incorporated only part of the primary auditory cortex and included parts of the belt regions as well, although the extent of the lesions with respect to the cortical fields described above was not defined. Similarly, large bilateral lesions that incorporate most or all of auditory cortex were also shown to give rise to localization deficits in ferrets.[33] In cats, small lesions restricted to the representation of a few frequencies in the primary auditory cortex result in sound localization deficits restricted to the frequencies represented in the lesioned zone.[34] Thus, in multiple non-human species, auditory cortical lesions result in pronounced deficits in the perception of acoustic space.

In humans, similar results have been found, although the extent of the lesions are less well documented and are inconsistent across patients and studies. An early study documented sound localization ability in normal subjects and patients with a variety of cortical and subcortical lesions.[35] This study found that patients with temporal lobe lesions, presumably incorporating part or all of auditory cortex, showed the most consistent and profound localization deficits in contralateral acoustic space. Patients with lesions in other, nonauditory, cortical areas generally had no sound localization deficits. These findings were extended more recently in patients who had undergone a hemispherectomy between the ages of 8 and 16 years and were tested 6 months to 25 years after surgery.[36] These patients had a deficit in localizing sounds in contralateral space, but not in ipsilateral space. In a more recent study, patients with partial auditory cortical lesions were examined with respect to sound localization as well as auditory recognition.[37] In this study, the two patients with more rostral auditory cortical lesions showed normal sound localization ability but were impaired in their ability to recognize environmental sounds. In contrast, the two patients with more caudal lesions were impaired in their ability to localize both static and moving sound stimuli but were able to recognize environmental sounds. These results indicate that different auditory cortical areas subserve these two auditory discrimination processes.

Taken together, these data show that the auditory cortex plays an essential role in sound location perception. The experiments in cats indicate that AI is essential for this discrimination,[34] which is consistent with the notion that auditory information is first processed in AI and then spatial information is further processed in other, more caudal, auditory cortical areas. More recent studies in humans show that the caudal regions of auditory cortex are critical for this perception, consistent with the anatomical and electrophysiological evidence (see below) observed in the macaque monkey. Thus, it is likely that the macaque monkey will serve as an invaluable animal model in studies of human sound localization ability.

V. ELECTROPHYSIOLOGICAL STUDIES

Experiments attempting to elucidate the cortical representation of acoustic space have concentrated mainly on anesthetized preparations, particularly in cats.[38-43] Fewer studies have investigated the spatial tuning properties of auditory cortical neurons in the awake monkey.[6,44,45] The results of these studies are converging on the notion that the acoustic space in auditory cortex does not have a spatially topographic organization. This is contrary to experimental evidence for spatially topographic representations of acoustic space in noncortical areas, the best example being the midbrain of the barn owl.[46-48] Spatially topographic representations are also seen in the superior colliculus of cats,[49] ferrets,[50] and monkeys.[51,52] Thus, topographical spatial representations, or "maps," can be constructed in the brain but are not constructed at the level of the auditory cortex.

A more recent study has taken advantage of the awake, behaving primate to investigate how spatial information is processed in the macaque auditory cortex.[6] In this study, the activity of single neurons in the AI and CM were recorded while monkeys performed a simple sound localization task. A restricted set of two different stimuli was presented to all recorded neurons: a tone that was near the characteristic frequency (the frequency that the neuron responded to at the lowest intensity) and either a band-passed noise that contained the characteristic frequency within its spectrum or a broadband noise. Representative examples of neuronal responses are shown in Figures 13.1A and B for stimulus locations presented directly ahead and at 15- and 30-degree eccentricity along both cardinal and oblique axes. Most neurons in both AI and CM were responsive to all stimulus locations, although neurons in CM generally had sharper spatial tuning compared to neurons in AI. The spatial tuning of both AI and CM neurons also showed the same stimulus dependence as the sound localization ability in the same monkeys. The spatial tuning of individual neurons in azimuth was weaker for tone stimuli compared to noise stimuli, consistent with the better localization performance for noise stimuli compared to tone stimuli

Cortical Mechanisms of Sound Localization and Plasticity in Primates

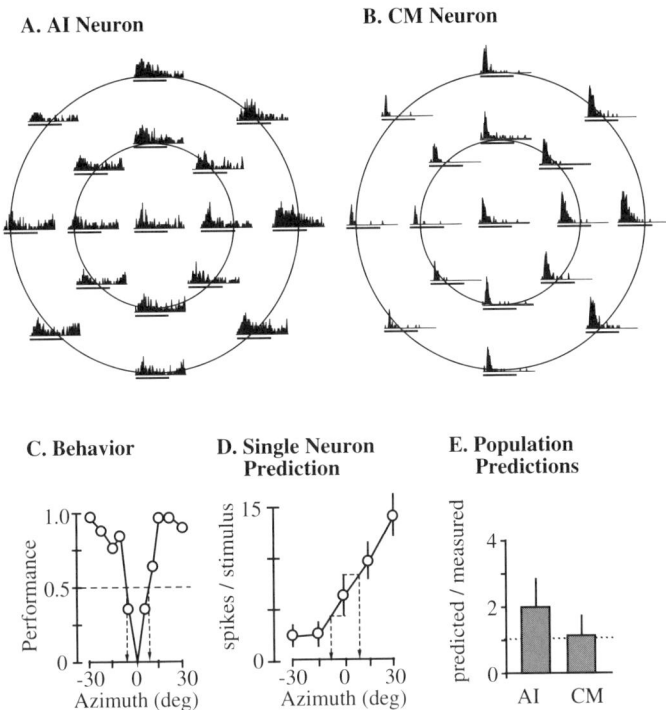

FIGURE 13.1 Responses from representative single neurons in the AI (A) and CM (B) to noise stimuli presented from 17 different locations in frontal space. The center PSTH shows the response of the neuron for stimuli presented directly in front of the monkey; the other PSTHs are positioned at 15- and 30-degree eccentricity along both cardinal and oblique axes. (C) Representative psychometric function showing localization in azimuth to a band-passed stimulus during a go/no-go paradigm. Threshold was taken at the point where the monkey detected this change at a 50% performance level (dashed lines). (D) Mean and standard deviation of activity of a single neuron as a function of stimulus eccentricity in azimuth. The neuronal threshold was taken as the eccentricity where the response differed from the response to the center location by one standard deviation (dashed line). (E) Pooled responses across neurons comparing the threshold predicted by the neuronal response to the behaviorally measured threshold across all stimulus types (tones and noise) in both azimuth and elevation. Only neurons that had statistically significant spatial tuning are shown. (Adapted from Reference 6.)

measured behaviorally (see above). Also, very few neurons were spatially tuned in elevation for tone stimuli or band-passed noise stimuli that contained only low frequencies, again consistent with the poor sound localization acuity in elevation for these stimuli.

Further analysis indicated that the population of neurons in CM, but not AI, had response properties that were consistent with the sound localization acuity for each animal. The eccentricity of a stimulus necessary for the neural response to be statistically significantly different from the response when the stimulus was directly in front of the monkey was calculated for each neuron (Figure 13.1D). This distance was then compared to the distance necessary for the monkey to correctly detect a change in location on 50% of the trials (Figure 13.1C). Overall, neurons in CM more accurately predicted the sound localization performance than neurons in AI. All neurons recorded in sessions using the same tone, band-passed, and/or broadband stimulus were then pooled and the analysis was repeated. The results (Figure 13.1E) showed that the pooled neuronal responses in CM were not significantly different than the behavioral measures, thus demonstrating that the firing rates of populations of cells in this cortical area, but not in AI, contain sufficient information to account for the monkey's behavior.

Less is currently known about the physiology of the human cerebral cortex underlying sound localization. Although functional imaging studies have provided valuable information about visual and somatosensory processing in humans, the auditory system has not been as extensively studied. This is undoubtedly due to the technical difficulties of overcoming the loud sounds (and constrained space) generated by the functional magnetic resonance imaging (fMRI) technique. However, several imaging studies have shown that regions of the auditory cortex are activated under different listening conditions, and a few studies have investigated sound localization in particular. Results from positron emission tomography indicate that the core and belt auditory areas generally do not show different levels of activation during sound localization tasks compared to passive listening to the same sounds (e.g., see Bushara et al.[53] and Weeks et al.[54,55]). However, there is significant activation in the inferior parietal lobule during sound localization tasks, with greater activation in the right hemisphere compared to the left.[54] Activation of the primary auditory cortex during active sound localization was revealed using magnetoencephalography, and again activation in the right hemisphere was greater than that measured in the left hemisphere.[56] This observation is consistent with some previous reports of greater deficits in sound localization following right hemisphere lesions compared to left hemisphere lesions in humans (e.g., see Haeske-Dewick et al.[57]), although other studies have reported no hemisphere difference (e.g., see Sanachez-Longo and Forster[35] and Poirier et al.[58]). Direct comparisons of deficits following left or right hemisphere lesions have not yet been performed on macaque monkeys, although no obvious differences were observed in squirrel monkeys.[32] A hemispheric dominance or lack thereof has yet to be definitively demonstrated in any non-human species.

VI. PLASTICITY OF SOUND LOCALIZATION

A. AUDITORY CORTICAL PLASTICITY

Several studies have shown that cortical activity can be altered in adults following peripheral or central lesions, behavioral training, or manipulation of neuromodulators.[59-62] In the auditory system, early studies showed that classical conditioning could induce changes in the responses of cortical neurons.[63,64] These changes can occur within only tens or scores of trials and can be specific to the paired frequency,[65] suggesting that such a mechanism could influence a widespread change in the tonotopic organization of auditory cortical fields.[66] Such changes in the tonotopic representation of the cochlea have been observed in AI following restricted cortical

Cortical Mechanisms of Sound Localization and Plasticity in Primates

lesions[67,68] and after pairing acoustic stimuli with basal forebrain stimulation,[69,70] which can also alter temporal processing of auditory cortical neurons.[70,71] The ability of central neural structures to alter their response properties is consistent with the ability to acquire new skills and behaviors, as well as the ability to adapt to changes in the animal's physical condition or in the environment.[59,72]

Plasticity of the primary auditory cortex has been directly related to changes in perception in owl monkeys.[73] Monkeys were trained to perform a frequency discrimination task using the same base frequency on consecutive days. Throughout several weeks of training, a steady improvement in performance at this task was observed, as reflected in progressively decreasing thresholds measured during each session (Figure 13.2A). Following this training, the animals were anesthetized, and the responses of neurons throughout AI were investigated. Figures 13.2B and C show the AI isofrequency bands encompassing neurons that have their thresholds in the 2- to 4-kHz range from two different monkeys. Figure 13.2B shows a normal monkey that was never trained and only a modest representation (black areas) of this narrow frequency range (approximately 2500 to 2700 Hz). In contrast, Figure 13.2C shows the representation of the same frequency range in a monkey trained to discriminate frequencies within this narrow range. This type of result was typical; monkeys trained at a particular frequency range had an enlarged representation of those frequencies within AI. Further, the behaviorally measured thresholds in the trained monkeys were significantly correlated with the area of representation of that frequency range. These results suggest that changes in perception, which commonly occur both during training and following periods without training, are likely manifest as changes in cortical representations of the stimulus features of the particular task.

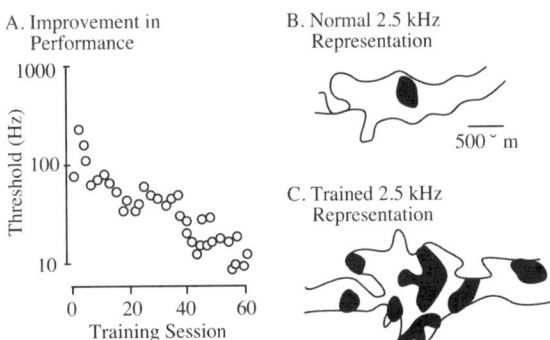

FIGURE 13.2 Plasticity of AI following behavioral training. Panel A shows the progressive improvement in performance with training. Each circle shows the threshold measured in the same monkey trained to discriminate changes in the frequency of a 2.5-kHz standard tone. Panel B shows the 2- to 4-kHz isofrequency band in the primary auditory cortex of a normal monkey that had not been trained at any task. The black region shows recording locations where the neurons responded to the narrow range of frequencies used in the training paradigm. Panel C shows the same isofrequency region in a monkey trained to perform the discrimination task. In this case, a much larger representation of the training frequencies was observed. (Adapted from Reference 73.)

B. CORTICAL PLASTICITY AND SOUND LOCALIZATION

Given the robust changes in cortical representations observed across sensory systems, it seems likely that similar changes in the cortical representation of acoustic space would take place during development and throughout adulthood. For example, increasing head size during development can cause changes in the interaural time and intensity cues. Studies in ferrets have shown that the spatial responses of auditory neurons are influenced during development in a manner consistent with the changes in both interaural and spectral cues.[7,74]

Studies on the ability of primates to adapt to changes in sound localization cues have been largely confined to humans. The sound localization acuity of humans that have been monaurally deaf since birth has been compared to normal humans, as well as humans wearing sound-attenuating ear plugs in one ear.[75] Two of five patients that were monaurally congenitally deaf had profound sound localization deficits similar to normal controls wearing a monaural earplug. Three patients, however, showed localization performance that was equivalent in the two hemifields and approached the level of the unplugged control subjects. In similar types of experiments, it has been found that raising owls with a frequency-specific monaural plug induces changes in the alignment of the auditory and visual receptive fields in the midbrain,[76] and ferrets raised with a monaural plug have sound localization ability similar to unplugged controls.[74] These results indicate that, at least in some instances, subjects forced to rely almost exclusively on monaural (spectral) cues can learn to accurately localize acoustic stimuli.

Humans can also adapt quickly to changes in sound localization cues. For example, changing the interaural timing differences initially results in sound localization errors, but subjects are able to compensate for these inappropriate cues within approximately one week.[77] Even more rapid adjustments have been observed when visual stimuli are simultaneously presented with acoustic stimuli, but at a spatially disparate location.[78] In these experiments, a visual stimulus is presented simultaneously with an acoustic stimulus, but at a consistent spatial disparity (e.g., 4 degrees to the left). Exposure to these disparities for as little as 20 minutes causes subjects to consistently localize stimuli incorrectly by approximately the same amount as the previous visual–auditory disparity (in this case, by 4 degrees to the left). This change in the internal representation of acoustic space has recently been shown to also occur in macaque monkeys.[79] Finally, manipulation of the localization cues has been shown to influence localization perception both within tens of minutes[80] and over the course of several days to weeks.[81] For example, Hofman et al.[81] showed that changing the spectral cues produced an initial deficit in localization in elevation, but most subjects slowly recovered to normal sound localization ability within several weeks. Interestingly, once the normal spectral cues were restored, the subjects were still able to localize sounds equally well, suggesting the presence of two independent representations of acoustic space in these subjects.

VII. SUMMARY

Humans and macaque monkeys have similar sound localization abilities, indicating that they likely share the same neuronal mechanisms to process acoustic space.

Electrophysiological and neuroanatomical experiments in monkeys have led to the hypothesis that spatial information is serially processed from the core areas through the caudal belt areas toward the parietal lobe. Imaging studies in humans are largely consistent with this idea, particularly with respect to the role of the parietal lobe in processing spatial information. Studies in humans that manipulate either the binaural or spectral cues, or both, have revealed some capacity to adapt to changing input conditions, suggesting a parallel neuronal plasticity that underlies this adaptation. Future studies in non-human primates may well reveal the neuronal locus of these plastic changes, providing further evidence of how auditory spatial information is processed in the primate cerebral cortex.

REFERENCES

1. Middlebrooks, J.C. and Green, D.M., Sound localization by human listeners, *Annu. Rev. Psychol.*, 42, 135, 1991.
2. Blauert, J., *Spatial Hearing: The Psychophysics of Human Sound Localization*, MIT Press, Cambridge, MA, 1997.
3. Carlile, S., Pleong, P., and Hyams, S., The nature and distribution of errors in sound localization by human listeners, *Hearing Res.*, 114, 179, 1997.
4. Brown, C.H. et al., Localization of noise bands by Old World monkeys, *J. Acoust. Soc. Am.*, 68, 127, 1980.
5. Brown, C.H. et al., Vertical and horizontal sound localization in primates, *J. Acoust. Soc. Am.*, 72, 1804, 1982.
6. Recanzone, G.H. et al., Correlation between the activity of single auditory cortical neurons and sound-localization behavior in the macaque monkey, *J. Neurophysiol.*, 83, 2723, 2000.
7. King, A.J., Parsons, C.H., and Moore, D.R., Plasticity in the neural coding of auditory space in the mammalian brain, *Proc. Natl. Acad. Sci. USA*, 97, 11821, 2000.
8. May, B.J. and Huang, A.Y., Sound orientation behavior in cats. I. Localization of broadband noise, *J. Acoust. Soc. Am.*, 100, 1059, 1996.
9. Populin, L.C. and Yin, T.C.T., Behavioral studies of sound localization in the cat, *J. Neurosci.*, 18, 2147, 1998.
10. Wightman, F. and Kistler, D., Of Vulcan ears, human ears and 'earprints,' *Nat. Neurosci.*, 1, 337, 1998.
11. Spezio, M.L. et al., Head-related transfer functions of the rhesus monkey, *Hearing Res.*, 144, 73–88, 2000.
12. Recanzone, G.H. et al., Comparison of relative and absolute sound localization ability in humans, *J. Acoust. Soc. Am.*, 103, 1085, 1998.
13. Phillips, D.P. and Brugge, J.F., Progress in neurophysiology of sound localization, *Annu. Rev. Psychol.*, 36, 245, 1985.
14. Merzenich, M.M. and Brugge, J.F., Representation of the cochlear partition on the superior temporal plane of the macaque monkey, *Brain Res.*, 50, 275, 1973.
15. Morel, A., Garraghty, P.E., and Kaas, J.H., Tonotopic organization, architectonic fields, and connections of auditory cortex in macaque monkeys, *J. Comp. Neurol.*, 335, 437, 1993.
16. Jones, E.G. et al., Subdivisions of macaque monkey auditory cortex revealed by calcium-binding protein immunoreactivity, *J. Comp. Neurol.*, 362, 153, 1995.

17. Molinari, M. et al., Auditory thalamocortical pathways defined in monkeys by calcium-binding protein immunoreactivity, *J. Comp. Neurol.*, 362, 171, 1995.
18. Hackett, T.A., Stepniewska, I., and Kaas, J.H., Subdivisions of auditory cortex and ipsilateral cortical connections of the parabelt auditory cortex in macaque monkeys, *J. Comp. Neurol.*, 394, 475, 1998.
19. Hackett, T.A., Stepniewska, I., and Kaas, J.H., Thalamocortical connections of the parabelt auditory cortex in macaque monkeys, *J. Comp. Neurol.*, 400, 271, 1998.
20. Rauschecker, J.P., Tian, B., and Hauser, M., Processing of complex sounds in the macaque nonprimary auditory cortex, *Science*, 268, 111, 1995.
21. Rauschecker, J.P. et al., Serial and parallel processing in rhesus monkey auditory cortex, *J. Comp. Neurol.*, 382, 89, 1997.
22. Kosaki, H. et al., Tonotopic organization of auditory cortical fields delineated by parvalbumin immunoreactivity in macaque monkeys, *J. Comp. Neurol.*, 386, 304, 1997.
23. Recanzone, G.H., Guard, D.C., and Phan, M.L., Frequency and intensity response properties of single neurons in the auditory cortex of the behaving macaque monkey, *J. Neurophysiol.*, 83, 2315, 2000.
24. Rauschecker, J.P. and Tian, B., Mechanisms and streams for processing of "what" and "where" in auditory cortex, *Proc. Natl. Acad. Sci. USA*, 97, 11800, 2000.
25. Tian, B. et al., Functional specialization in rhesus monkey auditory cortex, *Science*, 292, 290, 2001.
26. Woods, T.M., Su, T.K., and Recanzone, G.H., Spatial tuning as a function of stimulus intensity of single neurons in awake macaque monkey auditory cortex, *Soc. Neurosci. Abstr.*, 27, 2001.
27. Rauschecker, J.P., Parallel processing in the auditory cortex of primates, *Audiol. Neuro-Otol.*, 3, 86, 1998.
28. Kaas, J.H., Hackett, T.A., and Tramo, M.J., Auditory processing in primate cerebral cortex, *Curr. Opin. Neurobiol.*, 9, 164, 1999.
29. Ungerleider, L.G. and Mishkin, M., Two cortical visual systems, in *Analysis of Visual Behavior*, D.J. Ingle et al., Eds., MIT Press, Cambridge, MA, 1982, p. 549.
30. Ungerleider, L.G. and Haxby, J.V., "What" and "where" in the human brain, *Curr. Opin. Neurobiol.*, 4, 157, 1994.
31. Heffner, H.E. and Heffner, R.S., Effect of bilateral auditory cortex lesions on sound localization in Japanese macaques, *J. Neurophysiol.*, 64, 915, 1990.
32. Thompson, G.C. and Cortez, A.M., The inability of squirrel monkeys to localize sound after unilateral ablation of auditory cortex, *Behav. Brain Res.*, 8, 211, 1983.
33. Kavanagh, G.L. and Kelly, J.B., Contribution of auditory cortex to sound localization by the ferret (*Mustela putorius*), *J. Neurophysiol.*, 57, 1746, 1987.
34. Jenkins, W.M. and Merzenich, M.M., Role of cat primary auditory cortex for sound-localization behavior, *J. Neurophysiol.*, 52, 819, 1984.
35. Sanachez-Longo, L.P. and Forster, F.M., Clinical significance of impairment of sound localization, *Neurology*, 8, 119, 1958.
36. Poirier, P. et al., Sound localization in hemispherectomized patients, *Neuropsychologia*, 32, 541, 1994.
37. Clarke, S. et al., Auditory agnosia and auditory spatial deficits following left hemispheric lesions: evidence for distinct processing pathways, *Neuropsychologia*, 38, 797, 2000.
38. Eisenman, L.M., Neural encoding of sound localization: an electrophysiological study in auditory cortex (AI) of the cat using free field stimuli, *Brain Res.*, 75, 203, 1974.

39. Middlebrooks, J.C. and Pettigrew, J.D., Functional classes of neurons in primary auditory cortex of cat distinguished by sensitivity to sound location, *J. Neurosci.*, 1, 107, 1981.
40. Imig, T.J., Irons, W.A., and Samson, F.R., Single-unit selectivity to azimuthal direction and sound pressure level of noise bursts in cat high-frequency primary auditory cortex, *J. Neurophysiol.*, 63, 1448, 1990.
41. Rajan, R. et al., Azimuthal sensitivity of neurons in primary auditory cortex of cats. I. Types of sensitivity and the effects of variations in stimulus parameters, *J. Neurophysiol.*, 64, 872, 1990.
42. Brugge, J.F., Reale, R.A., and Hind, J.E., The structure of spatial receptive fields of neurons in primary auditory cortex of the cat, *J. Neurosci.*, 16, 4420, 1996.
43. Middlebrooks, J.C. et al., Codes for sound-source location in nontonotopic auditory cortex, *J. Neurophysiol.*, 80, 863, 1998.
44. Benson, D.A., Hienz, R.D., and Goldstein, M.H., Jr., Single-unit activity in the auditory cortex of monkeys actively localizing sound sources: spatial tuning and behavioral dependency, *Brain Res.*, 219, 249, 1981.
45. Ahissar, M. et al., Encoding of sound-source location and movement: activity of single neurons and interactions between adjacent neurons in the monkey auditory cortex, *J. Neurophysiol.*, 67, 203, 1992.
46. Knudsen, E.I. and Konishi, M., A neural map of auditory space in the owl, *Science*, 200, 795, 1978.
47. Knudsen, E.I. and Brainard, M.S., Creating a unified representation of visual and auditory space in the brain, *Annu. Rev. Neurosci.*, 18, 19, 1995.
48. Konishi, M., Study of sound localization by owls and its relevance to humans, *Comp. Biochem. Physiol. Part A*, 126, 459, 2000.
49. Middlebrooks, J.C. and Knudsen, E.I., A neural code for auditory space in the cat's superior colliculus, *J. Neurosci.*, 4, 2621, 1984.
50. King, A.J. and Hutchings, M.E., Spatial response properties of acoustically responsive neurons in the superior colliculus of the ferret: a map of auditory space, *J. Neurophysiol.*, 57, 596, 1987.
51. Jay, M.F. and Sparks, D.L., Auditory receptive fields in primate superior colliculus shift with changes in eye position, *Nature*, 309, 345, 1984.
52. Wallace, M.T., Wilkinson, L.K., and Stein, B.E., Representation and integration of multiple sensory inputs in primate superior colliculus, *J. Neurophysiol.*, 76, 1246, 1996.
53. Bushara, K.O. et al., Modality-specific frontal and parietal areas for auditory and visual spatial localization in humans, *Nature Neurosci.*, 2, 759, 1999.
54. Weeks, R. et al., A PET study of human auditory spatial processing, *Neurosci. Lett.*, 262, 155, 1999.
55. Weeks, R. et al., A positron emission tomographic study of auditory localization in the congenitally blind, *J. Neurosci.*, 20, 2664, 2000.
56. Palomäki, K. et al., Sound localization in the human brain: neuromagnetic observations, *NeuroReport*, 11, 1535, 2000.
57. Haeske-Dewick, H., Canavan, A.G.M., and Hömberg, V., Sound localization in egocentric space following hemispheric lesions, *Neuropsychologia*, 34, 937, 1996.
58. Poirier, P. et al., Sound localization in acallosal humans listeners, *Brain*, 116, 53, 1993.
59. Irvine, D.R. and Rajan, R., Injury- and use-related plasticity in the primary sensory cortex of adult mammals: possible relationship to perceptual learning, *Clin. Exp. Pharmacol. Physiol.*, 23, 939, 1996.

60. Buonomano, D.V. and Merzenich, M.M., Cortical plasticity: from synapses to maps, *Annu. Rev. Neurosci.*, 21, 149, 1998.
61. Kaas, J.H., The reorganization of somatosensory and motor cortex after peripheral nerve or spinal cord injury in primates, *Prog. Brain Res.*, 128, 173, 2000.
62. Jones, E.G., Cortical and subcortical contributions to activity-dependent plasticity in primate somatosensory cortex, *Annu. Rev. Neurosci.*, 23, 1, 2000.
63. Olds, J. et al., Learning centers of rat brain mapped by measuring latencies of conditioned unit responses, *J. Neurophysiol*, 38, 1114, 1972.
64. Oleson, T.D., Ashe, J.H., and Weinberger, N.M., Modification of auditory and somatosensory system activity during pupillary conditioning in the paralyzed cat, *J. Neurophysiol.*, 38, 1114, 1975.
65. Bakin, J.S. and Weinberger, N.M., Classical conditioning induces CS-specific receptive field plasticity in the auditory cortex of the guinea pig, *Brain Res.*, 536, 271, 1990.
66. Weinberger, N.M., Physiological memory in primary auditory cortex: characteristics and mechanisms, *Neurobiol. Learn. Memory*, 70, 226, 1998.
67. Robertson, D. and Irvine, D.R., Plasticity of frequency organization in auditory cortex of guinea pigs with partial unilateral deafness, *J. Comp. Neurol.*, 282, 456, 1989.
68. Rajan, R. et al., Effect of unilateral partial cochlear lesions in adult cats on the representation of lesioned and unlesioned cochleas in primary auditory cortex, *J. Comp. Neurol.*, 338, 17, 1993.
69. Kilgard, M.P. and Merzenich, M.M., Cortical map reorganization enabled by nucleus basalis activity, *Science*, 279, 1714, 1998.
70. Kilgard, M.P. et al., Sensory input directs spatial and temporal plasticity in primary auditory cortex, *J. Neurophysiol.*, 86, 326, 2001.
71. Kilgard, M.P. and Merzenich, M.M., Plasticity of temporal information processing in the primary auditory cortex, *Nat. Neurosci.*, 1, 727, 1998.
72. Irvine, D.R. et al., Specificity of perceptual learning in a frequency discrimination task, *J. Acoust. Soc. Am.*, 108, 2964, 2000.
73. Recanzone, G.H., Merzenich, M.M., and Schreiner, C.E., Plasticity in the frequency representation of primary auditory cortex following discrimination training in adult owl monkeys, *J. Neurosci.*, 13, 87, 1993.
74. King, A.J. et al., How plastic is spatial hearing?, *Audiol. Neuro-Otol.*, 6, 182–186, 2001.
75. Slattery, W.H. III and Middlebrooks, J.C., Monaural sound localization: acute versus chronic unilateral impairment, *Hearing Res.*, 75, 38, 1994.
76. Gold, J.I. and Knudsen, E.I., Hearing impairment induces frequency-specific adjustments in auditory spatial tuning in the optic tectum of young owls, *J. Neurophysiol.*, 82, 2197, 1999.
77. Javer, A.R. and Schwarz, D.W.F., Plasticity in human directional hearing, *J. Otolaryngol.*, 24, 111, 1995.
78. Recanzone, G.H., Rapidly induced auditory plasticity: the ventriloquism after effect, *Proc. Natl. Acad. Sci. USA*, 95, 869, 1998.
79. Woods, T.M. and Recanzone, G.H., Visually mediated plasticity of acoustic space perception in behaving macaque monkeys, *Soc. Neurosci. Abstr.*, 26, 1221, 2000.
80. Shinn-Cunningham, B., Models of plasticity in spatial auditory processing, *Audiol. Neuro-Otol.*, 6, 187, 2001.
81. Hofman, P.M., Van Riswick, J.G.A., and Van Opstal, A.J., Relearning sound localization with new ears, *Nat. Neurosci.*, 1, 417, 1998.

14 Anatomy and Physiology of Auditory–Prefrontal Interactions in Non-Human Primates

Lizabeth M. Romanski

CONTENTS

I. Introduction 259
II. Organization of Auditory Cortex 260
    A. Anatomy 260
    B. Functional Organization of Non-Primary Auditory Cortex 260
III. Frontal Lobe Organization 261
    A. Anatomical Organization 261
    B. Prefrontal Auditory Connections 261
    C. Cascades of Hierarchical Connections 266
    D. Functional Organization of the Prefrontal Cortex 266
    E. Auditory Domains in the Prefrontal Cortex 268
IV. Conclusions 271
References 274

I. INTRODUCTION

The frontal and temporal lobes have been implicated in language processing with lesion, imaging, and stimulation studies.[1,2] Certainly, language functions have been linked with the frontal lobes since the work of Paul Broca.[9] Early studies defined a role for the frontal lobe in the motor control of speech. Neuroimaging studies have extended the involvement of the ventrolateral frontal lobe to include comprehension, semantics, phonetics, and syntax, as well as auditory and visual working memory;[1–8] however, we still have little knowledge regarding the cellular processes and circuitry of the ventrolateral prefrontal cortex. This may be due, in part, to the fact that historically, few animal studies have examined auditory functions in the prefrontal cortex. Recent anatomical and electrophysiological studies have advanced our knowledge of auditory processes and organization both in the temporal and frontal lobes of primates. These studies have altered our view of the organization and function of auditory association areas[10,11] and have allowed us to pinpoint the regions of the frontal lobe that receive afferents from physiologically defined auditory association cortices. Anatomical localization of prefrontal auditory recipient zones has paved the way for electrophysiological analysis of prefrontal neurons and their receptivity to auditory stimuli. With renewed interest in auditory–prefrontal interactions and an animal model to study these interactions, we may be able to determine the cellular events that underlie auditory memory processes in primates and ultimately language processing in humans.

II. ORGANIZATION OF AUDITORY CORTEX

A. Anatomy

Early studies of the auditory cortical system distinguished a central "core" of primary auditory cortex (AI) on the supratemporal plane surrounded by a "belt" of non-primary association cortices based on cytoarchitectonic differences revealed by Nissl and myelin stains.[12–15] This core and belt organization has been updated by several groups[16–18] and most recently by Hackett et al.[19] These studies have established new criteria by which to view the primary auditory cortex and its surrounding belt regions.[10] For example, primary and non-primary auditory cortex can be distinguished on the basis of their differential staining for the calcium binding protein, parvalbumin.[17–19] AI is easily distinguished by its intense parvalbumin staining in layer IV and its appearance in Nissl stains, where a broad layer IV, densely packed with small size granule cells, is apparent. A more rostral field (R) resembles AI in architecture and has been included as a primary auditory cortical field on the basis of its anatomical[19] as well as electrophysiological[20,21] characteristics. An additional core field, RT, lies just anterior to R. The belt regions have a lighter pattern of parvalbumin staining which is nonetheless distinct when compared with more lateral regions of the parabelt — located ventral to the belt on the superior temporal gyrus (STG) — which have faint parvalbumin staining. In Nissl sections of the lateral belt cortex, layer IV is smaller and less densely packed than AI and R but is more dense than adjacent parabelt regions.

B. Functional Organization of Non-Primary Auditory Cortex

The recent anatomical parcellation of primary auditory cortex is consistent with prior physiological identification of AI by Merzenich and Brugge[22] and more recent electrophysiological analysis.[16,18,20,21] Physiological localization of the non-primary auditory cortex has also recently been achieved.[18,20] Rauschecker et al.[20] provided the first physiological evidence for three separate tonotopic regions (anterolateral, AL; middle lateral, ML; and caudolateral, CL) in the non-primary lateral belt association cortex. Compared to primary auditory cortical neurons, which readily respond to relatively simple acoustic elements such as pure tones, neurons of the lateral belt association cortex prefer complex stimuli including bandpassed noise and vocalizations.[20,21] The three lateral belt regions are physiologically distinguishable by the frequency reversals between them and by their differential responses to complex acoustic stimuli.[20,21,23,24] The anterior lateral belt region exhibits more selectivity for types of complex sounds, such as vocalizations, while the caudolateral field has neurons that show a greater spatial selectivity.[23,24] Together, the physiological and anatomical evidence supports a view of the auditory cortical system as a series of concentric circles with a central core. The primary auditory cortex is surrounded by a narrow belt of secondary association areas located laterally and medially, and a parabelt auditory association cortex is located on the ventral gyral surface of the STG just below the lateral belt (Figures 14.1 and 14.2A), with successively higher areas processing increasingly complex stimuli. The anterior temporal lobe and the dorsal bank of the superior temporal sulcus may represent later stages of possible auditory processing, as they receive afferents from the parabelt cortex.[19,25]

III. FRONTAL LOBE ORGANIZATION

A. ANATOMICAL ORGANIZATION

The organization of the frontal lobes has been described in terms of its anatomical structure and connections,[26–29] physiology,[30,31] and the effects of lesions.[31–33] The frontal lobes are composed of several regions that differ in both cytoarchitecture and connections. The most commonly used anatomical scheme for describing the organization of the frontal lobes is that of Walker,[26] who, like Brodmann, relied chiefly on Nissl stains to create his cytoarchitectonic parcellation of the frontal lobe. Barbas and Pandya[34] updated this organization, relying on both cytoarchitecture and connectional data to formulate their prefrontal cytoarchitectonic map. A more recent parcellation based on a variety of histological techniques including Nissl, myelin, acetylcholinesterase staining and others is that of Preuss and Goldman-Rakic.[28] In this extensive study, the authors clearly delineated the boundaries of granular prefrontal cortex, identifying subdivisions not previously recognized (Figure 14.1), and their nomenclature will be used in this chapter. An additional study[35] compared the cytoarchitecture of dorsolateral prefrontal cortex in the macaque and human brain with special attention to the parcellation of areas 46 and 8. This study argues for homologies between distinct macaque and human brain regions to allow for easier comparison of human functional imaging data with macaque behavioral and physiological data.

B. PREFRONTAL AUDITORY CONNECTIONS

In defining prefrontal–auditory interactions, it is necessary to have an understanding of the anatomical pathways by which auditory cortices can influence the frontal lobes, and vice versa. Research in the visual system has pointed to the importance of the prefrontal cortex as a recipient of visuo-spatial and visual object information via the dorsal and ventral visual streams.[36] In contrast to the visual connections, the prefrontal targets of central auditory pathways are less clear. One general principle that has been observed in most studies of prefrontal–auditory connections is that a rostrocaudal topography exists (Figure 14.1).[25,27,29,37–39] Direct, reciprocal connections are apparent between the caudal STG and caudal prefrontal cortex (PFC) (including periprincipalis

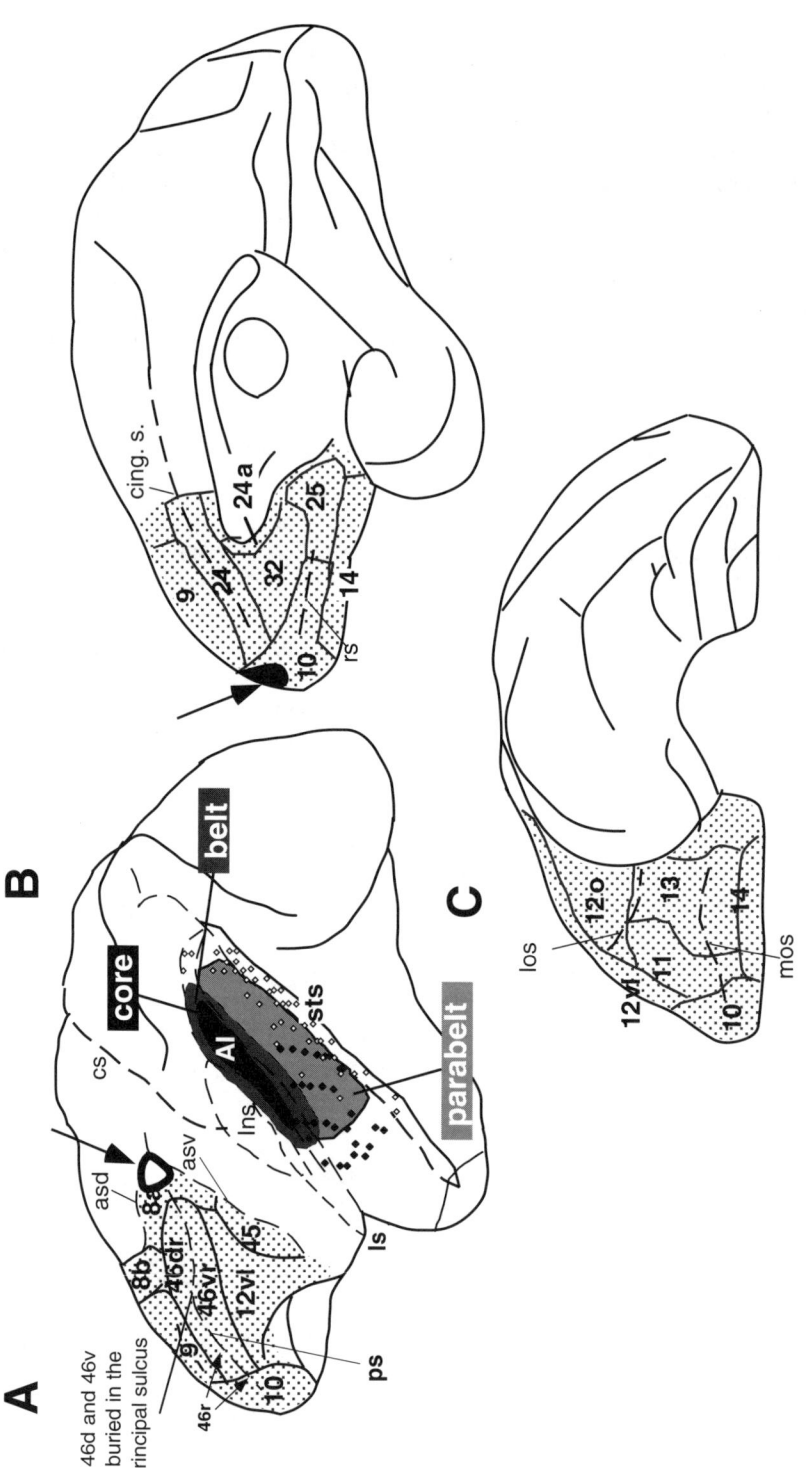

FIGURE 14.1 Cytoarchitectonic organization of prefrontal and auditory cortex and connections. Lateral (A), medial (B), and orbital (C) views of the macaque brain are illustrated with color-coded injections sites and retrograde cell labeling. (A) The parcellation of dorsolateral prefrontal cortex is based on Preuss and Goldman-Rakic.[28] The principal sulcus is divided into subregions including rostral, dorsal, and ventral regions. The depths of the sulcus are subdivided as 46d and 46v buried within the sulcus. The core and belt organization of the auditory cortex is depicted in A using the organizational scheme of Hackett et al.[19] The primary auditory cortex (AI) or core is surrounded by the belt region, and the parabelt lies adjacent to the lateral belt (gray). The medial and orbital frontal cytoarchitectonic boundaries are shown according to Carmichael and Price[43] in B and C. To illustrate the rostrocaudal topographic organization of frontotemporal connections, the results of two HRP injections in prefrontal cortex are shown. In A, an injection into caudal dorsolateral PFC (area 8) is shown in white (arrow) and the corresponding retrogradely labeled cells are shown in the caudal auditory cortex and in the dorsal bank of the superior temporal sulcus (as white squares). In B, an injection into rostral PFC (area 10) is colored black (arrow). This injection resulted in retrogradely labeled cells (black) in the rostral belt and parabelt depicted in A. Abbreviations: asd, arcuate sulcus, dorsal limb; asv, arcuate sulcus, ventral limb; cs, central sulcus; cing.s, cingulate sulcus; los, lateral orbital sulcus; ls, lateral sulcus; mos, medial orbital sulcus; ps, principal sulcus; rs, rostral sulcus; sts, superior temporal sulcus.

[area 46], periarcuate [area 8a], and the inferior convexity [area 45]).[27,29] Middle and rostral STG are reciprocally connected with rostral principalis (rostral 46 and 10) and orbitofrontal areas (areas 11 and 12).[37,38,40] In addition, a robust connection exists between the orbital and medial prefrontal cortices with the anterior temporal lobe (recently designated as STGr) and the temporal pole.[25,27,42–44]

Two recent studies have examined prefrontal–auditory interactions and utilized updated anatomical and physiological evidence to reiterate the rostrocaudal topography of auditory–prefrontal connections and to determine the precise prefrontal loci that are most densely innervated by belt and parabelt auditory association cortices.[25,44] Specifically, the rostral, orbital, and ventrolateral areas of the prefrontal cortex are reciprocally connected with rostral belt (areas AL and anterior ML) and parabelt association cortex (area RP, rostral parabelt), whereas caudal principalis and some ventrolateral regions of the prefrontal cortex are reciprocally connected with the caudal belt (caudal ML and CL) and parabelt (area CP, caudal parabelt) (Figure 14.2).[25] It has been noted that the parabelt was more strongly connected with the prefrontal cortex than the belt

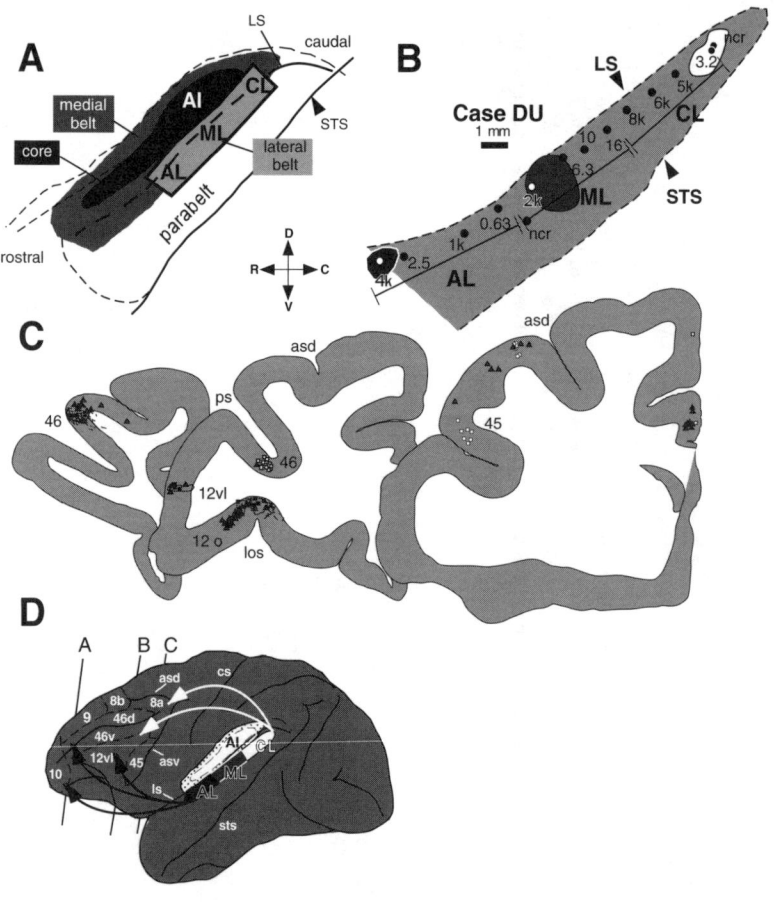

FIGURE 14.2

(Figure 14.2).[25,39,44] Furthermore, the dorsal bank of the superior temporal sulcus (STS) has a dense projection to several regions in dorso- and ventrolateral prefrontal cortex.[25]

While these anatomical studies suggest that auditory information is received by the prefrontal cortex, more direct evidence is obtained when anatomical and physiological methods are combined. In one such study, Romanski et al.[39] recorded from lateral belt auditory areas AL, ML, and CL and placed injections of anatomical tracers into these physiologically defined regions. The tracers included both anterograde and retrograde tracers that permitted visualization of whole axonal arbors as they entered different laminae in cortical targets. Five specific regions of the frontal lobes contained both retrogradely labeled cells and anterogradely labeled fibers. These frontal lobe targets included the frontal pole, principal sulcus, lateral inferior convexity, lateral orbital cortex, and dorsal periarcuate region (Figure 14.2). Moreover, these connections were topographically organized such that projections from AL typically involved the frontal pole (area 10), rostral principal sulcus (area 46), inferior convexity (areas 12vl and 45), and lateral orbital cortex (areas 11 and 12o; see Figure 14.2C). In contrast, projections from area CL targeted the dorsal periarcuate cortex (area 8a, frontal eye fields) and the caudal principal sulcus (area 46), as well as the caudal inferior convexity (areas 12vl and 45; see Figures 14.2C,D) and, in two cases, premotor cortex (area 6d). The frontal pole (area 10) and the lateral orbital cortices (areas 11 and 12) were devoid of anterograde labeling from injections into the caudal auditory region. Conversely, the frontal eye fields did not receive projections from anterior auditory area AL. The connections of area ML were a mix of the connections of AL and CL.

These highly specific rostrocaudal topographical frontal–temporal connections suggest the existence of separate streams of auditory information that target distinct domains of the frontal lobes (Figure 14.2D). One pathway, originating in CL, targets caudal

---

FIGURE 14.2 (See Color Figure 14.2 in color insert.) Auditory projections from physiologically identified lateral belt areas target specific regions of the prefrontal cortex. (A) Schematic of auditory cortex showing the medial and lateral belt, primary auditory cortex (AI), and parabelt cortex on the superior temporal gyrus. The portion of the lateral belt recorded is from Romanski et al.[39] (B) A physiological map of the lateral belt recording from one case. The best center frequency for each electrode penetration (black or white dots) is labeled in kHz. Injections of different anterograde and retrograde tracers (shaded regions) are shown. The boundaries of AL, ML, and CL are delineated by a bounded line and are derived from the frequency reversal points. In this example, the injection was centered in the 4- to 6-kHz region in all three lateral belt regions. (C) Three coronal sections (rostral-left to caudal-right) through the prefrontal cortex showing the anterograde (outlines) and retrograde (circles, triangles, and squares) labeling. Projections from AL (squares and dark lines) targeted the rostral principal sulcus and orbital cortex in the most rostral coronal section. In addition, the second coronal section shows projections from AL to the inferior convexity and the lateral orbital cortex. The most caudal coronal section shows evidence of projections from an injection into area CL (white squares). Projections from ML (triangles) overlapped with those from AL and CL. (D) Schematic summary of the dual streams connecting auditory and prefrontal cortex. Rostral auditory cortex projects to rostral and ventral prefrontal cortex (black arrows); caudal auditory cortex is connected with caudal and dorsal prefrontal cortex (white arrows). Abbreviations: LS, lateral sulcus; PS, principal sulcus; STS, superior temporal sulcus; D, dorsal; V, ventral; R, rostral; C, caudal. (Based on findings from Romanski et al.[39])

dorsolateral prefrontal cortex (DLPFC); the other pathway, originating in AL, targets rostral and ventral prefrontal areas. Previous studies have demonstrated that there are spatial and nonspatial *visual* streams that target dorsal–spatial and ventral–object prefrontal regions. Thus, it is possible that the pathways originating from the anterior and posterior *auditory* belt and parabelt cortices are analogous to the "what" and "where" streams of the visual system. The recent demonstration of separate functional specificity of anterior belt and caudal belt regions[23,24] adds further evidence to this notion.

C. Cascades of Hierarchical Connections

In addition to the existence of functionally distinct "what" and "where" auditory projections is the notion that auditory–prefrontal connections are a cascade of multiple streams or projections. Primary regions, AI and R, project to the lateral and medial belt regions of the auditory cortex.[16,19] It is here within the auditory belt that two auditory streams diverge.[18,21] One stream begins in the rostral belt (regions RTL, AL, and ML), sending small projections to the anterior and orbital prefrontal cortex and a dense projection to the rostral parabelt (area RP).[19,25] The rostral parabelt sends a somewhat larger projection to the frontal lobe and has further connections with the dorsal bank of the STS and the anterior temporal lobe. In turn, the STS and the anterior temporal lobe project to the frontal lobe and other association cortices.[27,38,45] The other auditory stream — which begins in the middle and caudal belt (areas ML and CL) — projects to the dorsal arcuate region, caudal principalis, and ventrolateral prefrontal cortex. It may also have an intermediate relay in the posterior parietal cortex that is densely connected with the caudal prefrontal cortex.[46]

Thus, a cascade of auditory afferents from two divergent streams targets the prefrontal cortex. Small streams enter the frontal lobe from early auditory processing stations (such as the lateral and medial belt) and successively larger projections originating in the parabelt, anterior temporal lobe, and dorsal bank of the STS join the flow of auditory information destined for the frontal lobe. The flow of visual information is known to follow a similar cascading pattern, with area TEO sending a restricted input to the frontal lobe and to the more rostrally located area TE, which provides a more widespread input to the prefrontal cortex than area TEO.[47] The role of multiple streams of auditory projections to the frontal lobe is unknown. It is possible, however, that a cascade of inputs from increasingly more complex auditory association cortex may be necessary to represent all qualities of a complex auditory stimulus. In other words, crudely processed information from early unimodal belt association cortex may arrive first at specific prefrontal targets, while complex information (such as voice identity, emotional relevance, etc.) from increasingly higher associative auditory areas may follow. This hierarchy of inputs allows for a rich representation of complex auditory stimuli to reach the prefrontal cortex in an efficient manner.

D. Functional Organization of the Prefrontal Cortex

The diverse anatomy of the prefrontal cortex gives rise to diverse function. Because both cytoarchitecture and connections of prefrontal areas differ, it is not surprising that function across these areas may differ as well; however, fractionation of functions within the prefrontal cortex has recently become a hotly debated topic. Although a variety of investigators have long argued for a segregation of functions based on lesion and electrophysiological data,[30–32,36,48,49] some recent work has focused on the homogeneity of working memory processes within the dorsolateral prefrontal cortex.[50,51] One classic theoretical construct invoked to describe the diversity of frontal lobe organization is that of *domain specificity*.[31,33] This hypothesis argues for a segregation of frontal lobe regions based on information domains. Goldman-Rakic[33] holds that "the dorsolateral prefrontal cortex as a whole has a generic function — 'on-line' processing ... or working memory" with autonomous subdivisions that process different types of information. Anatomical and physiological evidence supports the notion of separate information processing domains. A variety of studies have shown that the DLPFC has neurons involved in spatial processing, while the ventrolateral regions, including the inferior convexity (IFC), have recently been associated with object and face processing (Figure 14.3).[36,52,53] Various studies in primates have confirmed the visuo-spatial processing specialty of DLPFC, including electrophysiological recordings that have demonstrated the increased activity of neurons during the delay period of spatial, delayed-response tasks[54–56] (for review, see Goldman-Rakic[31,33]). In contrast, only a few studies have examined object- and pattern-specific responses in the IFC.[36,52,53,57–59] In these studies, neurons with receptive fields that included the fovea were shown to have robust and highly specific responses to pictures of objects and faces.[36]

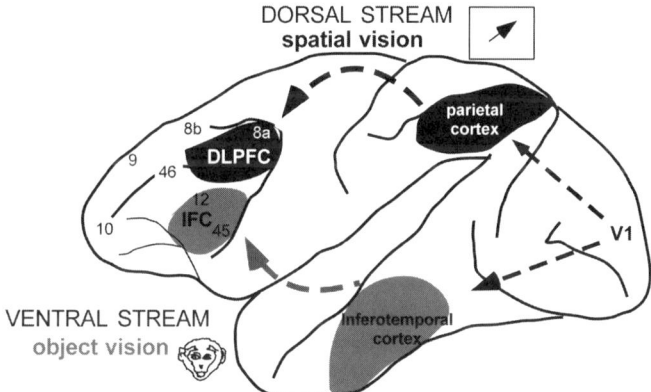

FIGURE 14.3 Visual "what" and "where" streams, previously proposed for transmission of visual information by Ungerleider and Mishkin,[86] are shown terminating in separate functional regions of the prefrontal cortex. The working memory domains in the PFC proposed by Goldman-Rakic[33] are shown on this lateral view of the macaque brain. The dorsal "spatial stream" (shown in black) continues from the posterior parietal cortex by a dense projection to the dorsolateral prefrontal cortex (DLPFC), including areas 8 and 46. Cells in DLPFC have delay activity related to the location of visual cues. A ventral or object stream (shown in gray) has been proposed which originates in inferotemporal cortex and projects to the ventrolateral prefrontal cortex, known as the inferior convexity (IFC; areas 12 and 45). Cells in the IFC are activated by pictures of patterns, objects, and faces.

E. AUDITORY DOMAINS IN THE PREFRONTAL CORTEX

Evidence for auditory processing within the prefrontal cortex of animals comes from anatomical (see above) and physiological studies. Involvement of the prefrontal cortex in auditory processing is further confirmed by clinical studies in humans showing deficits in auditory and language processing following lesions of frontal cortical areas. Large lesions of the lateral frontal cortex in non-human primates have been reported to disrupt performance of auditory discrimination tasks.[60-64] In many of these studies, ventrolateral prefrontal cortex or the principal sulcus was included in the lesion. The effects of prefrontal cortical lesions on auditory discrimination and response tasks have been interpreted to suggest that dorsolateral portions of the prefrontal cortex are mainly involved in response inhibition,[63,65] while more posterior portions of the prefrontal cortex may serve some role in conditional learning.[32] However, this conclusion is premature considering that no study has performed a careful analysis of the role of the prefrontal cortex in auditory working memory using tasks devoid of conditional, cross-modal, or spatial components to assess working memory. Clearly, more information on the effects of lesions of the prefrontal cortex in auditory processes is needed.

The essential role of the prefrontal cortex in complex auditory processes is also evident when the PFC is disconnected from the auditory association cortex. By lesioning first the prefrontal cortex and then the auditory cortex, Gaffan and Harrison[66] demonstrated that the left temporal lobe was specialized for auditory perception in the monkey and that the left prefrontal cortex was specialized for auditory–visual associations. Another example is provided by a study by Knight et al.,[67] where evoked potentials were analyzed in human patients with focal lesions of the DLPFC. It was found that the DLPFC lesions increased the amplitude of middle-latency auditory evoked potentials, suggesting that the prefrontal cortex may exert inhibitory modulation of input to the primary auditory cortex during normal auditory transmission.[67]

While numerous studies have demonstrated the proclivity of neurons in the macaque prefrontal cortex to respond to visual stimuli in memory and non-memory tasks, few studies have described the responses of prefrontal neurons to complex acoustic stimuli. Several physiological studies support a role for caudal prefrontal cortex in auditory spatial processing. For example, studies have found neurons responsive to simple auditory stimuli in the periarcuate region,[68-70] and lesions of the periarcuate cortex can impair auditory discrimination in primates.[32,60] Furthermore, neurons in the periarcuate region which respond to auditory stimuli are affected by the location of the sound source,[69,71,72] and the auditory responses of these neurons are affected by changes in gaze,[71] a finding that is in agreement with the role of the frontal eye field in saccadic eye movements to salient targets. Vaadia et al.[70] noted the increased response of neurons in the periarcuate region when non-human primates were engaged in an auditory localization task as compared to a passive listening task. These observations of DLPFC auditory–spatial function are complemented by the observation that neurons in the caudal belt region, which project to the DLPFC, are sensitive to the location of auditory stimuli.[24,73,74] Tian et al.[24] found that caudal auditory belt neurons were spatially selective (compared with anterior belt neurons, which were selective for different types of vocalizations). Thus, the caudal principalis and periarcuate region are the targets of a dorsal,

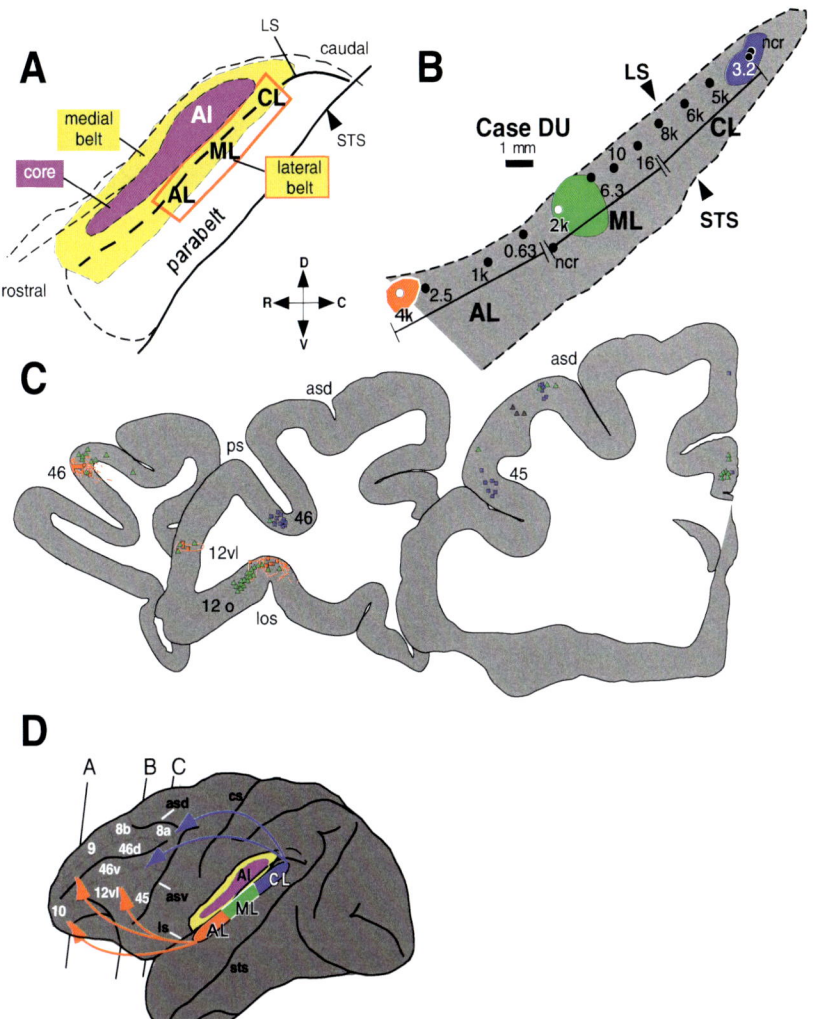

COLOR FIGURE 14.2 Auditory projections from physiologically identified lateral belt areas target specific regions of the prefrontal cortex. (A) Schematic of auditory cortex showing the locations of the medial and lateral belt (shaded in yellow), primary auditory cortex (AI, purple), and parabelt cortex on the superior temporal gyrus. The portion of the lateral belt recorded by Romanski et al.[39] is outlined in red. (B) A physiological map of the lateral belt recording from one case. The best center frequency for each electrode penetration (black or white dots) is labeled in kHz. Injections of different anterograde and retrograde tracers (red, green, and blue regions) are shown. The boundaries of AL, ML, and CL are delineated by a bounded line and are derived from the frequency reversal points. In this example, the injection was centered in the 4- to 6-kHz region in all three lateral belt regions. (C) Three coronal sections (rostral-left to caudal-right) through the prefrontal cortex showing the anterograde (outlines) and retrograde (colored circles and squares) labeling. Projections from AL (shown in red) targeted the rostral principal sulcus and orbital cortex in the most rostral coronal section. In addition, the second coronal section shows projections from AL (in red) in the inferior convexity and the lateral orbital cortex (area 12o). The most caudal coronal section shows evidence of projections from an injection into area CL (shown in blue). Projections from ML (shown in green) overlapped with those from AL and CL. (D) Schematic summary of the dual streams connecting auditory and prefrontal cortex. Rostral auditory cortex projects to rostral and ventral prefrontal cortex (red arrows); caudal auditory cortex is connected with caudal and dorsal prefrontal cortex. Abbreviations: LS, lateral sulcus; PS, principal sulcus; STS, superior temporal sulcus; D, dorsal; V, ventral; R, rostral; C, caudal. (Based on findings from Romanski et al.[39])

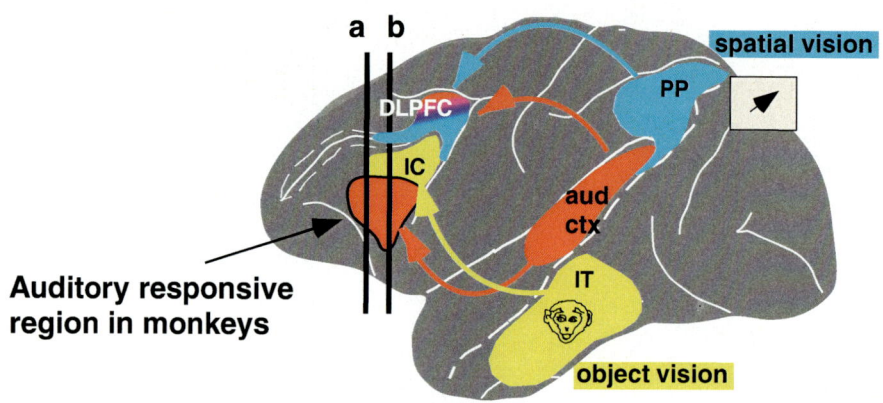

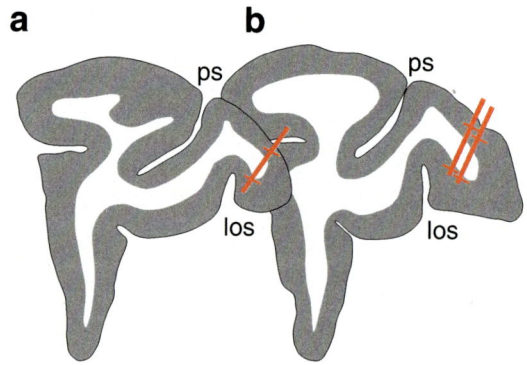

COLOR FIGURE 14.4 Location of auditory responsive neurons in the prefrontal cortex of the macaque. The locations of auditory responsive cells recorded from the ventrolateral prefrontal cortex are shown on a lateral view of the macaque prefrontal cortex in relationship to the visual domains previously described.[36,52,53] The visuo-spatial stream is shown in light blue, while the ventral-object stream is shown in yellow. Dual auditory streams targeting dorsal and ventral PFC regions are depicted with red arrows. The ventrolateral prefrontal auditory region is shown in red antero-lateral to a visual responsive domain (labeled IC and shaded yellow) and has neurons that respond to complex sounds including vocalizations. While visuo-spatial responses have been extensively demonstrated in DLPFC, recent experiments have also shown that there are auditory neurons in this same region that exhibit spatial tuning; thus DLPFC processes auditory and visual spatial information (shown with red and blue gradient). The lines a and b through the auditory prefrontal region refer to the coronal sections at right which depict the locations of auditory cells in the inferior convexity. a, b: The electrode penetrations in these coronal sections are depicted as black lines and the locations of cells are indicated by cross marks on these lines. Auditory cells were found on the lateral and orbital portions of the convexity, areas 12 and 45. Abbreviations: aud ctx, auditory cortex; IC, inferior convexity; other abbreviations as in Figure 14.1.

auditory–spatial processing stream, originating in the caudal belt and parabelt, similar to the dorsal visual stream, which terminates in the DLPFC.[36,47]

Although spatial localization of sounds has been examined physiologically in dorsolateral prefrontal cortex of non-human primates, comparable efforts to study cellular processes in the PFC (including the ventrolateral PFC, the inferior convexity) that mediate communication and nonspatial acoustic signaling are sparse. In fact, few recording studies have sampled neurons in ventrolateral and orbitofrontal regions with auditory stimuli. Responses to acoustic stimuli have been only sporadically noted in the frontal lobes of Old and New World monkeys.[75–81] Several of these studies used auditory stimuli in combination with visual stimuli as task elements and did not systematically explore auditory responsive cells.[68,80,81] One extensive mapping study of auditory, visual, and oculomotor stimuli in macaques failed to yield a significant number of auditory responsive cells or a distinct clustering of auditory cells.[78,79] More encouraging data have been noted in squirrel monkeys; one study noted that 20% of recorded cells in the frontal lobe responded to auditory stimuli, which included clicks, tones, and species-specific vocalizations. The auditory responsive neurons, however, were found over a fairly wide region of the frontal cortex, including dorsolateral, periarcuate, and premotor regions.[75] Thus, all previous studies of both Old and New World monkeys have failed to find discrete loci for the processing of auditory stimuli in the frontal lobe.

Building on the anatomical studies that predicted an auditory region in ventrolateral prefrontal cortex,[25,39] Romanski and Goldman-Rakic[82] have recently described an auditory responsive region in the macaque prefrontal cortex. In this study, the investigators demonstrated that a discrete region of the PFC has neurons that respond to complex acoustic stimuli, including species-specific vocalizations. They were able to establish the selectivity of these neurons for acoustic stimuli and rule out visual or saccadic activity as an alternative explanation for the responses. Most of the auditory neurons were localized to a 4-mm × 4-mm area in ventrolateral PFC (Figure 14.4). While visual responses in the same animals were noted over a wide region of the ventrolateral PFC, the auditory responsive cells were tightly clustered and were located anterolateral to visual neurons. Histological verification revealed that auditory neurons were localized to a region overlying the lateral surface of the inferior convexity (areas 12 lateral and 45) and extending to the lateral orbital cortex (area 12 orbital) (Figure 14.4). While most cells responded to both vocalization and nonvocalization stimuli (Figure 14.5A–C), a subset of cells responded stronger to vocalizations, and a small number of cells responded only to the category of vocalizations (Figure 14.5D). The localization of auditory responses to the inferior prefrontal convexity in the non-human primate is suggestive of some functional similarities between this region and the inferior frontal gyrus of the human brain (including Broca's area).[9]

Surprisingly, only one other electrophysiological study[76] has previously observed auditory responses in the macaque IFC, within the lateral orbital cortex near where auditory responsive neurons were found by Romanski and Goldman-Rakic.[82] In this study, light flashes and clicks were used to evoke responses in the orbital cortex.[76] The investigators found auditory, visual, and combined responses in this region. The sparse evidence for auditory responses in the ventrolateral PFC in earlier studies may be due to the fact that studies often confine electrode penetrations to caudal

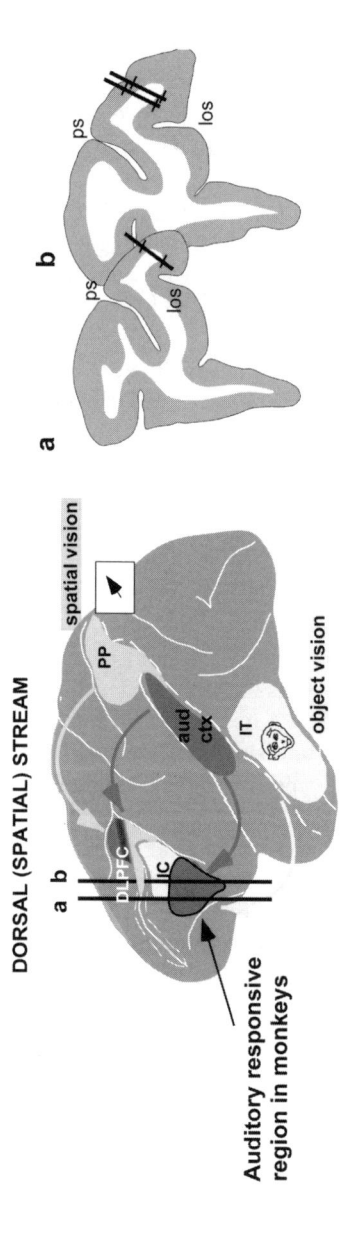

FIGURE 14.4 (See Color Figure 14.4 in color insert.) Location of auditory responsive neurons in the ventrolateralprefrontal cortex (vlPFC) of the macaque. The locations of auditory responsive cells recorded from the vlPFC (in dark gray) are shown on a lateral view of the macaque brain in relationship to visual domains previously described.[36,52,53] The visuo-spatial stream is shown terminating in DLPFC (from posterior parietal cortex, PP arrow) while the ventral-object stream is shown terminating in IC from inferotemporal cortex (IT, monkey face). Dual auditory streams targeting dorsal and ventral PFC regions are depicted from auditory cortex. The vlPFC auditory region (dark gray) is shown antero-lateral to the ventral visual object domain (IC) and has neurons that respond to complex sounds including vocalizations. While visuo-spatial responses have been demonstrated in DLPFC, recent experiments have also shown that auditory neurons in DLPFC exhibit spatial tuning thus DLPFC processes auditory and visual spatial information. The lines a and b through the auditory prefrontal region refer to the coronal sections at right depicting the locations of auditory cells in vlPFC. a, b: The electrode penetrations in these prefrontal coronal sections are depicted as dark lines and the locations of cells are indicated by cross marks on these lines. Auditory cells were found on the lateral and orbital portions of the convexity, areas 12 and 45. Abbreviations: aud ctx, auditory cortex; IC, inferior convexity; pp, parietal cortex; other abbreviations as in Figure 14.1.

and dorsolateral PFC.[69,78,79,81] In addition, the size of the auditory region described by Romanski and Goldman-Rakic[82] is small and difficult to find unless anatomical and physiological landmarks are utilized.[39,52,53]

A few studies have noted that salient environmental stimuli and, in particular, vocalizations are most effective in driving prefrontal neurons.[75,78,82] The preference of frontal lobe auditory neurons for vocalizations may be indicative of further specialization. First, the vocalization stimuli may represent the most complex and dynamic acoustic stimuli of the repertoire. In future studies, responses to vocalizations can be compared with responses to other complex acoustic stimuli (such as tone sequences and combinations, complex sweeps, etc.) which can be contrived to mirror the spectral and temporal richness of vocalizations but which will lend themselves to more precise spectral and temporal analysis.[83] Vocal stimuli may also represent a behaviorally relevant variant of acoustic stimuli that is most salient for monkeys. Finally, just as language-related sounds have been found to be more effective than tone or noise stimuli in activating the inferior frontal gyrus in human imaging studies,[84,85] vocalizations may be most effective in eliciting responses in the non-human primate ventrolateral frontal lobe due to their relevance as communication signals. As an example, studies of visual stimulus preference in an adjacent region of the ventrolateral prefrontal cortex indicate that pictures of faces, rather than simple patterns and colored squares, evoke responses in the "face" region of the prefrontal cortex.[52,53]

IV. CONCLUSIONS

The anatomical and physiological evidence reviewed is compatible with the existence of *at least two* anatomically and functionally distinct temporofrontal processing pathways for auditory information in non-human primates. Certainly the number of auditory processing streams is by no means limited to these two, but they may consist of several smaller streams involved in various aspects of auditory perception. However, the evidence does suggest a general dissociation between spatial and nonspatial processing. Just as spatial and nonspatial visual information generally flows via dorsal and ventral streams,[86] spatial and nonspatial streams of auditory information may emanate from caudal and rostral auditory cortex and terminate in distinct regions of prefrontal cortex that are specialized for spatial and object processing (Figure 14.4). Caudal auditory belt regions, which themselves exhibit spatial selectivity, target dorsolateral prefrontal cortex, which has been shown in the non-human primate to contain auditory neurons with spatial selectivity.[70,72] In addition, the rostral auditory belt cortex — which has been shown to be selectively responsive to monkey vocalizations[24] — projects to rostral, orbital, and ventrolateral prefrontal cortical regions. Moreover, the existence of an auditory responsive domain in the ventrolateral prefrontal cortex (also known as the inferior convexity), in an area that receives afferents from the rostral belt and parabelt, has recently been shown.[39] The localization of auditory responses to the inferior prefrontal convexity in the macaque is suggestive of some functional similarities between this region and the inferior frontal gyrus of the human brain (including Broca's area),[87] and its identification may provides us with an animal model by which to dissect the cellular mechanisms that underlie vocal communication in primates and, ultimately, language processing in humans.

272 Primate Audition: Ethology and Neurobiology

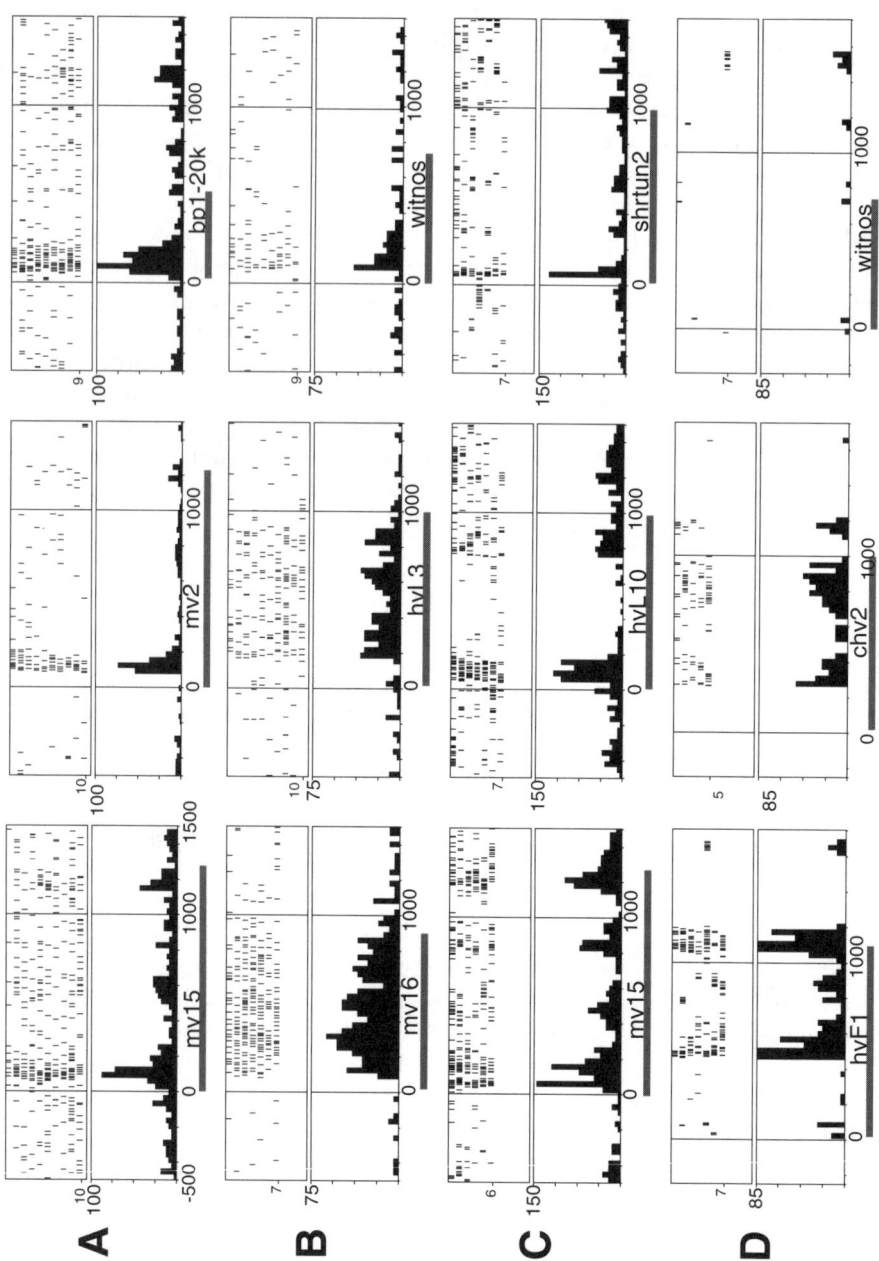

FIGURE 14.5 Auditory responsive neurons in the ventral prefrontal cortex of the rhesus macaque. The auditory responses of four cells (rows A to D) to three auditory stimuli (in columns) as raster (top panels) and PSTHs (bottom panels) are depicted. The gray bar below each histogram indicates the onset and duration of the auditory stimulus. Cell A gave a nonspecific phasic onset response to all auditory stimuli tested, while auditory stimuli elicited a tonic (sustained) response that lasted the length of the auditory stimuli in cell B. Cell C showed evidence of stimulus synchronized activity for some stimuli (mv15; shrtun). Although cells A to C responded to a variety of vocalization and nonvocalization stimuli, the response of cell D was significantly stronger for vocalization stimuli as evidenced by the lack of response to additional stimuli, including white noise (D, witnos), FM sweeps, and other complex stimuli. PSTHs are calculated as spikes/sec along the y-axis; bin width = 30 msec. Abbreviations: mv, monkey vocalization; hv, human vocalization; ch, chimp vocalization; bp1–20K, bandpassed noise range 1–20 kHz; shrtun, short music segment; witnos, white noise burst.

REFERENCES

1. Geschwind, N., The organization of language and the brain, *Science*, 170, 940–944, 1970.
2. Caramazza, A., Berndt, R.S., Basili, A.G., and Koller, J.J., Syntactic processing deficits in aphasia, *Cortex*, 17, 333–348, 1981.
3. Tramo, M.J., Baynes, K., and Volpe, B.T., Impaired syntactic comprehension and production in Broca's aphasia: CT lesion localization and recovery patterns, *Neurology*, 38, 95–98, 1988.
4. Demb, J.B., Desmond, J.E., Wagner, A.D., Vaidya, C.J., Glover, G.H., and Gabrieli, J.D., Semantic encoding and retrieval in the left inferior prefrontal cortex: a functional MRI study of task difficulty and process specificity, *J. Neurosci.*, 15, 5870–5878, 1995.
5. Petrides, M., Alivisatos, B., and Evans, A.C., Functional activation of the human ventrolateral frontal cortex during mnemonic retrieval of verbal information, *Proc. Natl. Acad. Sci. USA*, 92, 5803–5807, 1995.
6. Chao, L.L. and Knight, R.T., Prefrontal and posterior cortical activation during auditory working memory, *Cogn. Brain Res.*, 4, 27–37, 1996.
7. Smith, E.E., Jonides, J., and Koeppe, R.A., Dissociating verbal and spatial working memory using PET, *Cerebral Cortex*, 6, 11–20, 1996.
8. Gabrieli, J.D.E., Poldrack, R.A., and Desmond, J.E., The role of left prefrontal cortex in language and memory, *Proc. Natl. Acad. Sci. USA*, 95, 906–913, 1998.
9. Broca, P., Remarques sur le siege defaute de langage articule suivies d'une observation d'aphemie (perte de la parole), *Bull. Soc.'d'Anthropol.*, 2, 330–337, 1861.
10. Kaas, J.H. and Hackett, T.A., Subdivisions of auditory cortex and levels of processing in primates [review], *Audiol. Neuro-Otol.*, 3, 73–85, 1998.
11. Rauschecker, J.P., Parallel processing in the auditory cortex of primates, *Audiol. Neuro-Otol.*, 3, 86–103, 1998.
12. Pandya, D.N. and Sanides, F., Architectonic parcellation of the temporal operculum in rhesus monkey and its projection pattern, *Z. Anat. Entwicklungsgesch.*, 139, 127–161, 1973.
13. Galaburda, A. and Sanides, F., Cytoarchitectonic organization of the human auditory cortex, *J. Comp. Neurol.*, 190, 597–610, 1980.
14. Galaburda, A.M. and Pandya, D.N., The intrinsic architectonic and connectional organization of the superior temporal region of the rhesus monkey, *J. Comp. Neurol.*, 221, 169–184, 1983.
15. Cipolloni, P.B. and Pandya, D.N., Connectional analysis of the ipsilateral and contralateral afferent neurons of the superior temporal region in the rhesus monkey, *J. Comp. Neurol.*, 281, 567–585, 1989.
16. Morel, A., Garraghty, P.E., and Kaas, J.H., Tonotopic organization, architectonic fields, and connections of auditory cortex in macaque monkeys, *J. Comp. Neurol.*, 335, 437–459, 1993.
17. Jones, E.G., Dell'Anna, M.E., Molinari, M., Rausell, E., and Hashikawa, T., Subdivisions of macaque monkey auditory cortex revealed by calcium-binding protein immunoreactivity, *J. Comp. Neurol.*, 362, 153–170, 1995
18. Kosaki, H., Hashikawa, T., He, J., and Jones, E.G., Tonotopic organization of auditory cortical fields delineated by parvalbumin immunoreactivity in macaque monkeys, *J. Comp. Neurol.*, 386, 304–316, 1997.

19. Hackett, T.A., Stepniewska, I., and Kaas, J.H., Subdivisions of auditory cortex and ipsilateral cortical connections of the parabelt auditory cortex in macaque monkeys, *J. Comp. Neurol.*, 394, 475–495, 1998.
20. Rauschecker, J.P., Tian, B., and Hauser, M., Processing of complex sounds in the macaque nonprimary auditory cortex, *Science*, 268, 111–114, 1995.
21. Rauschecker, J.P., Tian, B., Pons, T., and Mishkin, M., Serial and parallel processing in rhesus monkey auditory cortex, *J. Comp. Neurol.*, 382, 89–103, 1997.
22. Merzenich, M.M., and Brugge, J.F., Representation of the cochlear partition on the superior temporal plane of the macaque monkey, *Brain Res.*, 50, 275–296, 1973.
23. Rauschecker, J.P. and Tian, B., Mechanisms and streams for processing of 'what' and 'where' in auditory cortex, *Proc. Natl. Acad. Sci. USA*, 97, 11800–11806, 2000.
24. Tian, B., Reser, D., Durham, A., Kustov, A., and Rauschecker, J.P., Functional specialization in rhesus monkey auditory cortex. *Science*, 292, 290–293, 2001.
25. Romanski, L.M., Bates, J.F., and Goldman-Rakic, P.S., Auditory belt and parabelt projections to the prefrontal cortex in the rhesus monkey, *J. Comp. Neurol.*, 403, 141–157, 1999.
26. Walker, A.E., A cytoarchitectural study of the prefrontal area of the macaque monkey, *J. Comp. Neurol.*, 73, 59–86, 1940.
27. Petrides, M. and Pandya, D.N., Association fiber pathways to the frontal cortex from the superior temporal region in the rhesus monkey, *J. Comp. Neurol.*, 273, 52–66, 1988.
28. Preuss, T.M. and Goldman-Rakic, P.S., Myelo- and cytoarchitecture of the granular frontal cortex and surrounding regions in the strepsirhine primate *Galago* and the anthropoid primate *Macaca*, *J. Comp. Neurol.*, 310, 429–474, 1991.
29. Barbas, H., Architecture and cortical connections of the prefrontal cortex in the rhesus monkey, *Adv. Neurol.*, 57, 91–115, 1992.
30. Fuster, J.M., *The Prefrontal Cortex*, Raven, New York, 1980.
31. Goldman-Rakic, P., Circuitry of primate prefrontal cortex and regulation of behavior by representational memory, in *Handbook of Physiology*, Section 1, *The Nervous System*, Vol. V, *Higher Functions of the Brain*, F., Plum, Ed., American Physiological Society, Bethesda, MD, 1987, pp. 373–418.
32. Petrides, M., The effect of periarcuate lesions in the monkey on the performance of symmetrically and asymmetrically reinforced visual and auditory go, no-go tasks, *J. Neurosci.*, 6, 2054–2063, 1986.
33. Goldman-Rakic, P.S., The prefrontal landscape: implications of functional architecture for understanding human mentation and the central executive, *Philos. Trans. Roy. Soc. London B, Biol. Sci.*, 351, 1445–1453, 1996.
34. Barbas, H. and Pandya, D.N., Architecture and intrinsic connections of the prefrontal cortex in the rhesus monkey, *J. Comp. Neurol.*, 286, 353–375, 1989.
35. Petrides, M. and Pandya, D.N., Dorsolateral prefrontal cortex: comparative cytoarchitectonic analysis in the human and the macaque brain and corticocortical connection patterns, *Eur. J. Neurosci.*, 11, 1011–1036, 1999.
36. Wilson, F.A., O'Scalaidhe, S.P., and Goldman-Rakic, P.S., Dissociation of object and spatial processing domains in primate prefrontal cortex, *Science*, 260, 1955–1958, 1993.
37. Pandya, D.N. and Kuypers, H.G., Cortico-cortical connections in the rhesus monkey, *Brain Res.*, 13, 13–36, 1969.
38. Chavis, D.A. and Pandya, D.N., Further observations on corticofrontal connections in the rhesus monkey, *Brain Res.*, 117, 369–386, 1976.

39. Romanski, L.M., Tian, B., Fritz, J., Mishkin, M., Goldman-Rakic, P.S., and Rauschecker, J.P., Dual streams of auditory afferents target multiple domains in the primate prefrontal cortex, *Nat. Neurosci.*, 2, 1131–1136, 1999.
40. Pandya, D.N., Hallett, M., and Kmukherjee, S.K., Intra- and interhemispheric connections of the neocortical auditory system in the rhesus monkey, *Brain Res.*, 14, 49–65, 1969.
41. Barbas, H. and Mesulam, M.M., Cortical afferent input to the principalis region of the rhesus monkey, *Neuroscience*, 15, 619–637, 1985.
42. Barbas, H., Organization of cortical afferent input to orbitofrontal areas in the rhesus monkey, *Neuroscience*, 56, 841–864, 1993.
43. Carmichael, S.T. and Price, J.L., Sensory and premotor connections of the orbital and medial prefrontal cortex of macaque monkeys, *J. Comp. Neurol.*, 363, 642–664, 1995.
44. Hackett, T.A., Stepniewska, I., and Kaas, J.H., Prefrontal connections of the parabelt auditory cortex in macaque monkeys, *Brain Res.*, 817, 45–58, 1999.
45. Seltzer, B. and Pandya, D.N., Frontal lobe connections of the superior temporal sulcus in the rhesus monkey, *J. Comp. Neurol.*, 281, 97–113, 1989.
46. Cavada, C. and Goldman-Rakic, P.S., Posterior parietal cortex in rhesus monkey. I. Parcellation of areas based on distinctive limbic and sensory corticocortical connections, *J. Comp. Neurol.*, 287, 393–421, 1989.
47. Webster, M.J., Bachevalier, J., and Ungerleider, L.G., Connections of inferior temporal areas TEO and TE with parietal and frontal cortex in macaque monkeys, *Cerebral Cortex*, 4, 470–483, 1994.
48. Fulton, J.F., *Functional Lobotomy and Affective Behavior*, Norton, New York, 1950.
49. Mishkin, M.M., Perseveraton of central sets after frontal lesions in monkeys, in *The Frontal Granular Cortex and Behavior*, J.K. Warren and K. Akert, Eds., McGraw-Hill, New York, 1964.
50. Rao, S.C., Rainer, G., and Miller, E.K., Integration of what and where in the primate prefrontal cortex, *Science*, 276, 821–824, 1997.
51. Rushworth, M.F.S. and Owen, A.M., The functional organization of the lateral frontal cortex: conjecture or conjuncture in the electrophysiology literature?, *Trends Cogn. Sci.*, 3, 46–53, 1998.
52. O'Scalaidhe, S.P., Wilson, F.A., and Goldman-Rakic, P.S., Areal segregation of face-processing neurons in prefrontal cortex, *Science*, 278, 1135–1138, 1997.
53. O'Scalaidhe, S.P.O., Wilson, F.A.W., and Goldman-Rakic, P.G.R., Face-selective neurons during passive viewing and working memory performance of rhesus monkeys: evidence for intrinsic specialization of neuronal coding, *Cerebral Cortex*, 9, 459–475, 1999.
54. Fuster, J.M. and Alexander, G.E., Neuron activity related to short-term memory. *Science*, 173, 652–654, 1971.
55. Funahashi, S., Bruce, C.J., and Goldman-Rakic, P.S., Visuospatial coding in primate prefrontal neurons revealed by oculomotor paradigms, *J. Neurophysiol.*, 63, 814–831, 1990.
56. Chafee, M.V. and Goldman-Rakic, P.S., Matching patterns of activity in primate prefrontal area 8a and parietal area 7ip neurons during a spatial working memory task, *J. Neurophysiol.*, 79, 2919–2940, 1998.
57. Pigarev, I.N., Rizzolatti, G., and Schandolara, C., Neurons responding to visual stimuli in the frontal lobe of macaque monkeys, *Neurosci. Lett.*, 12, 207–212, 1979.
58. Rosenkilde, C.E., Bauer, R.H., and Fuster, J.M., Single cell activity in ventral prefrontal cortex of behaving monkeys, *Brain Res.*, 209, 375–394, 1981.

59. Hoshi, E., Shima, K., and Tanji, J., Neuronal activity in the primate prefrontal cortex in the process of motor selection based on two behavioral rules, *J. Neurophysiol.*, 83, 2355–2373, 2000.
60. Gross, C.G. and Weiskrantz, L., Evidence for dissociation of impairment on auditory discrimination and delayed response following lateral frontal lesions in monkeys, *Exp. Neurol.*, 5, 453–476, 1962.
61. Gross, C.G., A comparison of the effects of partial and total lateral frontal lesions on test performance by monkeys, *J. Comp. Physiol. Psychol.*, 56, 41–47, 1963.
62. Goldman, P.S. and Rosvold, H.E., Localization of function within the dorsolateral prefrontal cortex of the rhesus monkey, *Exp. Neurol.*, 27, 291–304, 1970.
63. Iversen, S.D. and Mishkin, M., Perseverative interference in monkeys following selective lesions of the inferior prefrontal convexity, *Exp. Brain Res.*, 11, 376–386, 1970.
64. Mishkin, M. and Weiskrantz, L., Effects of delayed reward on visual-discrimination performance in monkeys with frontal lesions, *J. Comp. Physiol. Psychol.*, 51, 276–281, 1958.
65. Stamm, J.S., Functional dissociation between the inferior and arcuate segments of dorsolateral prefrontal cortex in the monkey, *Neuropsychologia*, 11,181–190, 1973.
66. Gaffan, D. and Harrison, S., Auditory-visual associations, hemispheric specialization and temporal–frontal interaction in the rhesus monkey, *Brain*, 114, 2133–2144, 1991.
67. Knight, R.T., Scabini, D., Woods, D.L., and Clayworth, C., The effects of lesions of superior temporal gyrus and inferior parietal lobe on temporal and vertex components of the human AEP, *Electroencephalogr. Clin. Neurophysiol.*, 70, 499–509, 1988.
68. Ito, S.I., Prefrontal unit activity of macaque monkeys during auditory and visual reaction time tasks, *Brain Res.*, 247, 39–47, 1982.
69. Azuma, M. and Suzuki, H., Properties and distribution of auditory neurons in the dorsolateral prefrontal cortex of the alert monkey, *Brain Res.*, 298, 343–346, 1984.
70. Vaadia, E., Benson, D.A., Hienz, R.D., and Goldstein, M.H. Jr., Unit study of monkey frontal cortex: active localization of auditory and of visual stimuli, *J. Neurophysiol.*, 56, 934–952, 1986.
71. Russo, G.S. and Bruce, C.J., Auditory receptive fields of neurons in frontal cortex of rhesus monkey shift with direction of gaze, *Soc. Neurosci. Abstr.*, 15, 1204, 1989.
72. Kikuchi-Yorioka, Y. and Sawaguchi, T., Parallel visuospatial and audiospatial working memory processes in the monkey dorsolateral prefrontal cortex, *Nat. Neurosci.*, 3, 1075–1076, 2000.
73. Leinonen, L., Hyvarinen, J., and Sovijarvi, A.R., Functional properties of neurons in the temporo-parietal association cortex of awake monkey, *Exp. Brain Res.*, 39, 203–215, 1980.
74. Benson, D.A., Hienz, R.D., and Goldstein, M.H., Jr., Single-unit activity in the auditory cortex of monkeys actively localizing sound sources: spatial tuning and behavioral dependency, *Brain Res.*, 219, 249–267, 1981.
75. Newman, J.D. and Lindsley, D.F., Single unit analysis of auditory processing in squirrel monkey frontal cortex, *Exp. Brain Res.*, 25, 169–181, 1976.
76. Benevento, L.A., Fallon, J., Davis, B.J., and Rezak, M., Auditory–visual interaction in single cells in the cortex of the superior temporal sulcus and the orbital frontal cortex of the macaque monkey, *Exp. Neurol.*, 57, 849–872, 1977.
77. Wollberg, Z. and Sela, J., Frontal cortex of the awake squirrel monkey: responses of single cells to visual and auditory stimuli, *Brain Res.*, 198, 216–220, 1980.

78. Tanila, H., Carlson, S., Linnankoski, I., Lindroos, F., and Kahila, H., Functional properties of dorsolateral prefrontal cortical neurons in awake monkey, *Behav. Brain Res.*, 47, 169–180, 1992.
79. Tanila, H., Carlson, S., Linnankoski, I., and Kahila, H., Regional distribution of functions in dorsolateral prefrontal cortex of the monkey, *Behav. Brain Res.*, 53, 63–71, 1993.
80. Watanabe, M., Frontal units of the monkey coding the associative significance of visual and auditory stimuli, *Exp. Brain Res.*, 89, 233–247, 1992.
81. Bodner, M., Kroger, J., and Fuster, J.M., Auditory memory cells in dorsolateral prefrontal cortex, *NeuroReport*, 7, 1905–1908, 1996.
82. Romanski, L.M. and Goldman-Rakic, P.S., An auditory domain in primate prefrontal cortex, *Nat. Neurosci.*, 5, 15–17, 2002.
83. Nelken, I., Prut, Y., Vaadia, E., and Abeles, M., In search of the best stimulus: an optimization procedure for finding efficient stimuli in the cat auditory cortex, *Hearing Res.*, 72, 237–253, 1994.
84. Zatorre, R.J., Evans, A.C., Meyer, E., and Gjedde, A., Lateralization of phonetic and pitch discrimination in speech processing, *Science*, 256, 846–849, 1992.
85. Chee, M.W., O'Craven, K.M., Bergida, R., Rosen, B.R., and Savoy, R.L., Auditory and visual word processing studied with fMRI, *Human Br. Map.*, 7, 15–28 1999.
86. Ungerleider, L.G. and Mishkin, M., Two cortical visual systems, in *Analysis of Visual Behavior*, D.J. Ingle, M.A. Goodale, and R.J.W. Mansfield, Eds., MIT Press, Cambridge, MA, 1982.
87. Deacon, T.W., Cortical connections of the inferior arcuate sulcus cortex in the macaque brain, *Brain Res.*, 573, 8–26, 1992.

15 Cortical Processing of Complex Sounds and Species-Specific Vocalizations in the Marmoset Monkey (*Callithrix jacchus*)

Xiaoqin Wang, Siddhartha C. Kadia, Thomas Lu, Li Liang, and James A. Agamaite

CONTENTS

I. Introduction 280
II. Temporal Processing 281
    A. Rate vs. Temporal Coding of Time-Varying Signals 281
    B. Processing of Acoustic Transients within the Temporal Integration Window 283
    C. Extraction of Temporal Profiles Embedded in Complex Sounds 284
III. Spectral Processing 287
    A. Multipeaked Tuning and Long-Range Horizontal Connections 287
    B. Spectral Inputs from Outside the Classical Receptive Field 287
IV. Representation of Species-Specific Vocalizations 290
    A. Historical Studies 290
    B. The Nature of Cortical Representations 291
    C. Dependence of Cortical Responses on Behavioral Relevance of Vocalizations 292
    D. Mechanisms Related to Coding of Species-Specific Vocalizations ...292
V. Summary 295
Acknowledgments 296
References 296

I. INTRODUCTION

Understanding how the brain processes vocal communication sounds remains one of the most challenging problems in neuroscience. Species-specific vocalizations of non-human primates are communication sounds used in intraspecies interactions, analogous to speech in humans. Primate vocalizations are of special interest to us because, compared with other animal species, primates share the most similarities with humans in the anatomical structures of their central nervous systems, including the cerebral cortex. Therefore, neural mechanisms underlying perception and production of species-specific primate vocalizations may have direct implications for those operating in the human brain for speech processing. Although field studies provide full access to the natural behavior of primates, it is difficult to combine them with physiological studies at the single neuron level in the same animals. The challenge is to develop appropriate primate models for laboratory studies where both vocal behavior and underlying physiological structures and mechanisms can be systematically investigated. This is a crucial step in understanding how the brain processes vocal communication sounds at the cellular and systems levels. Most primates have a well-developed and sophisticated vocal repertoire in their natural habitats; however, for many primate species such as macaque monkeys, vocal activities largely diminish under the captive conditions commonly found in research institutions, due in part to the lack of proper social housing environments. Fortunately, some primate species such as New World monkeys (e.g., marmosets, squirrel monkeys) remain highly vocal in properly configured captive conditions. These primate species can serve as excellent models to study neural mechanisms responsible for processing species-specific vocalizations.

The cerebral cortex is known to play an important role in processing species-specific vocalizations. Studies have shown that lesions of auditory cortex cause deficits in speech comprehension in humans and discrimination of vocalizations in primates.[1,2] It has been shown in humans, apes, and Old and New World monkeys that the sensory auditory cortex consists of a primary auditory field (AI) and surrounding secondary fields.[3-8] It has also been shown that they have similar cytoarchitecture in the superior temporal gyrus, where the sensory auditory cortex is located.[9] In the cerebral cortex, acoustic signals are first processed in the AI and then in the secondary fields. From there, the information flows to the frontal cortex and other parts of the cerebral cortex.

Progress in the study of neural mechanisms for encoding species-specific primate vocalizations has been relatively slow, compared with our understanding of other auditory functions such as echolocation and sound localization. This chapter discusses recent studies in our laboratory that have focused on cortical mechanisms for processing temporal and spectral features of species-specific vocalizations in a vocal primate model, the common marmoset (*Callithrix jacchus*). Our neurophysiological work has been based on single-unit recordings under the awake condition to avoid confounding factors of anesthetics on cortical responses.[10-12] The authors[13] have argued that three issues are central to our understanding of cortical mechanisms for encoding species-specific primate vocalizations: (1) neural encoding of the statistical structure of communication sounds, (2) the role of behavioral relevance in shaping cortical representations, and (3) sensory–motor interactions between vocal production and perception systems.

II. TEMPORAL PROCESSING

Temporal modulations are fundamental components of species-specific primate vocalizations, as well as human speech and music. Low-frequency modulations are important for speech perception and melody recognition, while higher frequency modulations produce other types of sensations such as pitch and roughness.[14,15] Both humans and animals are capable of perceiving the information contained in temporally modulated sounds across a wide range of time scales, from milliseconds to tens and hundreds of milliseconds. The neural representation of temporal modulations begins at the auditory periphery where auditory-nerve fibers faithfully represent fine details of complex sounds in their temporal discharge patterns.[16-19] At subsequent nuclei along the ascending auditory pathway (cochlear nucleus, CN; inferior colliculus, IC; and auditory thalamus, MGB), the precision of this temporal representation gradually degrades (CN,[20-22] IC,[23] MGB,[24,25]) due to the biophysical properties of neurons and temporal integration of converging inputs from one station to the next. In a modeling study of the transformation of temporal discharge patterns from the auditory nerve to the cochlear nucleus,[26] it was shown that the reduction of phase locking in stellate cells can result from three mechanisms: (1) convergence of subthreshold inputs on the soma, (2) inhibition, and (3) the well-known dendritic low-pass filtering.[27] These basic mechanisms may also operate at successive nuclei, leading to the auditory cortex, progressively reducing the temporal limit of stimulus-synchronized responses.

It has long been noted that neurons in the auditory cortex do not faithfully follow rapidly changing stimulus components.[10,28] A number of previous studies have shown that cortical neurons can only be synchronized to temporal modulations at a rate far less than 100 Hz,[28-34] compared with a limit of about 1 kHz at the auditory nerve.[17,18] The lack of synchronized cortical responses to rapid, but perceivable temporal modulation has been puzzling. Because most of the previous studies in the past three decades on this subject were conducted in anesthetized animals with a few exceptions,[10,24,28,33] it has been speculated that the reported low temporal response rate in the auditory cortex might be caused partially by anesthetics. For example, Goldstein et al.[10] showed that click-following rates of cortical evoked potentials were higher in unanesthetized cats than in anesthetized ones. Neural responses obtained under unanesthetized conditions are therefore of particular importance to our understanding of cortical representations of time-varying signals.

A. RATE VS. TEMPORAL CODING OF TIME-VARYING SIGNALS

We have systematically investigated responses of single neurons in the primary auditory cortex (AI) of awake marmoset monkeys to rapid sequences of acoustic transients.[35] One type of response is illustrated in Figure 15.1A. This neuron exhibited significant stimulus-synchronized discharges to sequences of acoustic events at long interstimulus intervals (ISIs) (greater than approximately 20 to 30 msec). The responses became nonsynchronized at medium ISIs and sustained discharges diminished at short ISIs (less than approximately 10 msec). The second type of response did not exhibit stimulus-synchronized discharges but showed monotonically changing discharge rate with increasing ISIs when the ISI was shorter than about 20 to

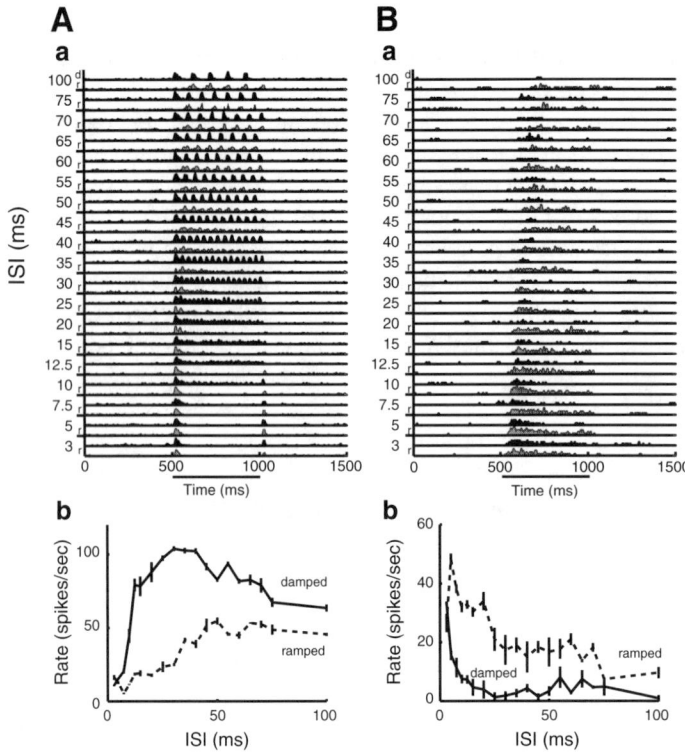

FIGURE 15.1 Stimulus-synchronized and nonsynchronized responses in the primary auditory cortex (AI) of awake marmosets (data based on Lu et al.[35]). (A) Response of a neuron with stimulus-synchronized discharges. The stimuli used were ramped or damped sinusoids with different ISIs (3–100 msec) and a fixed half-life.[36] This neuron responded more strongly to damped sinusoids across different interstimulus intervals (ISIs). (a) Post-stimulus time histograms (PSTH) are plotted in the order of increasing ISIs along the ordinate, with alternating responses to ramped (gray) and damped (black) sinusoids. The heights of the PSTHs are normalized to the maximum bin count over all stimulus conditions. Stimulus onset was at 500 msec, and duration was 500 msec. (b) Average discharge rates for ramped (dashed line) and damped (solid line) stimuli are plotted as functions of the ISI. Vertical bars represent standard errors of the means (SEM). (B) Example of a nonsynchronized neuron that responded more strongly to ramped sinusoids across the ISI. The format of the plots is the same as in A. (C) A combination of temporal and rate representations can encode a wide range of ISIs. Curves are the cumulative sum of the histograms representing response boundaries of two neural populations, one with stimulus-synchronized discharges ($N = 36$) and the other with nonsynchronized discharges ($N = 50$), respectively. The stimuli used in deriving these responses were click trains with different inter-click intervals (ICI, 3–100 msec). The dashed line shows the percentage of neurons with synchronization boundaries less than or equal to a given ICI. The solid line shows the percentage of neurons with rate-response boundaries greater than or equal to a given ICI. The total number of neurons analyzed is 94, including 8 neurons (not shown) with combined response characteristics. *(continued)*

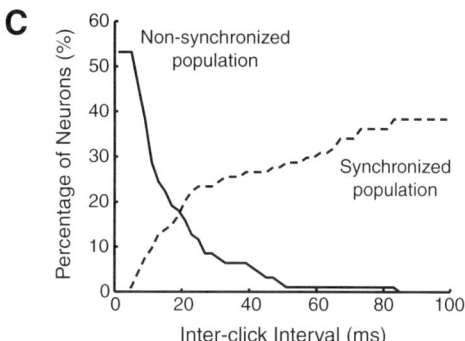

FIGURE 15.1 (continued).

30 msec, as illustrated by the example in Figure 15.1B. We have shown that the limit on stimulus-synchronized responses is on the order of 20 to 25 msec (median value of sampled population) in AI in the unanesthetized condition.[35] The observation that neurons are sensitive to changes of short ISIs indicates that a discharge-rate-based mechanism may be in operation when ISIs are shorter than 20 to 30 msec.

We have identified two populations of AI neurons displaying these two types of responses: synchronized and nonsynchronized.[35] The two populations appeared to encode sequential stimuli in complementary manners. Neurons in the synchronized population showed stimulus-synchronized discharges at long ISIs but few responses at short ISIs (Figure 15.1A). This population of neurons can thus represent slowly occurring temporal events explicitly using a temporal code. The nonsynchronized population of neurons did not exhibit stimulus-synchronized discharges at either long or short ISIs (Figure 15.1A). This population of neurons can implicitly represent rapidly changing temporal intervals by their average discharge rates. On the basis of these two populations of neurons, the auditory cortex can represent a wide range of time intervals in sequential, repetitive stimuli (Figure 15.1C). The large number of neurons with nonsynchronized and sustained discharges at short ISIs were observed in the AI of awake marmosets[35] but were not observed in the AI of anesthetized cats.[34]

B. Processing of Acoustic Transients within the Temporal Integration Window

The limited stimulus-synchronized responses observed in single cortical neurons suggest that a neuron integrates sequential stimuli over a brief time window that can be defined operationally as the temporal integration window. Two sequential acoustic events falling within the temporal integration window are not distinguished as separate events at the output of a neuron. The experiments discussed above suggest that AI neurons integrate stimulus components within a time window of about 30 msec and treat components outside this window as discrete acoustic events. Humans and animals are known to discriminate changes in acoustic signals at time scales shorter than the temporal integration window of AI neurons, suggesting that cortical neurons must be able to signal such rapid changes. We have investigated sensitivity of cortical neurons to rapid changes within the putative temporal integration window[36] using a class of temporally modulated signals termed ramped and damped sinusoids.[37] A damped sinusoid consists of a pure tone amplitude-modulated by an exponential function. It has a fast onset followed by a slow offset. The rate of amplitude decay is determined by the exponential half-life. A ramped sinusoid is a time-reversed damped sinusoid. Both types of sounds have identical long-term Fourier spectra. Most cortical neurons showed clear preference for either ramped or damped sinusoids.[36] Some neurons responded almost exclusively to one stimulus type. Figures 15.1A and B show the responses of two representative AI neurons to sequences of ramped and damped sinusoids at different ISIs, one with stimulus-synchronized discharges (Figure 15.1A) and the other with nonsynchronized discharges (Figure 15.1B). At ISIs longer than the presumed temporal integration window (>20–25 msec), discharges of the neuron shown in Figure 15.1A were synchronized to each period of ramped and damped sinusoids, while the neuron showed stronger responses to damped sinusoids (Figure 15.1A, part a). When ISIs were shorter than the presumed temporal integration window, synchronized discharges disappeared, but the overall preference for the damped sinusoids was maintained (Figure 15.1A, part b). Figure 15.1B shows a neuron of the nonsynchronized population that responded more strongly to ramped sinusoids across ISIs. These observations demonstrate that the sensitivity to temporal asymmetry is independent of the ability of a cortical neuron to synchronize to stimulus events, suggesting a rate-based coding mechanism for time scales shorter than the temporal integration window for AI neurons. These observations also show that temporal characteristics within the temporal integration window can profoundly modulate a cortical neuron's responsiveness.

C. Extraction of Temporal Profiles Embedded in Complex Sounds

Another important issue regarding cortical processing of time-varying signals is how cortical neurons represent similar temporal features that are introduced by different means. In the spectral domain, frequency tuning or filtering is a well-established concept for auditory cortical neurons. It is not clear, however, whether cortical neurons use a "temporal filter" in the time domain to extract common temporal features embedded in complex sounds. To answer these questions, it is necessary to test response properties of cortical neurons using temporally modulated signals with different spectral contents. In our experiments, temporal variations were introduced by sinusoidally modulating tone or noise carriers in amplitude or frequency domains.[38,39] These types of temporal modulations are commonly found in natural sounds such as species-specific vocalizations and environmental sounds. While these classes of stimuli differ significantly in their spectral contents, they share the same modulation waveform: a sinusoid with a particular temporal modulation frequency that gives rise to the perceived time-varying property of these sounds. Neurons in the auditory cortex of awake marmosets often responded to these temporally modulated stimuli with sustained firings that were much stronger than responses to unmodulated tones or noises.

A majority of AI neurons displayed bandpass-like modulation transfer functions when measured by average discharge rate or, in some cases, by discharge synchrony as well. Figure 15.2A shows an example of a neuron that responded with maximum discharge rates at 64-Hz modulation frequency for sinusoidal amplitude- or frequency-modulated tones (sAM or sFM), as well as sinusoidal amplitude-modulated noise (nAM). Data shown in Figure 15.2A (average discharge rate vs. modulation frequency) are conventionally referred to as the *rate-based modulation transfer function* (rMTF). The peak of an rMTF is referred to as the *rate-based best modulation frequency* (rBMF). For the neuron shown in Figure 15.2A, the rBMF is 64 Hz for all three stimuli. Because the spectral contents of amplitude-modulated noises

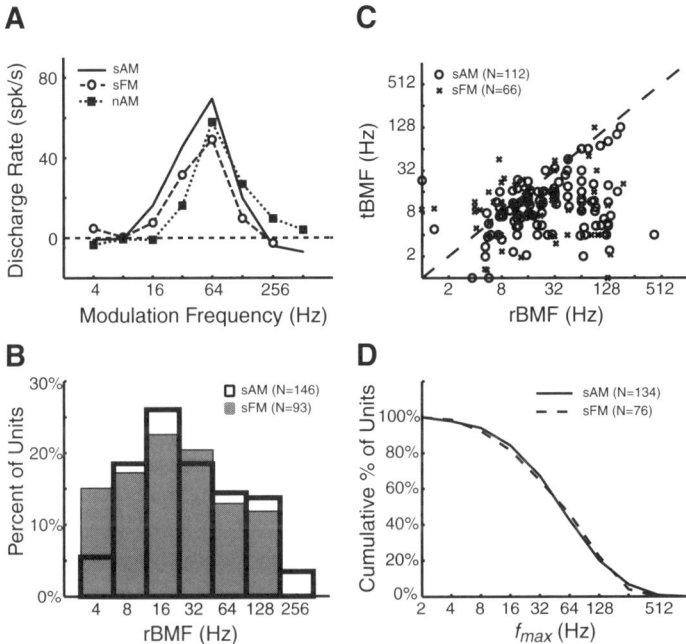

FIGURE 15.2 Cortical responses to temporally modulated stimuli recorded from AI of awake marmoset (data based on Liang et al.[38]). (A) Discharge-rate-based modulation transfer functions (rMTFs) measured in a single neuron using sAM (solid line), sFM (dashed line with open circles), and nAM (dotted line with filled squares). (B) Population properties for discharge-rate-based modulation frequency selectivity. Distribution of rBMF resulted from sAM (open) and sFM (gray) stimuli. The bin widths of the histograms are on a base-2 logarithmic scale. The distributions of $rBMF_{sAM}$ and $rBMF_{sFM}$ are not statistically different from each other (Wilcoxon rank sum test, $p = 0.17$). The medians of the rBMF distributions are 22.59 Hz (sAM) and 20.01 Hz (sFM). (C) Comparison between rate-based and synchrony-based modulation selectivity evaluated on the basis of individual neurons. tBMF is plotted vs. rBMF for sAM (circles) and sFM (crosses) stimuli. The distributions of tBMF and rBMF are statistically different from each other (Wilcoxon rank sum test, $p < 0.005$). The diagonal dashed line has a slope of 1. The medians of the tBMF distributions are 10.12 Hz (sAM) and 10.8 Hz (sFM). (D) Cumulative distributions of f_{max} for sAM (solid line) and sFM (dashed line) stimuli.

are spread much more broadly than that of the spectrum of the modulating waveform, this eliminates the possibility that the observed selectivity to modulation frequency was due to a spectral effect. These observations showed that individual neurons in the auditory cortex of awake marmosets have a preferred *temporal modulation frequency*, as measured by mean firing rate, which is largely independent of the spectral contents of modulated sounds. Furthermore, the rBMFs derived from sAM, sFM, and nAM stimuli (Figure 15.2A) match closely. The distributions of rBMFs across populations of AI neurons are shown in Figure 15.2B for sAM and sFM stimuli. Both distributions are centered between 16 and 32 Hz and are statistically indistinguishable.

Significant discharge synchronization was found in most, but not all, AI neurons. We used the vector strength (VS) to quantify stimulus-synchronized firing patterns and Rayleigh statistics to assess the statistical significance because low firing rates undermine the interpretation of the VS measure. The relationship between Rayleigh statistics and modulation frequency is referred to as the *temporal modulation transfer function* (tMTF) from which the *synchrony-based best modulation frequency* (tBMF) could be determined. AI neurons were typically tuned to a higher modulation frequency when measured by average discharge rate than by synchronized discharges. In Figure 15.2C, we compare rate-based and synchrony-based modulation frequency selectivity in individual neurons. For the vast majority of neurons, rBMF was greater than tBMF for both sAM and sFM stimuli. The distributions of tBMF and rBMF were statistically different from each other. This means that the average discharge rate reaches the maximum at modulation frequencies where stimulus-synchronized discharges are weaker than what could be optimally evoked in AI neurons. The distributions in Figure 15.2C again show the similarity between responses to sAM and sFM stimuli for each of the rBMF and tBMF measures. Another important measure of stimulus-synchronized discharges is the maximum synchronization frequency (f_{max}). This measure indicates the upper limit of stimulus-synchronized discharges in each neuron, whereas tBMF defines the modulation frequency at which the strongest discharge synchronization could be induced. Figure 15.2D shows the cumulative distributions of f_{max} for both sAM and sFM stimuli, which characterizes the upper boundary of stimulus-synchronized activities for the population of sampled neurons. These curves show how well AI neurons, as a whole, can represent temporal modulations by their temporal discharge patterns. The cumulative distributions of f_{max} for both sAM and sFM stimuli are highly overlapping, indicating the similarity in stimulus-synchronized discharges that resulted from these two classes of stimuli. The curves have low-pass shapes and begin to drop more rapidly above about 32 Hz. The medians of the cumulative f_{max} distributions are about 64 Hz for both types of stimuli (Figure 15.2D). Less than 10% of neurons were able to synchronize to modulation waveforms at 256 Hz.

These observations demonstrate that auditory cortical neurons can be selective to particular modulation frequencies by their mean firing rate or discharge synchrony and that this selectivity is similar for different temporally modulated stimuli. Because amplitude and frequency modulations are produced along different stimulus dimensions, the match between BMF_{sAM} and BMF_{sFM} suggests a *shared* representation of temporal modulation frequency by cortical neurons. The fact that auditory cortical neurons showed similar BMFs for tone or noise carrier further supports this notion; that is, it is the temporal modulation frequency, not the spectral contents, to which the neurons are selective. The selectivity for a particular temporal modulation frequency in terms of average discharge rate, in the absence of stimulus-synchronized discharge patterns, was often observed in our recordings from the auditory cortex of awake marmosets and represented a temporal-to-rate transformation by cortical neurons. These results are consistent with the notion that rate-coding may be used by the auditory cortex to encode temporal modulations at higher modulation frequencies than can be encoded by temporal discharge patterns.[35] Bieser and Mueller-Preuss[33] suggested that high amplitude-modulation frequencies may be encoded by a rate code. Our studies with sAM, sFM, and nAM stimuli have further extended this line of work by showing that rate coding is likely needed for encoding temporal modulations at high modulation frequencies, in both the amplitude and frequency domains.

III. SPECTRAL PROCESSING

A. Multipeaked Tuning and Long-Range Horizontal Connections

Like the primary visual cortex, neurons in AI are linked through a network of long-range horizontal connections.[40–44] Such anatomical structures suggest spectral integration over a large part of the acoustic spectrum. We found that single neurons in AI are influenced by spectral inputs outside the classical receptive field (i.e., an excitatory area surrounded by flanking inhibitory areas), as far as one to three octaves away from the characteristic frequency (CF) of a unit.[45] In a subpopulation of AI neurons, these distant spectral inputs were manifested in the form of multipeaked tuning (often harmonically related) as revealed by single tones. Figure 15.3A shows an example of a multipeaked neuron recorded from the auditory cortex of an awake marmoset. This neuron responded to tones at two harmonically related frequencies, 9.5 and 19 kHz (Figure 15.3A, part a). For neurons with multipeaked tuning, facilitation can be demonstrated by simultaneously stimulating multiple harmonically related CF. We used a two-tone paradigm to test the facilitation in the neuron shown in Figure 15.3A, part a. The response to two simultaneously presented tones was much stronger than the response to either tone presented alone (Figure 15.3A, part b). The distribution of all peak frequency ratios of a population of multipeaked neurons is shown in Figure 15.3B. A major peak at the ratio of 2 indicates that the most common relationship between the frequency peaks is that of an octave. It is possible that the harmonic structure of multipeaked neurons is created by widespread horizontal connections in the auditory cortex. In fact, our experiments have demonstrated that neurons in the supragranular layers of AI in cat make long-range horizontal connections with neurons of harmonically related frequency selectivity.[40]

B. Spectral Inputs from Outside the Classical Receptive Field

While a small population of AI neurons had multipeaked tuning, the majority of AI neurons exhibit well-known single-peaked tuning characteristics; however, when

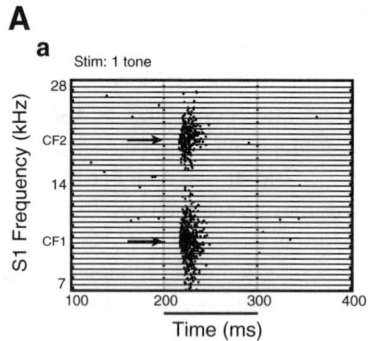
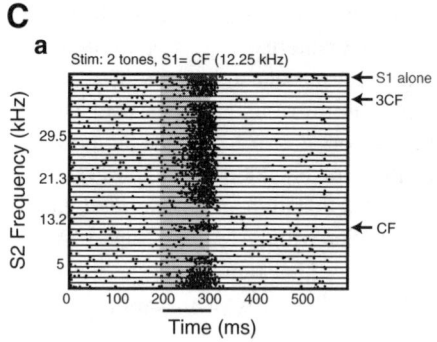
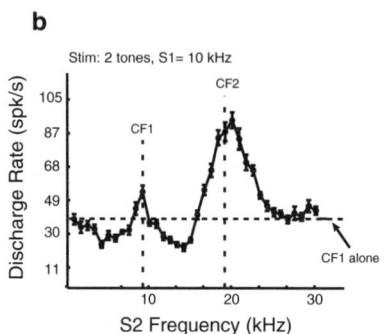
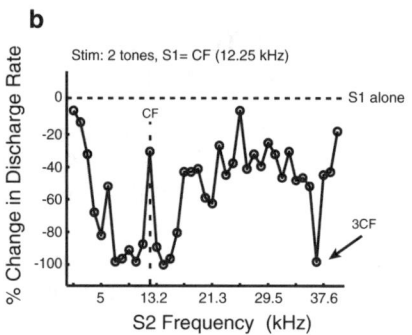
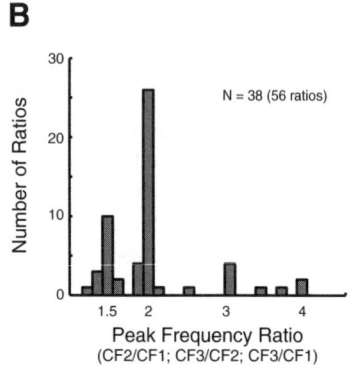
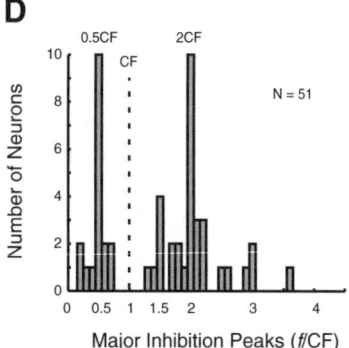

FIGURE 15.3

tested by simultaneous presentations of two tones, both facilitation and inhibition can be introduced by tones that are placed far away from the CF of a neuron.[45] In the example shown in Figure 15.3C, the second tone (S2) completely inhibited the response to the first tone (S1) at the frequency equal to 3 CF. This distant inhibition is distinctly different from inhibitory areas surrounding the CF. In contrast to neurons with multipeaked tuning, the distant inhibition rather than facilitation showed a harmonic structure (with respect to the CF) in the single-peaked neurons (Figure 15.3C). These findings indicate that AI neurons integrate spectral inputs from a broad range of frequencies, far beyond the region surrounding CF. The facilitation observed in AI neurons may serve as a basis for processing contextual information in acoustic scenes. The harmonically related inhibitory interactions may function to eliminate unwanted harmonic byproducts that are commonly produced by nonlinear mechanisms in acoustic generators in order to segregate a target signal from other concurrent sounds. Response enhancement and inhibition have also been reported using two nonsimultaneous tones in the auditory cortex of macaque monkeys.[46,47]

Our recent experiments further showed that the responsiveness of an AI neuron can be constrained by the temporal relationship between multiple spectral components. By temporally comodulating spectral components placed both inside and outside the excitatory receptive field of a neuron, the responsiveness of a neuron

FIGURE 15.3 Characteristics of multipeaked (A) and (B) and single-peaked (C) and (D) neurons recorded from AI of awake marmosets (data based on Kadia and Wang[45]). Each two-tone pair is defined by S1 (fixed tone) and S2 (variable tone). (A, part a) Dot raster showing single tone responses for a multipeaked neuron. Stimulus onset was at 200 msec, duration 100 msec (indicated by the bar under the horizontal axis). This neuron was responsive to two frequency regions, with peaks at 9.5 kHz (CF1) and 19 kHz (CF2). The most responsive frequency (the frequency evoking the strongest discharge rate) in each region was designated as a frequency peak. Note that CF1 and CF2 are harmonically related with a peak frequency ratio of 2. (A, part b) Responses to two simultaneously presented tones. Discharge rate (calculated for the duration of the stimuli) is plotted vs. S2 frequency; S1 frequency was 10 kHz. The error bars represent standard errors of the means (SEMs). The two-tone responses indicate maximum facilitation at the frequencies near ~20 kHz (CF2 of this neuron). The dotted line indicates the strength of response to CF1 tone alone. (B) Distribution of peak frequency ratios (CFn/CF1) in 38 multipeaked neurons. CFn is determined as the frequency with the strongest discharge rate at a sound level near threshold. Because a multipeaked neuron can have more than two peaks, a neuron may contribute to more than one ratio in this plot. (C, part a) Dot raster showing two-tone responses for a single-peaked neuron. (C, part b) Percent change in discharge rate, defined as $100 \times (R_{S1+S2} - R_{S1})/R_{S1}$, where R_{S1+S2} is the average discharge rate to two-tone pair, and R_{S1} is the average discharge rate to the S1 tone alone, is plotted vs. S2 frequency for the data shown in part a. The dotted line indicates the strength of response to the S1 tone alone. Notice the complete inhibition of discharges at a frequency equal to $3 \times $ CF, which can also be seen in part a. The drop of response magnitude at CF was due to the nonmonotonic rate-level function of this neuron. (D) Distribution of major inhibitory peaks in 51 single-peaked neurons. f/CF is the ratio of the inhibitory frequency to the CF of a unit. Single-peaked neurons in AI have inhibitory influences from a wide range of frequencies outside the receptive fields, predominantly from harmonically related frequencies ($0.5 \times $ CF, $2 \times $ CF).

can be significantly altered.[48] Together, these findings indicate that, while the auditory cortex is tonotopically organized globally, local integration of widespread spectral components is an important part of the information processing circuit.

IV. REPRESENTATION OF SPECIES-SPECIFIC VOCALIZATIONS

A. Historical Studies

Over the past several decades, a number of experimental attempts have been made to elucidate the mechanisms of cortical coding of primate vocalizations in the auditory cortex. The effort to understand the cortical representations of vocalizations reached a peak in the 1970s when a number of reports were published. The results of these studies were mixed, with no clear or consistent picture emerging as to how communication sounds are represented in the auditory cortex of primates.[49] This lack of success may be accounted for, in part, retrospectively, by expectations on the form of cortical coding of behaviorally important stimuli. For a time it was thought that primate vocalizations were encoded by highly specialized neurons, or *call detectors*;[50,51] however, individual neurons in the auditory cortex were often found to respond to more than one call or to various features of calls.[50-53] The initial estimate of the percentage of the call detectors was relatively high. Later, much smaller numbers were reported as more calls were tested and more neurons were studied. At the end of this series of explorations, it appeared that, at least in the initial stages of the auditory cortical pathway, the notion of highly specialized neurons was doubtful. Perhaps due to these seemingly disappointing findings, no systematic studies on this subject were reported for over a decade.

In these earlier studies, responses to vocalizations were not adequately related to basic functional properties of cortical neurons such as frequency tuning, temporal discharge patterns, etc. In addition, the responses to vocalizations were not interpreted in the context of the overall organization of a cortical field or the overall structure of the auditory cortex. Such information only became available in later years, and we are still defining much of the structure of the auditory cortex in primates. It is clear now that, until we achieve a good understanding of the properties of a neuron besides its responses to vocalizations, we will not be able to fully understand cortical coding of species-specific vocalizations. Another lesson learned from the previous studies is that one must consider the underlying statistical structure of a species' vocalizations. In earlier as well as some recent studies,[54] only vocalization tokens were used. This made it difficult, if not impossible, to accurately interpret cortical responses. We have previously argued[13] that it is not sufficient to study cortical responses to individual call tokens; one must study cortical representations of statistical structures of vocalizations (i.e., the distributions of acoustic features that distinguish call types and caller identity). Cortical coding of statistical structures of vocalizations can be systematically studied using accurately synthesized vocalizations, an approach currently being pursued in our laboratory.[55] In earlier days, before powerful computers and digital technology became available, it would have been a formidable task to quantify and model primate vocalizations. Nonetheless, the pioneer explorations of the primate auditory cortex in the 1970s using species-specific vocalizations served as an important stepping stone for current and future studies of this important problem in auditory neuroscience.

B. THE NATURE OF CORTICAL REPRESENTATIONS

The lack of convincing evidence of call detectors from earlier studies led us to consider other hypotheses regarding cortical coding of complex sounds such as vocalizations. An alternative coding strategy is to represent these complex sounds by spatially distributed neuronal populations. Recent work in the marmoset has provided evidence that such a population coding scheme may also operate at the input stage of the auditory cortex, but in a very different form from that observed at the periphery.[13,56] Compared with the auditory nerve, cortical neurons do not faithfully follow rapidly changing stimulus components within a complex vocalization, consistent with our recent studies of cortical coding of successive stimuli.[35] In other words, the responses of AI neurons are not faithful replicas of the spectrotemporal patterns of vocalizations, but have been transformed into a more abstract representation compared to that observed in the neural responses of the auditory periphery and brainstem.

It should be pointed out that the notion of population coding does not necessarily exclude the operation of highly specialized neurons. The questions, then, are to what extent do such specialized neurons exist and what are their functional roles? In earlier studies, it was expected that a large percentage of these call detector neurons would be found and that the coding of species-specific vocalizations would essentially be carried out by such neurons. Various studies since the 1970s have cast doubts on this line of thinking. In the anesthetized marmoset, a subpopulation of AI neurons appeared to be more selective to alterations of natural calls than did the rest of the population.[56] In our recent studies in awake marmoset, we identified neurons that were selective to specific call types or even individual callers in the auditory cortex.[57] Thus, it appears as though there are two general classes of cortical neurons: one responding selectively to call types or callers and the other responding to a wider range of sounds, including both vocalizations and non-vocalization signals. The latter group accounts for a larger proportion of cortical neurons in AI than in the lateral field (a secondary cortical area).[57] One may speculate that the task of encoding communication sounds is undertaken by the call-selective neurons alone, however small that population is. On the other hand, one cannot avoid considering the role of the nonselective population that also responds to communication sounds concurrently with the selective population. Our hypothesis is that the nonselective population serves as a general sound analyzer that provides the brain with a detailed description of a sound. The selective population, on the other hand, serves as a signal classifier, the task of which is to tell the brain that a particular call type or caller was heard.[13] One can imagine that the former analysis is needed when a new sound is heard or being learned, whereas the latter analysis is used when a familiar sound, such as a vocalization, is encountered. This and other theories concerning cortical coding of communication sounds[54,58] will be further tested by new experiments in coming years.

C. Dependence of Cortical Responses on Behavioral Relevance of Vocalizations

It has been shown that natural vocalizations of marmoset monkeys produced stronger responses in a subpopulation of AI neurons than did an equally complex but artificial sound such as a time-reversed version of the same call.[56] Moreover, the subpopulation of cortical neurons that were selective to a natural call had a more clear representation of the spectral shape of the call than did the nonselective subpopulation. When the reversed call was played, responses from the two populations were similar. These observations suggest that marmoset auditory cortex responds preferentially to vocalizations that are of behavioral significance as compared to behaviorally irrelevant sounds. To test this notion further, we have directly compared responses to natural and time-reversed calls in the auditory cortex of another mammal, the cat, whose AI shares similar basic physiological properties (e.g., CF, threshold, latency, etc.) to those of the marmoset.[59] Unlike the auditory cortex of the marmoset, however, the cat's auditory cortex does not differentiate natural marmoset vocalizations from their time-reversed versions.[60] The results of these experiments are summarized in Figure 15.4. Neurons in marmoset AI showed stronger responses to natural twitter calls than to time-reversed twitter calls (Figures 15.4A and B). This preference for natural marmoset twitter calls demonstrated in marmoset AI was absent in cat AI (Figure 15.4C).

The differential responses due to natural and reversed twitter calls observed in marmosets, but absent in cats, suggest that neural mechanisms other than those responsible for encoding acoustic structures of complex sounds are involved in cortical processing of behaviorally important communication sounds.[13] Because vocalizations are an integrated part of the marmoset's acoustic experience, it is likely that the differential neural responses observed in the marmoset's AI arise from structural and functional plasticity in sensory cortices formed through development[61] and adulthood.[62] Indeed, it has been demonstrated that repeated exposures to behaviorally important spatial-temporal sensory input patterns result in reorganization of response properties accordingly in the primary sensory cortices.[63,64] It is also possible that the preference to natural vocalizations observed in marmosets was partially due to a predisposed, specialized circuitry in the marmoset's auditory system that does not exist in cats. These findings have direct implications for interpreting cortical responses to natural and artificial sounds. They argue strongly that our exploration of cortical mechanisms responsible for encoding behaviorally important communication sounds must be based on biologically meaningful models.

D. Mechanisms Related to Coding of Species-Specific Vocalizations

A useful analysis in revealing coding mechanisms for species-specific vocalizations is to correlate neural responses with the statistical structures of a vocalization's spectrotemporal characteristics. Figure 15.5A shows an example of such a correlation. We analyzed two major types of species-specific vocalizations of the marmoset that display prominent frequency modulations, *trill* and *trillphee*.[65] Trills and trillphees are used in different

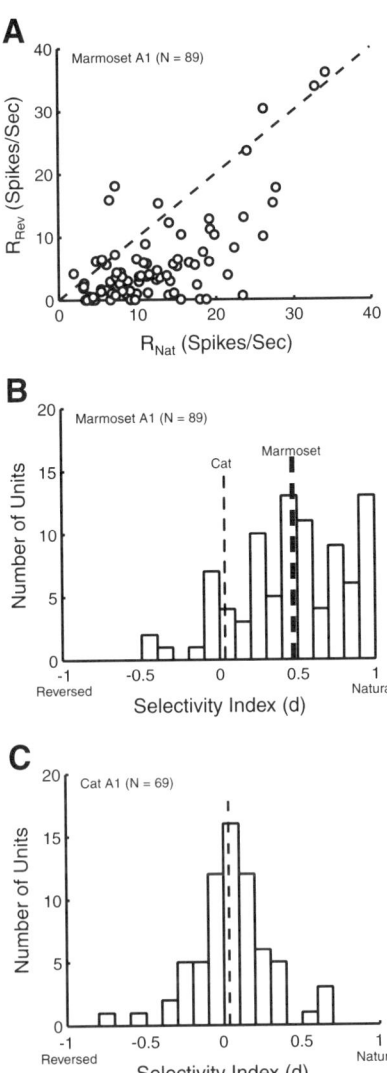

FIGURE 15.4 Comparison between cortical responses to natural and time-reversed marmoset twitter calls in the AI of two species, marmoset and cat (data based on Wang and Kadia[60]). (A) Mean firing rate of responses to a time-reversed twitter call (R_{rev}) is plotted vs. that of the natural twitter call (R_{nat}) on a unit-by-unit basis for marmosets. (B) The distribution of the selectivity index, $d = (R_{nat} - R_{rev})/(R_{nat} + R_{rev})$, is shown for marmosets. (C) The distribution of the selectivity index is shown for cats. The means of these two distributions, indicated by vertical dashed lines in part b, are as follows: cat, 0.047 ± 0.265; marmoset, 0.479 ± 0.361. The two distributions differ significantly (*t*-test, $p < 0.0001$).

social contexts by marmosets[66] and therefore must be discriminated by the animals. Our quantitative analysis showed that trill and trillphee calls were separable into clusters in a multidimensional space of particular acoustic parameters.[65] Vocalizations of each type

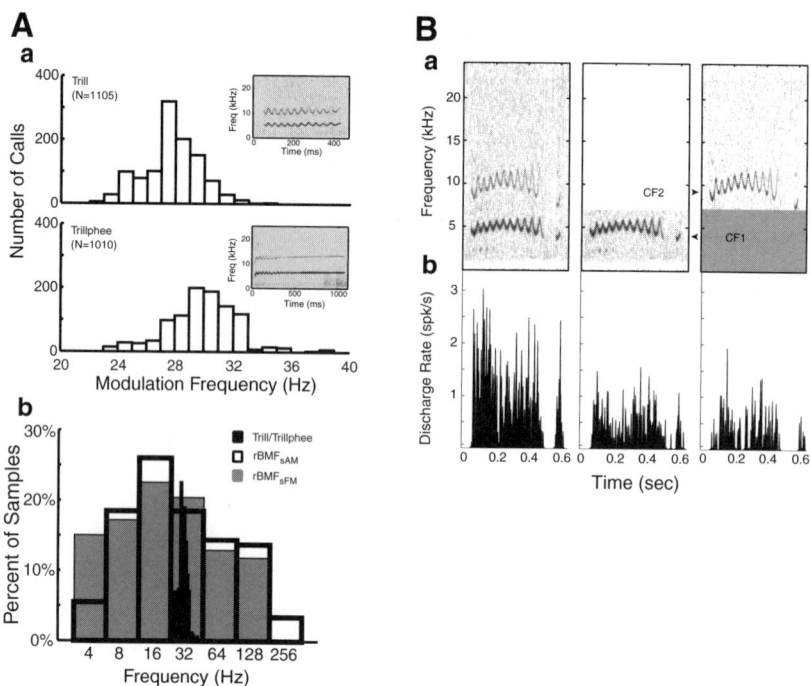

FIGURE 15.5 (A) Correlation between modulation frequency in marmoset vocalizations and temporal modulation selectivity of AI neurons. (a) Distribution of the modulation frequency measured from trill (top) and trillphee (bottom) calls.[65] The insets show spectra of the calls. Trill: $\mu \pm \text{STD} = 27.7 \pm 1.9$ Hz, $N = 1105$; trillphee: $\mu \pm \text{STD} = 29.8 \pm 2.4$ Hz, $N = 1010$. The two distributions are statistically different (Wilcoxon rank sum test, $p \ll 0.001$). The samples of each call type were recorded from four marmosets. (b) Comparison between the distribution of modulation frequency of trill and trillphee calls combined (black, $\mu = 28.7 \pm 2.4$ Hz, $N = 2115$) and those of rBMF_{sAM} (open) and rBMF_{sFM} (gray) obtained in awake marmoset.[38] (B) Nonlinear facilitation in the response of multipeaked neurons to marmoset vocalizations.[45] (a) Spectrogram of a natural trill call (left), its first harmonic component (middle), and its second harmonic component (right). (b) PSTHs of responses to the stimuli shown in part a. This neuron had two spectral peaks at approximately 4.8 kHz (CF1) and 9.6 kHz (CF2) which match two harmonics of the trill call. The average discharge rate to the natural trill call (30 spikes/sec) was greater than the sum of discharge rates to the first and second harmonics presented alone (12 and 9 spikes/sec, respectively).

produced by individual marmosets could be further separated into smaller clusters according to the identity of the caller.[13,65] A trill call has a sinusoid-like frequency modulation throughout the entire call duration, whereas a trillphee call has a sinusoid-like frequency modulation in the initial part of the call which transitions into a constant frequency in the later part of the call (Figure 15.5A, part a). In both types of calls, the frequency modulation was accompanied by amplitude modulation in the waveform of the calls. Interestingly, the distributions of call modulation frequencies are located near the center of the distributions of rBMFs of cortical neurons (Figure 15.5A, part b), but not near the center of tBMF distributions (Figure 15.2C). Because the width of the rMTF is much greater than the variations between modulation frequencies of trills or trillphee calls of individual marmosets, it is likely that a population of cortical neurons must be involved in discriminating individual identity embedded in the acoustic structure of the calls. Having a large number of cortical neurons available near behaviorally important modulation frequencies is certainly to the advantage of a population-based coding strategy. The importance of the amplitude modulation of vocalizations in evoking cortical responses has also been demonstrated in the auditory cortex of squirrel monkeys.[67]

A crucial step in understanding neural coding of species-specific vocalizations is to relate responses to vocalizations to functional structures of the auditory cortex. For example, the harmonic organization of auditory cortex neurons discussed above[40,45] has immediate implications for how marmoset vocalizations might be optimally processed. One key feature of most marmoset vocalizations is the presence of harmonic structures in the acoustic spectrum (Figure 15.5A, part a). The two-tone facilitation demonstrated in multipeaked neurons (Figure 15.3A) predicts that such a neuron will respond more strongly to an intact natural vocalization than to its components. Figure 15.5B shows one example of a marmoset trill call. Two main information-bearing elements are observed at harmonically related frequencies (CF1, CF2). We tested the response to this call in a neuron that had a matching two-peak receptive field. As can be predicted by its receptive field properties, the neuron responded more strongly when both harmonics were present than when either component was played alone (Figure 15.5B). This facilitation is nonlinear in that the response to the entire vocalization is greater than the sum of the responses to either harmonic component alone. Neurons such as the one shown in Figure 15.5B may function to detect the acoustic salience embedded in species-specific vocalizations.

V. SUMMARY

While much progress has been made in recent years in the study of the auditory cortex, our understanding of cortical mechanisms for encoding non-human primate vocalizations is still in its infancy. Much of our knowledge of the auditory cortex has been based on studies in anesthetized animals. An important step in moving this field forward is the use of awake and behaving preparations. Only then can issues such as the behavioral state of an animal and attentional modulation of cortical responses be adequately addressed. A limitation of existing neurophysiological techniques is that the vocal behavior of animals is substantially restricted or eliminated once an animal is restrained. Application of implanted and moveable electrodes attached to a telemetry device will significantly improve our ability to correlate natural vocal behavioral and brain activities. Finally, while the visual system has served as a good model system for understanding cortical coding mechanisms, we must bear in mind the fundamental difference between auditory and visual systems. Unlike the visual system, a significant portion of behaviorally important inputs to the auditory systems of vocal species such as humans and primates are actively produced by the species themselves (i.e., speech and species-specific vocalizations). The auditory cortical system in humans and non-human primates may therefore include a specialized pathway to process vocal communication sounds. Such a "vocal" pathway may or may not be involved when nonvocal acoustic signals (such as environmental sounds) are processed by the cerebral cortex. As we have suggested earlier,[13] this "vocal" pathway is likely a mutual communication channel between the superior temporal gyrus and other brain structures responsible for vocal productions (e.g., the frontal cortex). Modulations of neural responses in the auditory cortex resulting from voluntary vocalizations in primates provide supporting evidence for the interaction between vocal production and perception systems.[68,69] In this regard, the auditory cortical system may have pathways not only to determine the location ("where") and meaning ("what") of a sound, but also to determine "who" produces the sound.

ACKNOWLEDGMENTS

We thank Dr. Ross Snider and Steven Eliades for their work on the data acquisition system used in our studies and Ashley Pistorio for colony management, animal training, and proofreading the manuscript. Our research has been supported by NIH-NIDCD Grant DC03180 and by a Presidential Early Career Award for Scientists and Engineers (X.W.). Publications from our laboratory can be obtained at www.bme.jhu.edu/~xwang/papers.html.

REFERENCES

1. Heffner, H.E. and Heffner, R.S., Effect of unilateral and bilateral auditory cortex lesions on the discrimination of vocalizations by Japanese macaques, *J. Neurophysiol.*, 56, 683–701, 1986.
2. Penfield, W. and Roberts, L., *Speech and Brain Mechanisms*, Princeton University Press, Princeton, NJ, 1959.
3. Pandya, D.N. and Yeterian, E.H., Architecture and connections of cortical association areas, in *Cerebral Cortex*, Vol. 4, A. Peters and E.G. Jones, Eds., Plenum, New York, 1985, pp. 3–61.
4. Kaas, J.H., Hackett, T.A., and Tramo, M.J., Auditory processing in primate cerebral cortex, *Curr. Opin. Neurobiol.*, 9, 164–170, 1999.
5. Jones, E.G. and Powell, T.P.S., An anatomical study of converging sensory pathways within the cerebral cortex of monkey, *Brain*, 93, 793–820, 1970.
6. Morel, A. and Kaas, J.H., Subdivisions and connections of auditory cortex in owl monkeys, *J. Comp. Neurol.*, 318, 27–63, 1992.
7. Kaas, J.H. and Hackett, T.A., Subdivisions of auditory cortex and processing streams in primates, *Proc. Natl. Acad. Sci. USA*, 97, 11793–11799, 2000.
8. Hackett, T.A., Preuss, T.M., and Kaas, J.H., Architectonic identification of the core region in auditory cortex of macaques, chimpanzees, and humans, *J. Comp Neurol.*, 441, 197–222, 2001.
9. Brodmann, K., *Vergleichende Lokalisationslehre der Grobhirnrinde*, J.A. Barth, Leipzig, 1909.
10. Goldstein, M.H.J., Kiang, N.Y.-S., and Brown, R.M., Responses of the auditory cortex to repetitive acoustic stimuli, *J. Acoust. Soc. Am.*, 31, 356–364, 1959.

11. Zurita, P., Villa, A.E., and Rouiller, E.M., Changes of single unit activity in the cat's auditory thalamus and cortex associated to different anesthetic conditions, *Neurosci. Res.*, 19, 303–16, 1994.
12. Gaese, B.H. and Ostwald, J., Anesthesia changes frequency tuning of neurons in the rat primary auditory cortex, *J. Neurophysiol.*, 86, 1062–1066, 2001.
13. Wang, X., On cortical coding of vocal communication sounds in primate., *Proc. Natl. Acad. Sci. USA*, 97, 11843–11849, 2000.
14. Houtgast, T. and Steeneken, H.J.M., The modulation transfer function in room acoustics as a predictor of speech intelligibility, *Acustica*, 28, 66–73, 1973.
15. Rosen, S., Temporal information in speech: acoustic, auditory and linguistic aspects, *Philos. Trans. Roy. Soc. London B, Biol. Sci.*, 336, 367–373, 1992.
16. Johnson, D.H., The relationship between spike rate and synchrony in responses of auditory-nerve fibers to single tones, *J. Acoust. Soc. Am.*, 68, 1115–1122, 1980.
17. Palmer, A.R., Encoding of rapid amplitude fluctuations by cochlear-nerve fibers in the guinea-pig, *Arch. Otorhinolaryngol.*, 236, 197–202, 1982.
18. Joris, P.X. and Yin, T.C.T., Responses to amplitude-modulated tones in the auditory nerve of the cat, *J. Acoust. Soc. Am.*, 91, 215–232, 1992.
19. Wang, X. and Sachs, M.B., Neural encoding of single-formant stimuli in the cat. I. Responses of auditory-nerve fibers, *J. Neurophysiol.*, 70, 1054–1075, 1993.
20. Blackburn, C.C. and Sachs, M.B., Classification of unit types in the anteroventral cochlear nucleus: post-stimulus time histograms and regularity analysis, *J. Neurophysiol.*, 62, 1303–1329, 1989.
21. Frisina, R.D., Smith, R.L., and Chamberlain, S.C., Encoding of amplitude modulation in the gerbil cochlear nucleus. I. A hierarchy of enhancement, *Hearing Res.*, 44, 99–122, 1990.
22. Wang, X. and Sachs, M.B., Neural encoding of single-formant stimuli in the cat. II. Responses of anteroventral cochlear nucleus units, *J. Neurophysiol.*, 71, 59–78, 1994.
23. Langner, G. and Schreiner, C.E., Periodicity coding in the inferior colliculus of the cat. I. Neuronal mechanisms, *J. Neurophysiol.*, 60, 1799–1822, 1988.
24. Creutzfeldt, O., Hellweg, F.-C., and Schreiner, C., Thalamocortical transformation of responses to complex auditory stimuli, *Exp. Brain Res.*, 39, 87–104, 1980.
25. de Ribaupierre, F., Rouiller, E., Toros, A., and de Ribaupierre, Y., Transmission delay of phase-locked cells in the medial geniculate body, *Hearing Res.*, 3, 65–77, 1980.
26. Wang, X. and Sachs, M.B., Transformation of temporal discharge patterns in a VCN stellate cell model: implications for physiological mechanisms, *J. Neurophysiol.*, 73, 1600–1616, 1995.
27. Rall, W. and Agmon-Snir, H., Cable theory for dendritic neurons, in *Methods in Neuronal Modeling*, C.K.A.I. Segev, Ed., MIT Press, Cambridge, MA, 1998.
28. de Ribaupierre, F., Goldstein, M.H., and Yeni-Komshian, G., Cortical coding of repetitive acoustic pulses, *Brain Res.*, 48, 205–25., 1972.
29. Schreiner, C.E. and Urbas, J.V., Representation of amplitude modulation in the auditory cortex of the cat. II. Comparison between cortical fields, *Hearing Res.*, 32, 49–63, 1988.
30. Eggermont, J.J., Rate and synchronization measures of periodicity coding in cat primary auditory cortex, *Hearing Res.*, 56, 153–67, 1991.
31. Eggermont, J.J., Temporal modulation transfer functions for AM and FM stimuli in cat auditory cortex. Effects of carrier type, modulating waveform and intensity, *Hearing Res.*, 74, 51–66, 1994.
32. Gaese, B.H. and Ostwald, J., Temporal coding of amplitude and frequency modulation in the rat auditory cortex, *Eur. J. Neurosci.*, 7, 438–450, 1995.

33. Bieser, A. and Muller-Preuss, P., Auditory responsive cortex in the squirrel monkey: neural responses to amplitude-modulated sounds, *Exp. Brain Res.*, 108, 273–284, 1996.
34. Lu, T. and Wang, X., Temporal discharge patterns evoked by rapid sequences of wide- and narrow-band clicks in the primary auditory cortex of cat, *J. Neurophysiol.*, 84, 236–246, 2000.
35. Lu, T., Liang, L., and Wang, X., Temporal and rate representations of time-varying signals in the auditory cortex of awake primates, *Nat. Neurosci.*, 4, 1131–1138, 2001.
36. Lu, T., Liang, L., and Wang, X., Neural representation of temporally asymmetric stimuli in the auditory cortex of awake primates, *J. Neurophysiol.*, 85, 2364–2380, 2001.
37. R.D. Patterson, The sound of a sinusoid: spectral models, *J. Acoust. Soc. Am.*, 96, 1409–1418, 1994.
38. Liang, L., Lu, T., and Wang, X., Neural representations of sinusoidal amplitude and frequency modulations in the auditory cortex of awake primates, *J. Neurophysiol.*, 87, 2237–2261, 2002.
39. Liang, L., Lu, T., and Wang, X., Temporal encoding of amplitude modulated sounds with noise carrier in the lateral belt areas of the auditory cortex in awake marmoset monkeys, *Soc. Neurosci. Abstr.*, 29, 1999.
40. Kadia, S.C., Wang, X., Liang, L., and Ryugo, D.K., Harmonic structure in long-range horizontal connections within the primary auditory cortex of cat, *Soc. Neuroscience Abstr.*, 29, 1999.
41. Matsubara, J.A. and Phillips, D.P., Intracortical connections and their physiological correlates in the primary auditory cortex (AI) of the cat, *J. Comp. Neurol.*, 268, 38–48, 1988.
42. Ojima, H., Honda, C.N., and Jones, E.G., Patterns of axon collateralization of identified supragranular pyramidal neurons in the cat auditory cortex, *Cerebral Cortex*, 1, 80–94, 1991.
43. Wallace, M.N., Kitzes, L.M., and Jones, E.G., Intrinsic inter- and intralaminar connections and their relationship to the tonotopic map in cat primary auditory cortex, *Exp. Brain Res.*, 86, 527–544, 1991.
44. Reale, R.A., Brugge, J.F., and Feng, J.Z., Geometry and orientation of neuronal processes in cat primary auditory cortex (AI) related to characteristic-frequency maps, *Proc. Natl. Acad. Sci. USA*, 80, 5449–5453, 1983.
45. Kadia, S.C. and Wang, X., Spectral integration in the primary auditory cortex of the awake primate: neurons with single-peaked and multi-peaked tuning curves, *J. Neurophysiol.*, in press.
46. Brosch, M., Schulz, A., and Scheich, H., Processing of sound sequences in macaque auditory cortex: response enhancement, *J. Neurophysiol.*, 82, 1542–1559, 1999.
47. Brosch, M., Schulz, A., and Scheich, H., Neuronal mechanisms of auditory backward recognition masking in macaque auditory cortex, *NeuroReport*, 9, 2551–2555, 1998.
48. Barbour, D. and Wang, X., The sensitivity of auditory cortical neurons to comodulated tones, *Assoc. Res. Otolaryngol. Abstr.*, 23, 2000.
49. Pelleg-Toiba, R. and Wollberg, Z., Discrimination of communication calls in the squirrel monkey: call detectors or cell ensembles?, *J. Basic Clin. Physiol. Pharmacol.*, 2, 257–272, 1991.
50. Winter, P. and Funkenstein, H.H., The effect of species-specific vocalization on the discharge of auditory cortical cells in the awake squirrel monkeys (*Saimiri sciureus*), *Exp. Brain Res.*, 18, 489–504, 1973.

51. Newman, J.D. and Wollberg, Z., Responses of single neurons in the auditory cortex of squirrel monkeys to variants of a single call type, *Exp. Neurol.*, 40, 821–824, 1973.
52. Wollberg, Z. and Newman, J.D., Auditory cortex of squirrel monkey: response patterns of single cells to species-specific vocalizations, *Science*, 175, 212–214, 1972.
53. Manley, J.A. and Mueller-Preuss, P., Response variability of auditory cortex cells in the squirrel monkey to constant acoustic stimuli, *Exp. Brain Res.*, 32, 171–180, 1978.
54. Tian, B., Reser, D., Durham, A., Kustov, A., and Rauschecker, J.P., Functional specialization in rhesus monkey auditory cortex, *Science*, 292, 290–293, 2001.
55. DiMattina, C. and Wang, X., Virtual vocalization stimuli for systematic investigation of cortical coding of vocal communication sounds, *Assoc. Res. Otolaryngol. Abstr.*, 25, 2002.
56. Wang, X., Merzenich, M.M., Beitel, R., and Schreiner, C.E., Representation of a species-specific vocalization in the primary auditory cortex of the common marmoset: temporal and spectral characteristics, *J. Neurophysiol.*, 74, 2685–2706, 1995.
57. Wang, X., Cortical processing of communication sounds in nonhuman primates, *Assoc. Res. Otolaryngol. Abstr.*, 22, 1999.
58. Suga, N., Processing of auditory information carried by species-specific complex sounds, in *The Cognitive Neuroscience*, M.S. Gazzanica, Ed., MIT Press, Cambridge, MA, 1994, PP. 295–313.
59. Schreiner, C.E., Read, H.L., and Sutter, M.L., Modular organization of frequency integration in primary auditory cortex, *Annu. Rev. Neurosci.*, 23, 501–529, 2000.
60. Wang, X. and Kadia, S.C., Differential representation of species-specific primate vocalizations in the auditory cortices of marmoset and cat, *J. Neurophysiol.*, 86, 2616–2620, 2001.
61. Wiesel, T.N. and Hubel, D.H., Comparison of the effects of unilateral and bilateral eye closure on cortical unit responses in kittens, *J. Neurophysiol.*, 28, 1029–1040, 1965.
62. Merzenich, M.M., Kaas, J.H., Wall, J.T., Sur, M., Nelson, R.J., and Felleman, D.J., Progression of change following median nerve section in the cortical representation of the hand in areas 3b and 1 in adult owl and squirrel monkeys, *Neuroscience*, 10, 639–665, 1983.
63. Recanzone, G.H., Schreiner, C.E., and Merzenich, M.M., Plasticity in the primary auditory cortex following discrimination training in adult owl monkeys, *J. Neurosci.*, 13, 87–104, 1992.
64. Wang, X., Merzenich, M.M., Sameshima, K., and Jenkins, W.M., Remodeling of hand representation in adult cortex determined by timing of tactile stimulation, *Nature*, 378, 71–75, 1995.
65. Agamaite, J.A. and Wang, X., Quantitative classification of the vocal repertoire of the common marmoset (*Callithrix jacchus jacchus*), *Assoc. Res. Otolaryngol. Abstr.*, 20, 144, 1997.
66. Epple, G., Comparative studies on vocalization in marmoset monkeys (*Hapalidae*), *Folia Primatol.*, 8, 1–40, 1968.
67. Bieser, A., Processing of twitter-call fundamental frequencies in insula and auditory cortex of squirrel monkeys, *Exp. Brain Res.*, 122, 139–148, 1998.
68. Muller-Preuss, P. and Ploog, D., Inhibition of auditory cortical neurons during phonation, *Brain Res.*, 215, 61–76, 1981.
69. Eliades. S.J. and Wang, X., Suppression of neural activities prior to and during vocalization suggests sensory-motor interactions in the primate auditory cortex., *Soc. Neuroscience Abs.*, 30, 2000.

Index

A

Acoustic distortion mechanisms, 128
Acoustic features
    amplitude-based cues in communication calls, 66–67, 78–79
    frequency-based cues in communication calls, 66
    functionally relevant, 49–53
    phase-based cues in communication calls, 67
Acoustic gradation, infants' response to, 117–120
Acoustic impediments, 129
Acoustic processing vs. communicative processing, 65–66
Acoustic scatter, 130
Acoustic stimulation, 179
    complex, 271
    MDS used to examine perception of, 65
    reliance of monkeys in rain forest on, 16
Acoustic streams, 55
Acoustic structure, 43
Acoustic transients, 281
    processing of within the temporal integration window, 283–284
Acoustic variation, 128
Acoustically responsive region, 200
Acoustics
    habitat, 115, 127–128, 134
    linear, 87
    physical, 87
    vocalization, 66
Active theory of phonation, 89–90
Adaptive specializations, 1
Afferents, 266
AI. *See* primary auditory cortex
Air sacs, 95–96, 99
Alarm calls, 2
    guinea fowl, 22–24
    response of Diana monkeys to chimpanzees', 17–18
    semantic specificity of, 19–22
Alarms calls, response of Diana monkeys to Campbell's monkeys', 18–22
Alouatta. *See* howler monkeys
Amplitude decay, 284
Amplitude fluctuation, 130
    insensitivity of perceptual systems to, 134

Amplitudes
    acoustic cues in communication calls based on, 66–67, 73–75, 78–79
    species-typical, 8–10
Answer chucks, 230
Antecedent-consequent sequences, primate understanding of, 14–15
Anterior cingulate cortex, 114
Anterior temporal lobe, 261
Antiphonal calling, 8–9
    isolated, 45–46
        cotton-top tamarins and, 47
    social, 45–46, 50
    tamarin response to, 52–53
Antiresonances, 99
Anurans, auditory systems of, 1
Aotus. *See* owl monkey
Aperiodic sources, 143
Arched scream vocalizations, 77
ARMA models, 99
Articulatory apparatuses, 62
Articulatory positions, 102
Ateles. *See* spider monkeys
Attractors, 137–140
Auditory belt, 266
Auditory continuity, 47
Auditory core, 204
Auditory cortex, 280. *See also* primary auditory cortex
    anatomy, 186, 260
    combination sensitivity of neurons in, 9
    connections of, 206–208
    core architecture of, 204–206
    early studies of squirrel monkey, 232–233
    electrophysiological studies of, 250–252
    functional organization of non-primary, 260–261
    functional structures of related to vocalization responses, 295
    interaction between excitatory and inhibitory acoustic inputs on, 236
    lesion studies, 186, 249–250
    modulations of neural responses in, 296
    neuroanatomy and neurophysiology of squirrel monkey, 233
    neuronal ensembles in, 166–169
    organization of, 202–204

primate, 248–249
  representation of sound location in, 187
  representation of sound pattern in, 155–169
  role of behavioral relevance in shaping
    representations of, 280
  studies of auditory structures below the,
    239–240
  tuning properties of cells of, 237
Auditory cortical mechanisms, relevant to vocal
  communication, 233–239
Auditory cortical plasticity, 252–253
  sound localization and, 254
Auditory cues, food detection and, 4
Auditory discrimination, 1, 152
Auditory feedback, 112
Auditory function, synthesis as a method to
  understand, 97–98
Auditory nerve, physiological properties in
  squirrel monkeys, 232
Auditory neuroscience, 88
Auditory pattern-recognition system
  of monkeys, 153–154
  role of auditory cortex in, 154–155
Auditory perception, study of in cotton-top
  tamarins, 45
Auditory processing, 65
  in the prefrontal cortex, 268–271
Auditory spatial process, in caudal prefrontal
  cortex, 268
Auditory temporal integration. *See also* temporal,
  integration of information
  leaky integrator model of, 30
  phylogenetic differences in, 39–40
  spectral processing differences and, 40
Auditory-prefrontla connections, rostrocaudal
  topography of, 264
Auditory-responsive cortex in squirrel monkeys,
  232
Avian species, auditory systems of, 1
Aye-aye, food detection by, 4
Azimuth selective neurons of the inferior
  colliculus, 182

B

Baboons
  infant response to barks of, 118–120
  recognition of cause-and-effect by, 15
  subhyoid air sacs in, 95
  vocalizations of, 46
Barbary macaques
  alarm calls of, 115
  copulation calls of females, 6–7
  shrill barks of, 117

Barks
  alarm, 118
  contact, 118
  fear, 15
  shrill, 8, 117
Behavioral context, responsiveness of auditory
  cortex neurons to, 187
Behavioral founder effect, 114
Behavioral relevance, role of in shaping cortical
  representations, 280, 292
Belt regions of the auditory cortex, 186, 200, 203,
    208–210, 249, 260
  architecture of, 209–210
  connections of, 210
Best modulation frequency, 159
Bifurcations, 93–94, 128, 140, 144–145
Binaural cues, 248
Bioacoustics
  linear prediction in, 98–100
  nonlinear dynamics and, 136
Biomechanics of vocal folds, 144–146
Biomechanics of voicing, 127–128
Biphonation, 93, 140–142, 144
Bird song, vocal tract filtering in, 91
Black and white colobus monkeys, roar of, 135
Bloch's law, 29
Blue monkey, temporal integration studies of, 38
Bonnet macaque, 3
Boom vocalization of Campbell's monkeys,
    19–22
Brain, auditory-responsive areas of, 189
Broad tuning, 180
Broca's area, 269, 271
Bronx cheer, 116

C

Cackle element of the chuck call, 230
Cackling, 111
Call detector, 237, 290
Callichtrichids, dietary classifications of, 4
Callithrix. *See* Callichtrichids
Callithrix jacchus. *See* marmosets
Calls
  acoustically distinct, 140
  alarm, 2–4, 63
    chimpanzee, 17–18
    of Barbary macaques, 115
    response of infant vervet monkeys to, 117
    superb starling, 117
  caregiver, 231
  chuck, 230–231
  clear, 16

compensation of perceptual systems for
 duration/frequency of stereotypic,
 135
convergence of, 114–117
coo, 49–50, 134–135
 discreteness of SEH/SLH, 72–73
 field observation of, 67–68
 harmonic amplitude of, 73–75
 laboratory experiments on discrimination
 of, 68–72
 copulation, 6–8
 facial postures during, 102
 food, 5
 interaction between production and perception
 of, 49–53
 long, 8–9
 mating, 1, 6–7
 species-specific, MDS used to examine
 perception of, 65
 tonal, 143
 trill, 292
 trillphee, 292
 usage and comprehension of, 110
Campbell's monkeys
 alarm calls of, 16
 response of Diana monkeys to, 18–22
 boom vocalization of, 19–22
Caregiver calls of squirrel monkeys, 231
Categorical perception, 50, 54, 72–73, 98,
 117–118
Category identification, 63–64
Caudal belt fields, connection with primary
 auditory complex, 206
Caudal dorsolateral prefrontal cortex. See
 dorsolateral prefrontal cortex
Caudal prefrontal cortex, 261, 266
 auditory spatial processing in, 268
Caudal principalis, 268
Caudal superior temporal cortex, 212. See also
 auditory cortex
Caudal superior temporal polysensory region, 186
Caudolateral field, 249
Caudomedial field, 249
 sound localization and the, 252
Causal reasoning, 13
 in humans, 14
Causality concept, 14
 understanding of, 24
Causation, 15
Cebuella. See marmosets
Central auditory mechanisms, of squirrel
 monkeys, 233
Cercocebus albigena. See grey-cheeked
 mangabey
Cercopithecinae, cheek pouches of, 4

Cercopithecus aethiops. See vervet monkeys
Cercopithecus albogularis. See Syke's monkeys
Cercopithecus campbelli. See Campbell's
 monkeys
Cercopithecus diana. See Diana monkeys
Cercopithecus mitis. See blue monkey
Cercopithecus neglectus. See DeBrazza's monkey
Cerebral cortex, role of in processing species-
 specific vocalizations, 280
Cervus unicolor. See sambar deer
Cetacean clicks, vocal tract filtering in, 91
Chaos, 93, 137
Characteristic delay, 182
Characteristic frequency, 287
Chemoarchitecture of belt regions of the auditory
 cortex, 210
Chimpanzees
 as predators of Taï monkeys, 16
 auditory cues used for food detection by, 5
 pant-hoots of, 115–116
 SOS screams of, 17
 vocal membranes of, 96
 vocalizations of, 46
Chirps, 45
Chucks, 44
 acoustic attributes of in squirrel monkeys,
 230–231
CL. See caudolateral field
Classical receptive field, spectral inputs from
 outside the, 287–290
CLC, 45
 functional classes of exemplars of, 50
 functionally relevant acoustic features, 49–53
 structure of cotton-top tamarin, 47
Clear call, 16
Click trains, 155, 159
CM. See caudomedial field
Cochlea
 physiological properties in squirrel monkeys,
 232
 topographic organization of, 178
Combination long call. See CLC
Communication, 43
 neural encoding of the statistical structure of
 sounds, 280
Communicative processing vs. acoustic
 processing, 65–66
Comparative anatomy and physiology, 87
Complex acoustic stimuli, 271
 extraction of temporal profiles embedded in,
 284–287
Complex sweeps, 271
Conspecifics
 alarm calls of, 3
 distinguishing utterances of, 129

temporally manipulated vocalizations of, 50
    vocalizations of, 1
Coo calls, 134–135
    discrimination of in auditory cortex lessened
        animals, 154
    field observations of, 67–68
    in rhesus macaques, 111–112
    of Japanese macaques, 49–50, 64
Copulation calls, 6–7, 8
Core regions of the auditory cortex, 186, 200, 203,
        248–249, 260
    architecture of, 204–206
    connections of, 206–208
Cortical connections, 206–208
Cortical lesions, effect of on sound perception,
        154–155
Cortical neurons
    classes of, 291
    ensembles of, 166–169
    representation of sequences of simple stimuli,
        161–166
    representation of simple stimuli by, 155–159
    representation of the superposition of simple
        stimuli by, 161
    response patterning of, 159–161
Cortical representations, nature of, 291
Cotton-top tamarins
    food calls of, 5
    natural signals of, 46–53
    vocalizations of, 45, 99
Crested guinea fowl, response of Diana monkeys
        to alarm calls of, 22–24
Critical bands, 97
Crowned-hawk eagles, as predators of Taï
        monkeys, 16
Cytoarchitectonic parcellation of the frontal lobe,
        261
Cytoarchitecture
    of the core, 204–205
    of the parabelt region, 211

D

Daubentonia madagascarensis. *See* aye-aye
DeBrazza's monkey, 95
Deterministic chaos, 93–94, 137
Dialects, 114–117
Diana monkeys
    alarm calls of, 16
    response to alarm calls of Campbell's
        monkeys, 18–22
    response to alarm calls of chimpanzees, 17–18
    response to alarm calls of guinea fowl, 22–24

Dichotic stimulus, 179
    characteristic delay of IC neuron response to,
        182
    sensitivity of medial superior olives to, 180
Dietary classes of primates, 4
Digital sound synthesis, 97
Direct auditory cues, 4
Directionally sensitive neurons in the inferior
        colliculus, 182
Discrimination enhancement, 72
Distortion, as an impediment to vocal
        communication, 130–133
DLPFC. *See* dorsolateral prefrontal cortex
Domain specificity, 267
Dorsal periarcuate region, 265, 268
Dorsolateral prefrontal cortex, 236, 268
Duet songs, 7–8
Duplex perception, 98

E

Eagles
    as predators, 2–3
    crowned-hawk, as predators of Taï monkeys,
        16
Ears, anatomy of, 4
Echolocation, 1
Effective path-length fluctuation, 133
Ensemble formation, 166
Equal-loudness contours, 97
Errs of squirrel monkeys, 231
Euoticus elegantulus. *See* galago
Exponential model of temporal integration, 30–31
    scaling prameter of, 32
    slope parameter of, 33–35
Eye position
    acoustic stimuli modulated as a function of,
        183, 188
    modulation of auditory responses in the
        auditory cortex by, 187
    relation of superior colliculus and frontal eye
        fields to, 188–189

F

F_0. *See* first harmonic
Facial configurations, 67
    spectral peaks and, 92
Facilitation, 289
Female copulation calls, 6–7
Ferrier, David, 200
First harmonic, 66
First temporal field, 238
Flag element of the chuck call, 230–231

Foliovores, 4
Food-associated vocalizations, 5. *See also* indirect auditory cues
Formants, 90–92, 134–135, 151
  frequencies as an indication of body size, 96
  transition, 98
Fourier analysis, relationship of quality of information and time, 27
Fractionation, of functions within the prefrontal cortex, 266–267
Free-field stimuli, 179, 185
Frequency
  acoustic cues in communication calls based on, 66, 73–75
  discrimination, 154
  inflection, 70
  perceptually salient components of, 134–135
  resonant, 92
  sensitivity of primates to harmonically related, 152–153
Frequency difference limens, 152
Frequency modulations, sensitivity of perceptual systems to rapid, 135
Frequency response area, 155
Frequency tuning curve, 155
Frequency-specific attenuation, 130–131
Frogs
  mating calls of, 6
  vocal communication systems of Tungara, 44
  vocalizations of, 92
Frontal eye fields, 188–189
Frontal lobe
  anatomical organization of, 261
  auditory responses within, 240
  responses to acoustic stimuli in monkeys, 269
Frontal pole, 265
Frugivores, 4
Functional magnetic resonance imaging (fMRI), 178
Fundamental frequency of phonation, 66, 89–90, 137–140
Fuzzy logic, 77

G

Γ-oscillations, 160–161
Galago, food detection by, 4–5
Galago demidovii. *See* galago
Gaps, detection of in noise, 154
Gelada, vocalizations of, 46
Germany, auditory communication studies in, 237–239
Gibbons
  songs of, 7–8
  vocal membranes of, 96
  vocalizations of, 46
Go/no-go test procedure, 63–64
Gramnivores, 4
Granular prefrontal cortex, 261
Granulous cytoarchitecture of the core, 204
Grey-cheeked mangabey
  grunt vocalizations of, 143
  temporal integration studies of, 38
  whoop-gobble call of, 135
Growth of sensation, 30
Grunts, 8, 15
Guinea fowl, predator alarm call of, 22–24
Gummivores, 4
Guttera pulcheri. See guinea fowl

H

H1. *See* first harmonic
Habitat acoustics, 115, 127–128, 134
Habituation discrimination, 45, 50
  speech perception and, 54–55
Hanuman langurs, 3
Harmonics, 89
  amplitude variation of, 73–75
  in vocalizations of marmosets, 295
  multiples of fundamental frequency of phonation, 66
Head-orienting bias, 8, 183
Head-related transfer function, 248
Helmeted guinea fowl, response of Diana monkeys to alarm calls of, 22–24
Hemisphere-specific specialized processing centers, 69–70, 252
Hierarchical connections, cascades of, 266
High-frequency scatter, 131
Homotypical cortex of the superior temporal gyrus, 211
Howler monkeys, vocal adaptations of, 95
Humans
  causal reasoning in, 14
  infant babbling, 113–114
  poachers as predators of Taï monkeys, 16
  speech perception of, 54
  temporal integration studies of, 38
  understanding the language of, 109
Hyla versicolor. See frogs
Hylobates agilis. See gibbons
Hylobates syndactylus. See siamangs

I

IC. *See* inferior colliculus
Identity-related acoustic cues, 76

Indirect auditory cues, 4. *See also* food-associated vocalizations
Inferior colliculus, 232, 238, 239–240
　anatomy of, 180–181
　lesion studies of, 181–182
　representation of sound location in, 182–184
Inferior convexity, 267
Information theory, 27
Inhibition, 289
Insectivores, 4
Integration rates, 33
Intensity differences, 248
Intention
　impulse causality and, 14, 24
　primate recognition of, 15
Interaural level differences, sensitivity of lateral superior olives to, 180
Interstimulus intervals, 281–283
Isolated antiphonal calling, 45–46
　cotton-top tamarins and, 47
Isolation peep, 111, 235
　acoustic attributes of in squirrel monkeys, 229–230
Israel, auditory communication studies in, 237

J

Japanese macaque, 114
　auditory cortex lesion studies of, 186
　coo calls of, 49–50, 64, 67–68, 76
　laboratory experiments on vocal discrimination of, 68–72
Jürgens, Uwe, 234

K

Koniocellular cytoarchitecture of the core, 204
Koniocortex, 204–205

L

Labeled line coding, 157
Language mode vs. psychophysical mode of processing, 65
Language processing, 259
Langurs, 3
Laryngeal anatomy
　air sacs, 95–96
　mammalian, 89
Laryngoceles, 95
Lateral belt region of the auditory cortex, 209
Lateral inferior convexity, 265
Lateral intraparietal cortex, 188

Lateral orbital cortex, 265
Lateral superior olives, 180
Leaky integrator, 30
Lemurs, vocalizations of, 46, 88
Leontopithecus. *See* tamarins
Leopards, as predators, 2
　Taï monkeys, 16
Limit-cycle attractors, 137
Linear acoustics, 87
Linear prediction
　analysis and synthesis using, 100–102
　in bioacoustics, 98–100
Linear predictive coding. *See* linear prediction
Long calls, 8–9
Long-range horizontal connections, multipeaked tuning and, 287
LP. *See* linear prediction
LPC. *See* linear prediction

M

Macaca mulatta. *See* rhesus macaques
Macacca. *See* macaques
Macaques
　architecture of parabelt regions of the auditory cortex of, 211
　auditory cortex architecture of, 209
　auditory cortex lesion studies of, 186, 249–250
　Barbary
　　alarm calls of, 115
　　shrill barks of, 117
　copulation calls of female, 6–7
　food calls of, 5
　Japanese, 49–50, 64, 115 (See also Japanese macaque)
　koniocortex of, 205
　laboratory experiments on vocal discrimination of, 68–72
　neurophysiological responses to behavior and perception of, 247–248
　pigtailed, 76
　recognition of pitch contours by, 153
　rhesus, 8, 50, 76–78 (See also rhesus macaques)
　　pant-threat vocalization of, 101
　scream vocalizations of, 76–78, 145
　spectral discrimination of, 40
　stumptail, call acoustics of, 76
　subhyoid air sacs in, 95
　Sulawesi crested black, 76
Male copulation calls, 6
Mammalian laryngeal anatomy, 89

Index

Marmosets
    anatomical adaptations for food detection, 4
    trill calls of, 292–294
    twitter calls of, 292
Mast element of the chuck call, 230–231
Mating calls, 1, 6–7
Matlab, 98, 100
Matriline relationships, in rhesus macaques, 70
MDS. *See* multidimensional scaling
Medial belt region of the auditory cortex, 209
Medial geniculate body, 181, 185, 206, 232
Medial geniculate nucleus, 239
Medial intraparietal cortex, 188
Medial prefrontal cortex, connection with orbital prefrontal and anterior temporal lobe, 264
Medial superior olives, 180
Memory demands, 73
Memory processing, 65
Midbrain reticular formation, 236
Mismatch negativity, 160
Missing fundamental, 97
Modulation transfer function, 155
Monkeys
    auditory discrimination of, 152–154
    auditory perceptual capabilities of, 169
    black and white colobus, roar of, 135
    blue, temporal integration studies of, 38
    Campbell's, 16
    DeBrazza's, 95
    Diana, alarm calls of, 16
    functional organization of cerebral cortex of, 200
    howler, vocal adaptations of, 95
    macaque, 67 (See also Japanese macaque; rhesus macaques)
    red colobus, 2
    responses to acoustic stimuli in frontal lobes of, 269
    squirrel, ontogenetic development of calls of, 110–111
    Syke's, vocalizations of, 66
    vervet, 2–3, 63
        response of infants to alarm calls, 117
Monotonic response-location functions, 189
Motor theories of perception, 94
Mouse lemurs, ultrasonic vocalizations of, 88
MRF. *See* midbrain reticular formation
Müller-Preuss, Peter, 234
Multidimensional scaling, 62, 64–65, 70, 73
Multipeaked tuning, long-range horizontal connections and, 287
Myeloarchitecture of the core, 204
Myoelastic-aerodynamic theory of phonation, 89–90

N

National Institutes of Health, auditory communication studies at, 234–237
Natural vocal communication, 44
Neural encoding, of communication sounds, 280
Neural selectivity, 9
Neural synchrony, 167–169
Neuroanatomy of the squirrel monkey auditory cortex, 233
Neuroimaging techniques, sound localization and, 178
Neuronal code, 155
Neuronal ensemble, 155
Neurophysiology of the squirrel monkey auditory cortex, 233
New World primates, 228
    air sacs of, 95
    auditory cortical organization in, 209
    food-associated vocalizations produced by, 5
    vocal membranes of, 96
    vocalizations of in captivity, 280
Nilgiri langurs, 3
Noise, as an impediment to vocal communication, 129–130
Noisy scream vocalizations, 77
Non-human primates. *See* primates
Non-primary auditory cortex, functional organization of, 260–261
Nondirectional cells of the inferior colliculus, 182
Nonlinear dynamics, 87
    application of to the analysis of voicing, 136
    in vocal production, 93–94
Nonmnotonic response-location functions, 189
Numida meleagris. *See* helmeted guinea fowl
Nyquist frequency, 101

O

Observing response, 65
Old World primates
    air sacs of, 95
    auditory cortical organization in, 209
    copulation calls of, 6
    food-associated vocalizations produced by, 5
    sound power of, 129
    vocal membranes of, 96
Omnidirectional cells of the inferior colliculus, 182
Operant reporting response, 65
Orbital prefrontal cortex, connection with medial prefrontal and anterior temporal lobe, 264
Organ pipe formation, 211
Owl monkey, auditory cortex of, 232

P

Pan troglodytes. *See* chimpanzees
Pant-hoots of chimpanzees, 115–116
Panthera pardus. *See* leopards
Papio cynocephalus ursinus. *See* baboons
Parabelt regions of the auditory cortex, 186, 203, 210–212, 249
    architecture of, 211–212
    connections of, 212
Parainsular auditory field, 238
Parietal cortex, 187–188
Parieto-occipital area, 188
Path-length-fluctuation distortion process, 135
Peak tracing, 237
Perceptual analyses of acoustics in primates, 44, 128
Perceptual predisposition, 65
Perceptual systems, sensitivity of, 134
Perceptual units, 46. *See also* units of perception problem
Periarcuate region, 265, 268
PFC. *See* caudal prefrontal cortex; dorsolateral prefrontal cortex
Phase-based acoustic cues in communication calls, 67
Phase-vocoding, 98
Phonation
    fundamental frequency of, 66
    harmonic multiples of fundamental frequency of, 66
    myoelastic-aerodynamic theory of, 89–90
Phonetic processing, 65
Phonotaxis, selective, 46, 51–52
Physalaemus pustolosus. *See* frogs
Physalaemus pustulosus. *See* frogs
Physical acoustics, 87
Physical causation, 15
Pigtailed macaque, scream vocalizations of, 76
Place coding, 157–158
Planum temporale, 209
Plasticity, auditory cortical, 252–253
Play peep of squirrel monkeys, 231
Ploog, Detlev, 233–234
Poles, 99
Population coding, 291
Population vector, 166
Positional information, 177
Positron emission tomography (PET), 178
Power model of temporal integration, 30–31
Praat, 100
Predation, behavioral causation in the context of, 13, 15–24
Predation rates, 2

Predators
    avoidance of, 2–4
    understanding behavior of, 16–18
Prefrontal auditory connections, 261–266
Prefrontal cortex, 259
    auditory domains in the, 268–271
    functional organization of, 266–267
    specialization for auditory-visual associations of the left, 268
Prefrontal neurons, salient environmental stimuli and, 271
Premotor cortex, 265
Primary auditory cortex, 158, 200, 204, 232, 280. *See also* auditory cortex
    inferior colliculus, 180–181
    long-range horizontal connections in, 287
    plasticity of, 252–253
    sound location processing of, 250, 252
    synchronized and nonsynchronized neurons of, 283
Primates
    acoustic features of vocalizations of, 66–67
    acoustic perception in, 44
    alarm calls of, 2–4
    architecture of the core in, 204–206
    auditory cortex of, 248–249
    auditory pathways of, 178
    auditory pattern recognition in, 152–154
    behavioral and neurobiological studies of vocalizations of, 134–136
    copulation calls of, 6–7
    dietary classes of, 4
    identification of auditory-related fields in cortex of, 200
    mechanics of vocal production of, 136–146
    New World
        air sacs of, 95
        auditory cortical organization in, 209
        food vocalizations produced by, 5
        mechanisms mediating vocal communication of, 228
        vocal membranes of, 96
    nonlinear dynamical theory of vocalizations of, 93–94
    Old World
        air sacs of, 95
        auditory cortical organization in, 209
        copulation calls of, 6
        food vocalizations produced by, 5
        sound power of, 129
        vocal membranes of, 96
    organization of auditory cortex of, 202–204
    rates of temporal integration of non-human, 40
    songs of, 7–8
    speech perception of, 53–56

understanding of intention in non-human, 24–25
units of perception in communication of, 49
vocal adaptations of, 94–97
vocal apparatus control limitations of, 114–117
vocal communication systems of, 88
vocal development of, 109–110
vocal production, 121
  principles of, 89–94
Principal sulcus, 265
Processing, classification of, 65
Processing asymmetry, 69–70
Procolobus badius. *See* red colobus monkeys
Production mechanisms, understanding signaling systems by, 44
Prosimians. *See also* aye-aye; galago
  food detection by, 4
Prototype magnet effects, 98
PSOLA, 98
Psychoacoustics, 152
Psychoacoustics, of squirrel monkeys, 233
Psychophysical mode vs. language mode of processing, 65
Pulsed scream vocalizations, 77
Punctate call elements, sensitivity of perceptual systems to brief, 135
Purrs of squirrel monkeys, 231
Pygmy marmoset, babbling of, 113–114
Pythons, as predators, 3

Q

Question chucks, 230

R

Rain forests, ecological impediments to vocal communication in, 129
Rainshower formation, 205
Rate coding, 155, 157–158, 281–283, 287
Rate-based best modulation frequency, 285
Rate-based modulation transfer, 285
Receptive field of neurons, 155, 157
Red colobus monkeys
  auditory detection of by chimpanzees, 5
  predation of, 2
Red-bellied tamarins, food calls of, 6
Reporting response, 65
Reproduction, copulation calls and, 6–7
Resonances, 99
Resonant frequencies, 92
Response attenuation, 165
Response enhancement, 165

Response modifications, 166
Response patterning of cortical neurons, 159–161
Response selectivity, 235
Retroactive response attenuation, 165
Reverberation, 130–131
  minimization of, 135
Rhesus macaques
  coo calls of, 68, 70–72, 111–112
  copulation calls of, 6
  facial postures during call types of, 102
  food calls of, 5
  functionally relevant acoustic features of vocalizations of, 50
  pant-threat vocalization of, 101
  scream vocalizations of, 76–78
  temporal cues in calls of, 8
  temporal integration studies of, 31, 38
Ripple stimuli, 161
Riverine forests, ecological impediments to vocal communication in, 129
Rostral belt fields, 260
  connection with primary auditory complex, 207
  lesions of, 250
Rostral parabelt, 266
Rostrocaudal topography of the prefrontal cortex, 261, 264

S

Saccadic eye movements, 188–189
Saguinus. *See* tamarins
Saimiri sciureus. *See* squirrel monkeys
Sambar deer, 3
Same/different test procedure, 63
Savanna, ecological impediments to vocal communication in, 129
Scaling parameter, 32
  rate of integration vs., 33
  species specificity of, 37–38
Scream vocalizations, 76–78
  deterministic chaos and, 94
SEH. *See* smooth-early-high (SEH)/smooth-late-high (SLH) coo calls
Selection, effect on vocalizations of, 136
Selective attention, 134
Selective phonotaxis, 46
Semnopithecus entellus. *See* Hanuman langurs
Sensory adaptation, 28
Sensory auditory cortex, 280. *See also* auditory cortex
Sensory processing, 65
Sensory receptors, topographic maps of, 199
Sensory specializations, 1

Sensory stimuli, parietal cortex control of orientation to, 187
Sensory thresholds, 28
Sensory-motor, interactions between vocal production and perception systems, 280
Shrill bark, 8, 117
Siamangs, duet songs of, 8
Signal distortion, as an impediment to vocal communication, 130–133
Signal intensity, temporal patterns of, 53
Signal production, identification of parallels among primates, 62
Singing, 7–8
Single-peaked tuning characteristics, 287–290
Sinusoids, 284
Size exaggeration hypothesis, 96–97
SLH. *See* smooth-early-high (SEH)/smooth-late-high (SLH) coo calls
Slope parameter, temporal integration correlation with, 33–35
Smooth-early-high (SEH)/smooth-late-high (SLH) coo calls, 49, 79
   discreteness of, 72–73
   of Japanese macaques, 64, 68–70, 76
   of stumptail macaques, 76
Snakes, as predators, 2
SOA. *See* stimulus onset asynchrony
SOC. *See* superior olivary complex
Social antiphonal calling, 45–46, 50
Social causation, 15
SOS screams, 17
Sound localization, 1, 178
   auditory cortical function studies and, 247–248
   coding for in the superior colliculus, 188–189
   cues and perception, 248
   lateralization of by superior olivary complex, 180
   neural basis of, 178
   plasticity of, 252–254
   spatial tuning and, 250–251, 269
Sound location, coding for in inferior colliculus neurons, 182–184
Source-filter theory of speech production, 88
Source-filter theory of vocal production, 90–92
Spatial attention, parietal cortex control of, 187
Spatial information, processing of throughout the auditory cortex, 249
Spatial sensitivity, of neurons in the auditory cortex, 187
Spatial tuning, 180
   of inferior colliculus neurons, 183
Specialized processing centers, hemisphere-specific, 69–70

Species-specific calls, MDS used to examine perception of, 65
Spectral inputs, outside the classical receptive field, 287–290
Spectral peaks, 92
Spectral processing differences, 40
Speech accommodation, 116
Speech perception, 44, 53–56, 98
   MDS used to examine, 65
Speech production, 88
Spider monkeys
   koniocortex of, 205
   vocalizations of, 46
Spontaneous firing rate, 235
Spreo superbus. *See* superb starling
Squirrel monkeys, 228
   auditory cortex lesion studies of, 186
   auditory cortex studies of, 232–233
   dichotic stimuli response of, 185
   Gothic-arch form, 229
   ontogenetic development of calls of, 110–111
   Roman-arch forms, 229
   vocal behavior of, 228–232
Stephanoaetus coronatus. *See* crowned-hawk eagles
Stereotypic calls, compensation of perceptual systems for duration/frequency of, 135
Stimulus intensity, 29
Stimulus location, 177
Stimulus onset asynchrony, 163–166
Stumptail macaques, call acoustics of, 76
Subharmonics, 93
Subhyoid air sacs, 95
Subject training, 73
Submission, scream vocalizations within the context of, 77
Subregio temporalis transversa prima, 206
Sulawesi crested black macaques, scream vocalizations of, 76
Superb starling, alarm calls of, 117
Superior colliculus, 188–189
Superior olivary complex, 180
Superior temporal cortex, 204
Superior temporal gyrus, 237, 260, 296
Superior temporal region, 200–202
Superior temporal sulcus, 266
   dorsal bank of, 261
Supralaryngeal vocal tract, 66
Supratemporal regions of the auditory cortex, 209
Supratemporalis granulosa, 205
Suprathreshold stimulus manipulations, 73
Syke's monkeys
   electroglottograph waveform of grunt of, 137
   squeals of, 144–145

Index 311

vocalizations of, 66
Syllable order, 50
Sylvian fissure, 202, 204
Symmes, David, 234
Sympatric species, 3
   distinguishing utterances of, 129
Synchronized firing of neurons, 167–169
Synchrony-based best modulation frequency, 286
Syntactic modification of alarm calls, 19–22, 24

T

Taï National Park, primates of, 16
Tamarins
   anatomical adaptations for food detection, 4
   antiphonal calling response of, 52–53
   cotton-top
      food calls of, 5
      vocalizations of, 45
   individual identity information extracted from acoustical parameters by, 51–52
   red-bellied, food calls of, 6
   temporal cues in long calls of, 8–9
Temporal
   coding of neurons, 155, 158–159, 281–283
   cue sensitivity of primates, 152–153
   envelope shape, 67
   features of vocal signals, 72–73
   filter, 284
   integration of information, 28 (See also auditory temporal integration)
      effect of species factors on, 33
      exponential model of, 30–31
      frequency correlation with, 33
      leaky integrator model of, 30
      power model of, 30–31
   modulation frequency, 286
   modulation transfer functions, 152, 155, 165, 281, 285, 286
   patterns of signal intensity, 53
   properties of CLC, 47
Temporal integration window, processing of acoustic transients within the, 283–284
Temporal lobe
   anterior, 261, 266
   as primary auditory cortex in squirrel monkeys, 232
   connection with orbital and medial prefrontal cortices, 264
   lesions of, 249
   regions of, 209
   specialization for auditory function of the left, 268

Temporal profiles, embedded in complex sounds, 284–287
Temporoparietal association cortex, 186
Thalamus, 185
Time constraints, studies of temporal integration and, 38–39
Time-varying signals, rate vs. temporal coding of, 281–283
Tonal calls, 143
Tonal scream vocalizations, 77
Tone sequences, 271
Tonotopic organization, 157, 235
   physiological evidence for, 260–261
Topographic organization of spatial tuning, 183, 250
Toque macaques, food calls of, 5
Trachypithecus johnii. See Nilgiri langurs
Trading relations, 98
Trill, 292
Trillo, 143
Trillphee, 292
Tuning characteristics
   multipeaked, 287
   single-peaked, 287–290
Turbulent noise, 90, 136–137
Twitters of squirrel monkeys, 231

U

Ultrasonic vocalizations of mouse lemurs, 88
Undulated scream vocalizations, 77
Units of perception problem, 44, 46–47

V

Ventral parietal cortex, 188
Ventrolateral frontal lobe, 259
   auditory responses in, 269
Ventrolateral prefrontal cortex. *See* prefrontal cortex
Vervet monkeys
   alarm call response of infant, 117
   category identification testing used with, 63
   functionally relevant acoustic features vocalizations of, 50
   predation of, 2–3
   subhyoid air sacs in, 95
Virtual space stimuli, 179
Visual cues of danger, 3
Visual localization, 178
Vocal accommodation, 116
Vocal communication systems, 43–44
   auditory cortical mechanisms relevant to, 233–239

brain processing of, 280
Vocal convergence, 114–117
Vocal development, 109–110
Vocal discrimination, laboratory experiments on macaque, 68–72
Vocal folds, 89, 94
  biomechanics of, 136–146
  oscillation of, 137
Vocal imitation, 121
Vocal lips. *See* vocal membranes
Vocal matching, 116
Vocal membranes, 96, 114
Vocal perception, neurobiological and behavioral studies of, 144–146
Vocal production, 87–89
  nonlinear dynamics in, 93–94
  principles of primate, 89–94
  source-filter theory of, 90–92
Vocal recognition, 1
Vocal tract
  elongation, 96–97
  harmonic energy distribution correlation with configuration of, 73
  morphology, 62, 66
  resonant frequencies of, 91
  transfer function, 66
Vocalizations
  acoustic features of primate, 66–67
  anterior lateral belt region of the non-primary auditory cortex and, 261
  behavioral and neurobiological studies of primate, 134–136
  categorical testing to determine perceptual classifications of, 64
  conspecific, 1
  dependence of cortical responses on behavioral relevance of, 292
  distortion measurements for, 146
  functionally relevant acoustic features of primate, 49–53
  infants' response to, 117–120
  preference of frontal lobe auditory neurons for, 271
  scream, 76–78
  species-specific, 127
    mechanisms related to coding of, 292–295
    representations of, 290
  species-specific primate, 280
  squirrel monkey, 228–232
  temporal modulations in, 284
  variations within the range of a species', 114
Vocalizer identity, 68
Voice onset time, 98, 151, 153
Voicing biochmechanics, 127–128, 136–146

W

Waveforms, 127–128
Waves, interference between direct and reflected, 131–133
Western Ivory Coast, primates of the Taï National Park of, 16
Whine, 44
Whistles, 45
Winter, Peter, 234
Wollberg, Zvi, 234

Z

Zeros, 99